Edited by **Holger Stark**

Histamine H_4 Receptor:
A Novel Drug Target in Immunoregulation and Inflammation

VERSITA

Versita Discipline: Chemistry, Life Sciences

Published by Versita, Versita Ltd, 78 York Street, London W1H 1DP, Great Britain.

ISBN (paperback): 978-83-7656-054-0

ISBN (hardcover): 978-83-7656-055-7

ISBN (for electronic copy): 978-83-7656-056-4

Managing Editor: Anna Rulka

Language Editors: Brent Paulter, Michael Jones

Cover illustration: Holger Stark with support from E. Tiligada and son

www.versita.com

Contents

Chapter 7
Non-professional Histamine Producing Cells, Immune Responses and Autoimmunity 201

Preface

Some fields of research are tremendously growing within a few years due to high scientific interest. The histamine H_4 receptor field is such a scientific area with huge input from various sources due to high therapeutical interest and unmet medical need in many related diseases. To cover many of these aspects we have put together this book covering a variety of research approaches surrounding the histamine H_4 receptor. One of the main intentions was to attract young researchers going into this field with enhanced background knowledge. Therefore, we have put together some basic aspects for the beginners, some more advanced material for the advanced researchers as well as some more detailed aspects on points of discussion for the experts. The following nine chapters are the result of decades of research on the histamine area as well as of the recent COST Action of the European Union BM0806 "Recent advances in histamine receptor H_4R research" from 2009 to 2013 having Ekaterini Tiligada, Athens/Greece, as Chair. This COST Action succeeded in putting together different research groups of different topics and different countries to join and to work together. I deeply thank all authors from all over the world for their highly valuable contributions.

I also deeply thank Versita with Anna Rulka and co-workers for the continuous support during the publication process as well as the extraordinary possibility to make this book available for everyone by the open access publication form.

It is our hope that the visibility and the use of this book may support and accelerate the introduction of new histamine H_4 receptor-based drugs for the patients.

Holger Stark, Germany

Chapter 1

The histamine H_4 receptor story

Holger Stark

Heinrich-Heine-University Duesseldorf, Institute for Pharmaceutical and Medicinal Chemistry, Universitaetsstr. 1, 40225 Duesseldorf, Germany; E-Mail: stark@hhu.de

The long-lasting story of histamine and the receptors of this biogenic amine started in the beginning of the last century with the detection of the biological activity of the decarboxylation product of L-histidine and followed later with the introduction of the so-called "anti-histamines". The story has been refreshed several times with the innovation of histamine H_2 receptor blockers giving the classical pharmacological characterization of this new receptor subtype, with the neurotransmitter function based on the histamine H_3 receptor and soon after its cloning the detection of the histamine H_4 receptor subtype by molecular biology methods.

In different chapters of this book, the authors go into some details on the historic importance of the "old" histamine. These historic views may explain some of the blind alleys and one-ways that have been followed during different steps of knowledge as well as some new and very recent findings. Due to the medicinal chemistry background of the editor, this book has not started with the detection of the receptor but with the identification and development of some early lead compounds used for the characterization of function and cross-actions of the histamine H_4 receptor (Figure 1.1).

Schreeb et al. (2013) gave a general description of the structural relationships on agonist and antagonist of the different histamine receptor subtypes. They went into further details with the narrative of structure-affinity and structure-activity relationships of different lead structures on histamine H_4 receptor. Based on some histamine-related imidazole-containing compounds they transiently went over to benzimidazoles and pyrimidine derivates. With the structural descriptions, they show the overlap of pharmacophoric moieties within the different classes stressing the possibilities as well as the limitations for chemical modifications in these developments. It is interesting to see in which way the position of some ring nitrogen moieties or amino substituents can greatly change or maintain the binding properties within highly related scaffolds.

The formation and fate of histamine is the topic of the chapter by Schwelberger et al. (2013). Recent X-ray studies on the human L-histidine

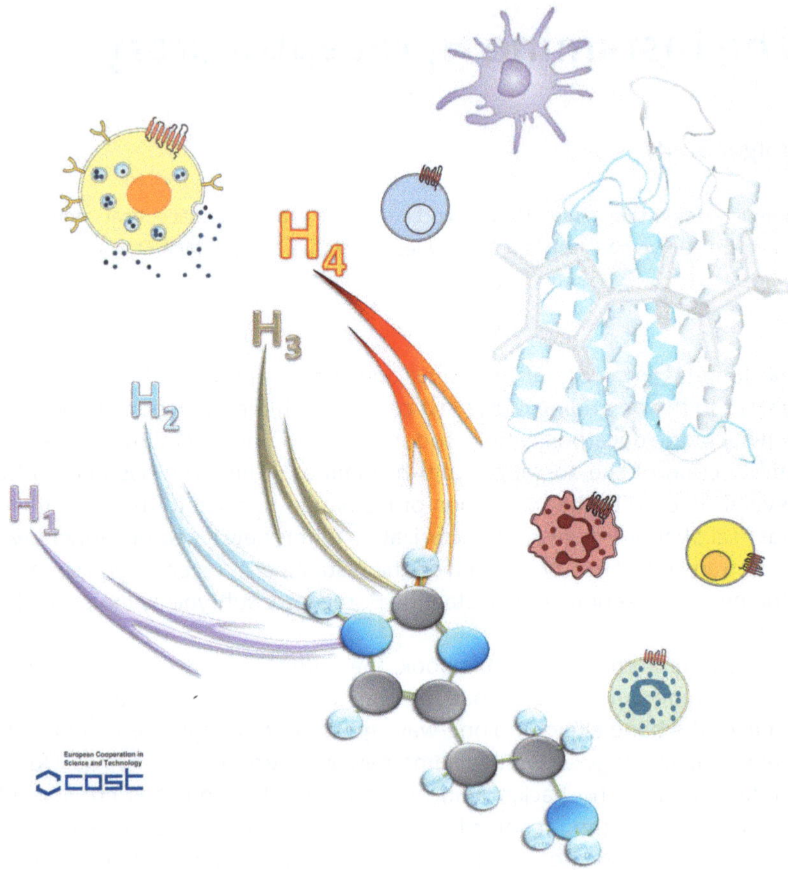

Figure 1.1 Illustration from the COST Action BM8006 hinting the transition from molecules to function and the complex network in histamine H_4 receptor research.

decarboxylase greatly contributed to the understanding of the binding domains. The not fully understood transport mechanisms are mainly organized by organic cation transporters with different tissue distribution and membrane localization. Other monoamine transporters of different families with related influences on the transport in different cells, e.g. neuronal and endocrine cells, make these uptake and secretion processes even more complex. Beyond storage and release, the metabolic pathway highlights large differences depending on the localization in the periphery or the central nervous system. The central histamine *N*-methyltransferase leads to a different product than the peripherally localized diamine oxidase, which products can then undergo further catabolic and coupling reactions. Although these main pathways are well worked-out, the exact fine-tuning in different tissues is still under investigation.

Some of these signaling mechanisms in mast cells, basophils, eosinophils, neutrophils, monocytes, macrophages and dendritic cells are explained by Ennis et al. (2013) in more detail. With the focus on the histamine H_4 receptor and inflammatory processes, they show the orchestration within the innate immune response and adaptive responses. Since the mast cells and the dendritic cells are the first defense line of the immune system responding to antigen, toxins and other pathogenic elements, it is clear that here the histamine H_4 receptor must have an important regulatory function due to its prominent localization. Not only are allergic and immunomodulatory reactions on the haematopoetic system are influenced by this histamine receptor subtype, but it may also be the case that the polymorphonuclear neutrophils are the line of defense against bacteria and fungi which is also influenced by this G-protein coupled receptor. In many fields we have already obtained large knowledge on the exact function, while others are clearly in their infancy.

This can also be seen when we go from the cellular function to some disease-related pathological states and potential therapeutic treatments. Bäumer et al. (2013) have put their emphasis the influence of histamine H_4 receptor modulation in inflammatory skin diseases. The different cell subpopulations and their expression level of the histamine H_4 receptor with varying functions on chemotaxis or numerous pro- and anti-inflammatory cytokines. The Th cell polarization seems to be a major regulatory process linking innate and adaptive immune pathways. The relevant differences on the miscellaneous animal models on atopic dermatitis and pruritus with different species will be helpful on early preclinical evaluations.

Going from the skin to the respiratory system, the inflammatory problems remain constant but the models and the potential therapeutic approaches change. Masini et al. (2013) report on animal models of allergic asthma and pulmonary fibrosis in connection to other signaling mechanisms in which NO as well as some prostaglandines from the arachidonic pathway play a major role. Th2-mediated diseases can be positively influenced by histamine H_4 receptor antagonists leading to the promising hypothesis that a combination of H_1 receptor antagonist, H_4 receptor antagonist and a mast cell stabilisator and/or a leukotriene inhibitor (or a related approach) may be used for the treatment of asthma.

The transitions from Th1 to Th2 cells with different disease states raises the need for further input in immune responses and immunological regulations. Konttinen et al. (2013) provide an overview on the historic developments of the different histamine receptor subtypes as well as the immunological processes involved in non-allergic and autoimmune diseases. Some of the authors have recently shown that Sjögren´s syndrome, a female dominant autoimmune disease of exocrine glands, have high histamine H_4 receptor overexpression in salivary glands. Some results obtained on models on allergic conjunctivitis have shown large potential for therapeutic benefit. If the same is true on rheumatoid arthritis, systemic lupus erythematosus, ulcerative colitis or Chrohn´s disease is still a matter of debate. The T_{reg} and Th17 cells along with their influence on the histamine H_4 receptors requires further evaluation.

The effect of histamine is not only on differentiation of immune competent cells but also on cell proliferation and migration. Histamine itself is already in use as an adjuvant orphan drug for cancer therapy in combination with IL-2 on treatment of acute myelonic leukemia. Medina et al. (2013) summarize the effects of histamine H_4 receptor ligands on melanoma, breast, colon, pancreatic and lung cancer and discuss the effects obtained in leukemia as well as in malignant lymphomas. The picture of the histamine H_4 receptor on cancer diseases is more heterogeneous and less sharp that that for many other diseases. The effects of agonists and antagonists cannot always be taken as opposing players. The xenograft tumor model shows promises in some breast cancer cell lines whereas it failed with numerous biomarker characterizations. As reported before on the inflammatory and autoimmune diseases in the digestive system some gastrointestinal tumors are potentially (co-)regulated by histamine H_4 receptors. It appears that the regulation of cell growth and immune processes are not well balanced in different cancers. Mast cells may play an important role on the drive of these pathophysiological processes.

It is clear that histamine is initiating and regulating much more processes than that by interacting with its G-protein coupled receptor subtypes. To highlight these possibilities Kyriakidis and Tiligada (2013) discuss the binding of histamine to DNA as well as that to the glutamine moiety in proteins. The effects of histamine in chemotaxis in pathogens are once again discussed. This is not taken from the human defense line as before, but from the bacterial utilization of histamine on survival mechanism. The dibasic histamine shares some functional polyamine elements and influences numerous non-receptor mediated effects.

In the final chapter of this book, Tiligada (2013) summarize the recent progress in the identification of the therapeutic potential of histamine H_4 receptor ligands in different diseases, mainly in inflammatory and immunological states

This book is covering numerous aspects on histamine and the histamine H_4 receptor. With the given status on drug development with some clinical candidates one may be optimistic that some so far unmet inflammatory and immune diseases with high medical need are approached. The histamine H_4 receptor ligands are not expected to revolutionize pharmacotherapy, but to give new possibilities and chances as well as hopefully largely improved treatments in the near future.

References

Bäumer, W., Glatzer, F., Roßbach, K., Ohtsu, H., Seike, M., Mommert, S., Werfel, T., Gutzmer, R. (2013) Histamine in atopic disorders: Atopic dermatitis and pruritus. In: Histamine H_4 receptor: A novel drug target in immunoregulatory and inflammatory diseases (Ed. Stark, H.), Versita, London/UK, pp 173-199.

Ennis, M., Ciz, M., Dib, K., Friedman, S., Singh Gangwar, R., Gibbs, B. F., Levi-Schaffer, F., Lojek, A., Migalovich-Sheikhet, H., O'Mahony, L., Perecko, T., Sánchez-Jiménez, F., Urdiales, J. L., Vasicek, O. (2013) Histamine receptors and inflammatory cells. In: Histamine H_4 receptor: A novel drug target in immunoregulatory and inflammatory diseases (Ed. Stark, H.), Versita, London/UK, pp 103-143.

Konttinen, Y. T., Husu, H., Han, X., Passani, M. B., Ballerini, C., Stegaev, V., Sillat, T., Mackiewicz, Z. (2013) Non-professional histamine producing cells, immune responses and autoimmunity. In: Histamine H_4 receptor: A novel drug target in immunoregulatory and inflammatory diseases (Ed. Stark, H.), Versita, London/UK, pp 201-258.

Kyriakidis, D. A., Tiligada, E. (2013) Non receptor-mediated histamine actions. In: Histamine H_4 receptor: A novel drug target in immunoregulatory and inflammatory diseases (Ed. Stark, H.), Versita, London/UK, pp 309-332.

Masini, E., Lucarini, L., Sydbom, A., Dahlén, B., Dahlén, S.-E., (2013) Histamine in asthmatic and fibrotic disorders. In: Histamine H_4 receptor: A novel drug target in immunoregulatory and inflammatory diseases (Ed. Stark, H.), Versita, London/UK, pp 145-171.

Medina, V. A., Coruzzi, G., Martinel Lamas, D. J., Massari, N., Adami, M., Levi-Schaffer, F., Ben-Zimra, M., Schwelberger, H., Rivera, E. S. (2013) Histamine in cancer. In: Histamine H_4 receptor: A novel drug target in immunoregulatory and inflammatory diseases (Ed. Stark, H.), Versita, London/UK, pp 259-308.

Schreeb, A., Dove, S., Lazewska, D., Buschauer, A., Kiec-Knonowiecz, K. Stark, H. (2013) Histamine H_4 receptor ligands. In: Histamine H_4 receptor: A novel drug target in immunoregulatory and inflammatory diseases (Ed. Stark, H.), Versita, London/UK, pp 21-61.

Schwelberger, H. G., Ahrens, F., Fogel, W. A., Sánchez-Jiménez, F. (2013) Histamine metabolism. In: Histamine H_4 receptor: A novel drug target in immunoregulatory and inflammatory diseases (Ed. Stark, H.), Versita, London/UK, pp 63-101.

Tiligada, E. (2013) Perspectives in H_4R research and therapeutic exploitation. In: Histamine H_4 receptor: A novel drug target in immunoregulatory and inflammatory diseases (Ed. Stark, H.), Versita, London/UK, pp 333-351.

Chapter 2

Histamine H_4 Receptor Ligands

A. Schreeb[1], D. Łażewska[2], S. Dove[3], A. Buschauer[3], K. Kieć-Kononowicz[2] and H. Stark[1,4*]

[1]Johann Wolfgang Goethe University, Institute of Pharmaceutical Chemistry, Biozentrum, ZAFES/CMP/ICNF, Max-von-Laue-Str. 9, 60438 Frankfurt am Main, Germany

[2]Jagiellonian University, Collegium Medicum, Department of Technology and Biotechnology of Drugs, Medyczna 9 st., Pl 30-688 Krakow, Poland

[3]University of Regensburg, Institute of Pharmacy, Universitätsstr. 31, 93040 Regensburg, Germany

[4]Heinrich-Heine-University Duesseldorf, Institute of Pharmaceutical and Medicinal Chemistry, Universitaetsstr. 1, 40225 Duesseldorf, Germany; E-mail: stark@hhu.de

Abstract

Since identification and cloning of the histamine H_4 receptor (H_4R) in 2000 by several groups, the development of H_4R ligands has been boosted by different drug development programs. The newest member of the histamine receptor family is considered a promising drug target (as described in this book in different chapters). Highly potent and selective H_4R agonists and antagonists have been published by several groups. The effort to improve the pharmacokinetic properties of the currently available H_4R ligands is reflected in a steadily growing number of scientific publications and patent applications. Preclinical data strongly confirms the need for novel potent H_4R ligands to explore their potential therapeutic value in treating allergies, inflammation, autoimmune disorders and, possibly, cancer. The main structural classes of H_4R ligands are (benz)imidazoles and six-membered nitrogen-containing heterocycles with numerous variations. The objective of this review is to compile currently available H_4R ligands and to present noticeable structure-activity and structure-selectivity relationships as well as some selected functional and (pre)clinical data.

2.1. Introduction

Histamine is an important mediator and neurotransmitter that is involved in a broad spectrum of central and peripheral physiological as well as

pathophysiological processes and exerts its specific effects by activation of four receptor subtypes (H_1R – H_4R) (Parsons & Ganellin, 2006; Walter & Stark, 2012). The histamine receptors belong to the family of rhodopsin-like (class A) G-protein coupled receptors and differ in receptor distribution, ligand binding properties, signaling pathways and functions. Histamine is a more potent agonist at the hH_4R than at the other histamine receptor subtypes (Leurs et al., 1994; Oda et al., 2000; Liu et al., 2001; Morse et al., 2001; Zhu et al., 2001; Bakker et al., 2001).

H_1R and H_2R have been targets of blockbuster drugs for treatment of allergic diseases and gastric/duodenal ulcer, respectively (Lacy et al., 2012; Yanai et al., 2012). hH_3R antagonists/inverse agonists, mainly designed for the treatment of neuronal diseases, are on the way to market (Sander et al., 2008; Leurs et al., 2011). Although preclinical data strongly suggest that the hH_4R may play a role as a drug target in the therapy of allergies, inflammation, autoimmune disorders and, possibly, cancer, no clinical studies of hH_4R ligands have been published yet.

Among the four histamine receptor subtypes, only the H_3R and H_4R are closely related (35% and 58% sequence identity overall and in the transmembrane domains, respectively; Parsons and Ganellin, 2006). The early pharmacological and physiological characterization of the H_4R was based on H_3R ligands containing an imidazole heterocycle, e.g. the H_3R/H_4R agonist imetit and the H_3R antagonist/H_4R agonist clobenprobit (**10**) shown in figure 2.1. Disadvantages of such imidazole-containing compounds, in particular a lack of receptor subtype selectivity and inhibition of cytochrome P450, stimulated the search for non-imidazoles as selective H_4R ligands. This process has been advanced by screening programs strongly supported by computational chemistry. The indolylpiperazine JNJ-7777120 (**11**; pK_i 8.4; Jablonowski et al., 2003) resulted from in-house high-throughput screening (HTS) followed by optimization at Johnson & Johnson (J&J).

Although JNJ-7777120 has been widely used as reference antagonist to investigate the H_4R, unfavourable pharmacokinetic properties and questions arising from partial agonist activity with low to moderate intrinsic activity in certain pharmacological models, underline the urgent need for new bioactive compounds.

A rough H_4R antagonist blueprint can be deduced from different known antagonist structures, consisting of a lipophilic residue, a (conjugated) hydrogen-bond acceptor/donor system and a basic moiety, mostly an *N*-alkylated tertiary cycloaliphatic amine (Figure 2.2). Remarkable for the H_4R is that rather small structural variations in the ligands may result in largely changed functional properties (Sander et al., 2009). Many hH_4R ligands identified among or derived from known H_xR ligands show only low H_4R selectivity (Seifert et al., 2013). In particular, selectivity for the hH_4R over the hH_3R is difficult to achieve because of the high degree of sequence homology. A recent orthogonal analysis based

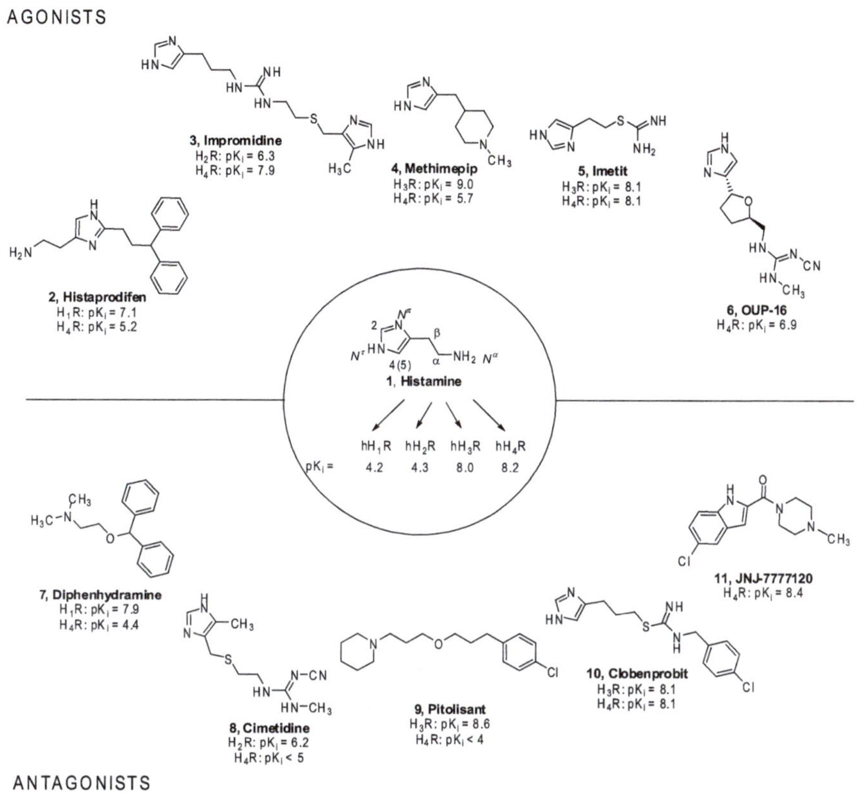

Figure 2.1 Structures and receptor binding data of histamine and representative $hH_{1-4}R$ ligands.

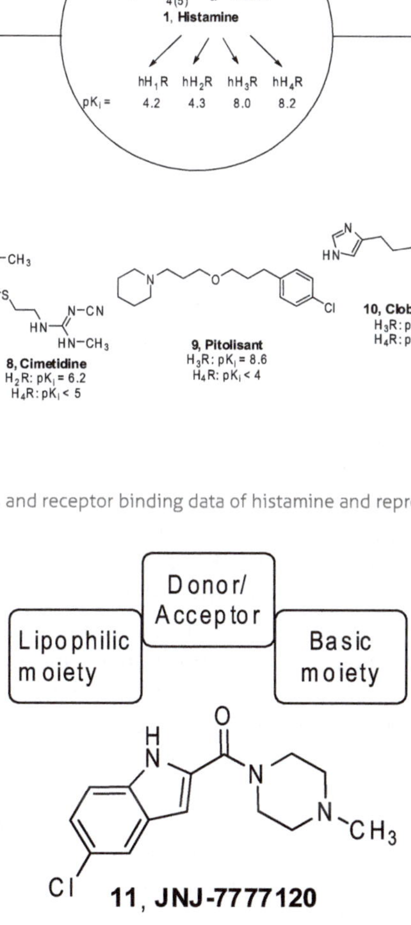

Figure 2.2 General structural blueprint for H_4R antagonists deduced from JNJ-7777120.

on ligand structures and enabling the identification of GPCR similarities beyond receptor subfamilies also confirmed the relationships between H_4R and H_3R (Lin et al., 2013). In addition, with respect to translational animal models, marked differences between H_4R species orthologues have to be taken into account (Lim et al., 2010; Liu et al., 2001b; Strasser et al., 2013). Affinities and potencies of many ligands, especially at rodent H_4Rs, differ significantly from those at the hH_4R.

2.2. Imidazole Derivatives

It is the aim of this section to review hH_4R affinities and functional data of various imidazole-type ligands. The compounds **12-48** were selected to present the major structural classes of such hH_4R ligands and to provide a basis for the discussion of structure-activity relationships. The imidazolyl moiety is essential for the agonistic activity of histamine, and the vast majority of the title compounds are (full or partial) agonists, too (for a recent review on hH_4R agonists, see Igel et al., 2010).

2.2.1. Analogues of Histamine

Histamine (**1**) is a full hH_4R agonist with pK_i (binding) and pEC_{50} (potency) values ~8. Studies on H_4R species orthologues revealed that the affinity of histamine is comparable at human, guinea pig and monkey H_4R, but about one order of magnitude lower at mouse, rat and dog H_4Rs (Lim et al., 2010). When measured in steady-state GTPase assays at H_4Rs expressed in Sf9 cell membranes, the potency of histamine at canine, rat and murine H_4Rs was even more than two orders of magnitude lower than that at hH_4R (Schnell et al., 2011b). Systematic studies with chimeras revealed that the region between $V141^{4.51}$ and $E182^{5.46}$ (superscripts according to the Ballesteros and Weinstein numbering (Ballesteros & Weinstein, 1995)) involving the second extracellular loop (ECL2) may account for the different histamine affinity at the human and mouse H_4R (Lim et al., 2008). Among point mutations in this region, the hH_4R-F169V mutant indicated that F169 in ECL2 is the key amino acid for differential interactions of certain agonists with the hH_4R and the mH_4R, respectively (Lim et al., 2008).

Based on the alignment with two corresponding consecutive phenylalanines in the β_2-adrenoceptor structure (Cherezov et al., 2007), Lim et al. (2008) suggested that a network of hydrophobic interactions involving F169 stabilizes ECL2 in a conformation which positions the neighboring F168 towards the binding pocket.

However, homology models of the hH_4R in its active state suggest possible ECL2 conformations enabling direct interactions of F169 with agonists

(Schneider et al., 2010). Figure 2.3 presents the putative binding mode of histamine at the hH_4R, supported by in-vitro mutagenesis studies (Shin et al., 2002; Lim et al., 2008) and previous docking approaches (Jongejan et al., 2008). The protonated amino group forms a salt bridge with $D94^{3.32}$, and the imidazolylethyl moiety ring is surrounded by $Y95^{3.33}$, F169 (ECL2) and $Y319^{6.51}$. Bidentate hydrogen bonds of the imidazole nitrogens with $T178^{5.42}$ and $E182^{5.46}$ account for the hH_4R agonism. A suggested alternative, "reverse" binding mode of histamine is based on the assumption that the imidazole NH interacts with $D94^{3.32}$ and that the amino group forms a salt bridge with $E182^{5.46}$ (Jojart et al., 2008).

2-Methylhistamine (**12**), reported as a H_1R-preferring agonist in the 1970s (Durant et al., 1975), proved to be an almost full agonist at the hH_4R with a potency similar to that at the hH_1R (Lim, 2005; Deml et al., 2009) (Figure 2.4). As expected from the low sequence homology between the hH_1R and hH_4R

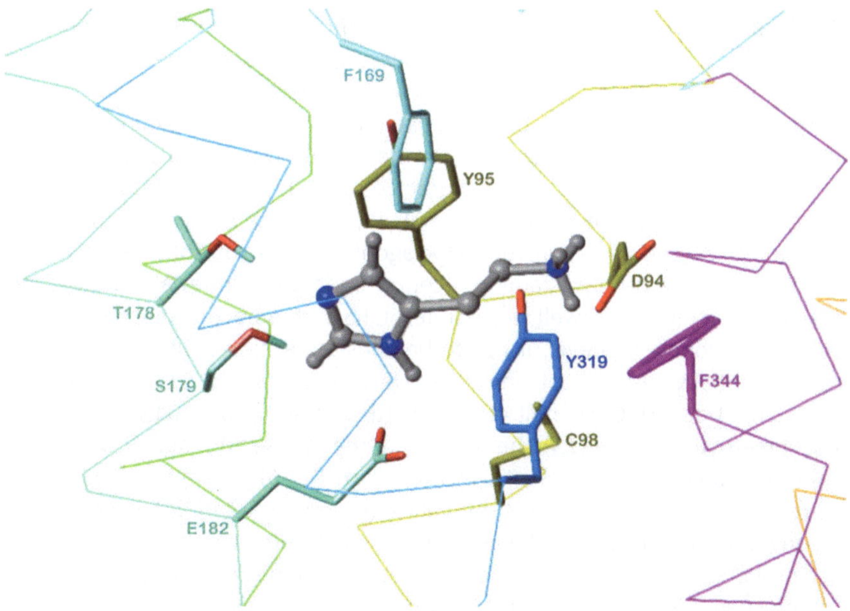

Figure 2.3 Putative binding mode of histamine (**1**) at the hH_4R, based on a model of the active hH_4R state (for details of generation, cf. Igel et al., 2009a; Schneider et al., 2010). The backbone (Cα trace, lines), C and some essential H atoms of side chains are drawn in spectral colours: TM2, orange; TM3, yellow; TM4, green; TM5, green-blue; TM6, blue; TM7, magenta; nitrogen, blue; oxygen, red; sulfur, yellow. Histamine (C and essential H atoms, grey) is drawn as ball-and-stick model. The model was generated and the docking and energy minimization were performed with the modeling suite Sybyl 8.0 (Tripos L.P., St. Louis, MO, USA) on a Silicon Graphics Octane workstation.

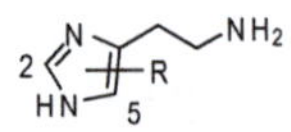

12, 2-Methylhistamine
pK_i = 6.1
13, 5-Methylhistamine
pK_i = 7.3

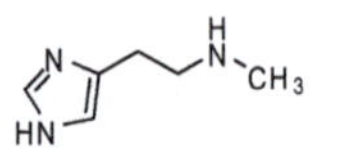

14, *N*$^{\alpha}$-Methylhistamine
pK_i = 6.5

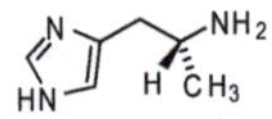

15, (*R*)-α-Methylhistamine
pK_i = 6.6

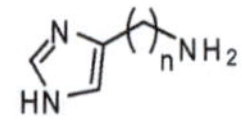

16, Homohistamine: n = 3
pK_i = 7.5
17, Imbutamine: n = 4
pK_i = 8.0

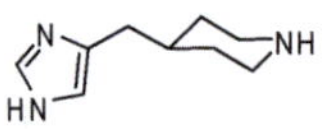

18, Immepip
pK_i = 7.7

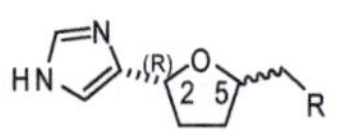

19, Imifuramine: (2R,5R); R = NH_2
pK_i = 6.1
20: (2R,5S); R = O-benzyl
pK_i = 6.4

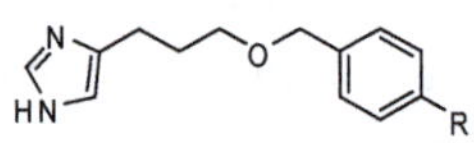

21, Proxyfan: R = H
pK_i = 7.3
22, Iodoproxyfan: R = I
pK_i = 7.9

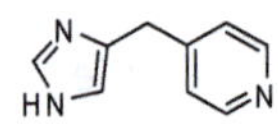

23, Immethridine
pK_i = 6.6

Figure 2.4 H_4R ligands related to histamine.

(Oda et al., 2000), most H_1R ligands had low affinity for the hH_4R, and only few compounds, e.g. some 2-phenylhistamines originally developed as selective H_1R agonists (Pertz et al., 2004), show partial agonistic activity.

The structural isomer of 2-methylhistamine, 5-methylhistamine (**13**; also referred to as 4-methylhistamine in literature), was originally described as H_2R agonist (Black et al., 1972; Durant et al., 1975). Regardless of the low sequence identity of the hH_2R and the hH_4R (22%), the compound turned out to be a potent and selective full hH_4R agonist with >100-fold selectivity for the hH_4R over the other histamine receptor subtypes (Lim et al., 2005). Since 5(4)-methylhistamine is well accessible, it has become the most frequently used hH_4R agonist. However, it has to be considered that many hH_4R agonists are substantially less potent and selective in rodents (Liu, 2001b; Lim et al, 2008; Lim et al., 2010; Schnell et al., 2011a; Strasser et al., 2013). In the case of 5(4)-methylhistamine (**13**), the potency at the mouse and rat H_4R was almost 100 fold reduced (pEC_{50} hH_4R: 7.4, mH_4R: 5.8, rH_4R: 5.6) (Lim et al., 2005). Again, the replacement of F169 (hH_4R) by valine (rodent H_4Rs) in ECL2 may be the reason for these differences. Unlike agonists, most H_4R antagonists are devoid of such pronounced species differences (Liu et al., 2001b). For example, the potent and selective H_4R antagonist JNJ-7777120 binds with almost identical affinity to the human, rat and mouse H_4Rs (Thurmond et al., 2004).

Since the hH_4R shares the highest sequence homology with the hH_3R among all histamine receptor subtypes (Morse et al., 2001), it is not surprising that many hH_3R ligands bind to the hH_4R as well. Most H_3R (partial) agonists were reported to have also (partial) agonistic activity at the hH_4R (Lim et al., 2005). However, the potencies at the hH_3R were 5- to 15,000-fold higher than at the hH_4R (Lim et al., 2005). The reference hH_3R agonists *N*$^{\alpha}$-methylhistamine (**14**) and (*R*)-α-methylhistamine (**15**) were 80- and 40-fold selective for the hH_3R, respectively. The (*S*)-enantiomer of **15** had about 10-fold lower affinity to both subtypes. The histamine homologues homohistamine (**16**) and imbutamine (**17**) were hH_3R and hH_4R agonists with similar affinity to both subtypes (Lim et al., 2005). With extension of the alkyl chain to pentamethylene (impentamine) the agonistic potency at the hH_3R was retained, whereas the agonistic activity at the hH_4R was lost. However, conformational constraints are obviously tolerated. The hH_3R agonist immepip (**18**) (Vollinga et al., 1995), an imidazolylmethylpiperidine, was comparable to imbutamine (**17**) regarding affinity and potency at the hH_4R (Lim et al., 2005).

Imifuramine (**19**), a chiral imidazolyltetrahydrofuranylmethanamine with (2*R*,5*R*)-configuration, was described as a potent full H_3R agonist with 50-fold selectivity for the hH_3R versus the hH_4R and about the same affinity to both subtypes (Hashimoto et al., 2003). The stereomers of **19** were less active at the hH_4R (stereogenic centres numbered according to (Hashimoto et al., 2003), see Figure 2.4). In particular, (2*R*)-configuration proved to be favourable for activity at the hH_4R, whereas both configurations in 2-position were tolerated by the hH_3R. Replacement of the basic amino group by a benzyloxy moiety resulted in compound **20**, an agonist at both hH_3R and hH_4R and devoid of subtype selectivity. In this case, the (2*R*,5*S*)-configuration was most favourable at the hH_4R. Compared to imifuramine, the hH_4R affinity and potency of **20** was slightly increased (Hashimoto et al., 2003). Taken together, these data indicate a binding mode of compound **20** different from that of compound **19**, which has two basic groups.

A benzyl ether instead of a basic group is also characteristic of the hH_3R agonists proxyfan (**21**) and iodoproxyfan (**22**). Both compounds were characterized as partial agonists with submicromolar affinities and potencies at the hH_4R, but 10- to 100-fold selectivity for the hH_3R (Lim et al., 2005). The hH_4R potency of the imidazolylmethylpyridine immethridine (**23**) was one order of magnitude lower, resulting in the highest hH_3R selectivity within this group of compounds (Lim et al., 2005).

2.2.2. Carbamidothioates (Isothioureas)

Imetit (**5**, Figure 2.1), VUF-8328 (**24**, Figure 2.5) and clobenpropit (**10**, Figure 2.1) are characterized by a basic carbamidothioates moiety (pK_a ~ 8.5) in place of the amino group in histamine or homohistamine. The compounds **5**, **10** and **24** were potent hH_4R partial agonists, but showed 10- to 100-fold selectivity for the hH_3R (Lim et al., 2005). Neither the length of the connecting chain (two or three methylene groups), nor a *p*-Cl-benzyl substituent had significant influence on hH_4R affinity and potency. However, clobenpropit was identified as an inverse agonist at the hH_3R and is one of the few compounds that activate the hH_4R, but not the hH_3R.

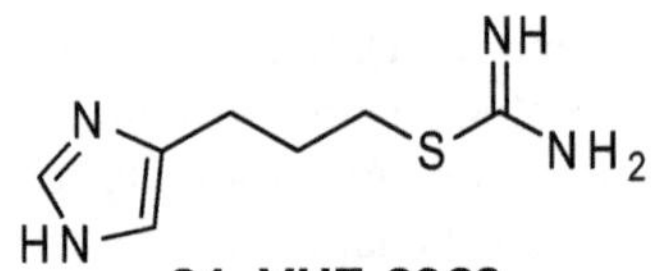

24, VUF-8328
$pK_i = 8.0$

Figure 2.5 Chemical structure of the H_4R partial agonist VUF-8328.

2.2.3. Alkylthioureas

In 1995, imidazolylalkylthioureas were described as potent H_3R antagonists (investigated at the guinea-pig ileum) (Vollinga et al., 1995). This series represents a further example of histamine receptor ligands derived from the endogenous agonist and characterized by a non-basic moiety instead of the amino group. Burimamide (**25**, Figure 2.6), a very weak H_2R antagonist, initially used by Black et al. (1972) to pharmacologically define the H_2R, has considerably higher antagonistic activity at the H_3R (Vollinga et al., 1995). Moreover, **25** and the analogues **26** – **29** proved to be hH_4R partial agonists with affinities and potencies in the two- to three-digit nanomolar range (Lim et al., 2005). Significant selectivity for one of the subtypes, the guinea pig H_3R or the hH_4R, was not observed. A tetramethylene spacer between the imidazole ring and the thiourea moiety turned out to be essential for hH_4R agonistic activity, whereas the size of the alkyl substituent at the thiourea group, methyl (**25**), ethyl (**26**), n-propyl (**27**) or isopropyl (**28**), was of minor relevance. Benzyl (**29**) and phenethyl analogues were much weaker partial hH_4R agonists. With a pentamethylene chain, hH_4R antagonism resulted in the case of an ethyl and inverse agonism in the case of a *p*-chlorobenzyl (**30**) substituent at the thiourea group.

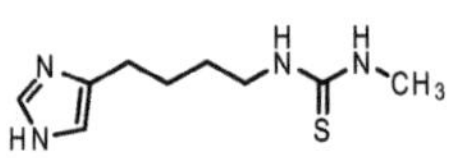

25, burimamide
$pK_i = 7.4$

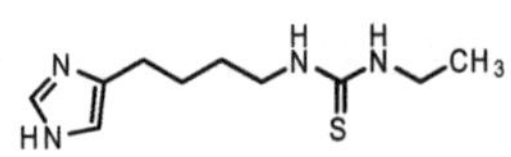

26
$pK_i = 7.6$

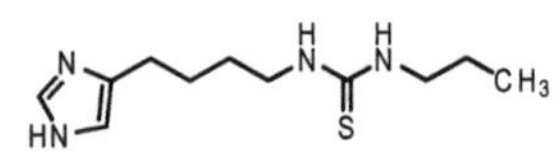

27
$pK_i = 8.0$

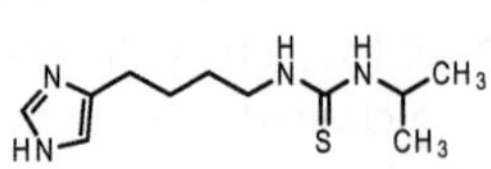

28
$pK_i = 8.1$

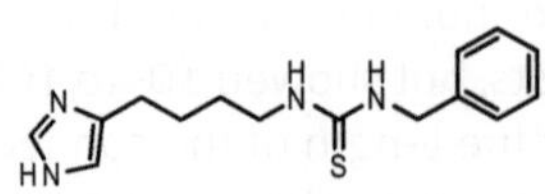

29
$pK_i = 7.3$

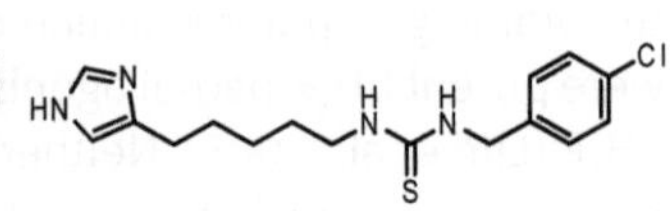

30
$pK_i = 6.9$

Figure 2.6 Influence of different alkyl substituents at the thiourea group on the H_4R affinity.

2.2.4. Alkylguanidines

Strongly basic imidazolylpropylguanidines such as impromidine (**31**, Figure 2.7) and arpromidine-like compounds belong to the most potent H_2R agonists (Durant et al., 1985; Buschauer et al., 1989; Buschauer et al., 1992; Dove et al., 2004). Moreover, many of these compounds showed moderate H_1R antagonistic and potent H_3R antagonistic activities (Dove et al., 2004). At the hH_4R, impromidine proved to be a potent partial agonist (Lim et al., 2005). A special compound to be included in this chapter is VUF-8430 (**32**), combining a guanidine via an ethylene spacer with an isothiourea group, which is capable of forming bidentate hydrogen bonds like imidazole, i.e., both the amine and the imidazole in histamine are replaced by bioisosteric moieties. VUF-8430 was initially designed as H_2R agonist (Sterk et al., 1986), but proved to be a potent hH_4R agonist (Lim et al., 2006; Lim et al., 2009) with ~30-fold selectivity over the hH_3R (full agonism) and negligible affinity for hH_1R and hH_2R (partial agonism). In spite of higher potency at the hH_4R, full agonistic activity of **32** at the hH_3R has to be taken into account in pharmacological studies on the H_4R, unless investigations are performed in models devoid of H_3R. For example, in contrast to the H_4R, the H_3R is not expressed in immune cells such as mast cells or eosinophils (Lippert et al., 2003; Ling et al., 2004; Thurmond et al., 2008).

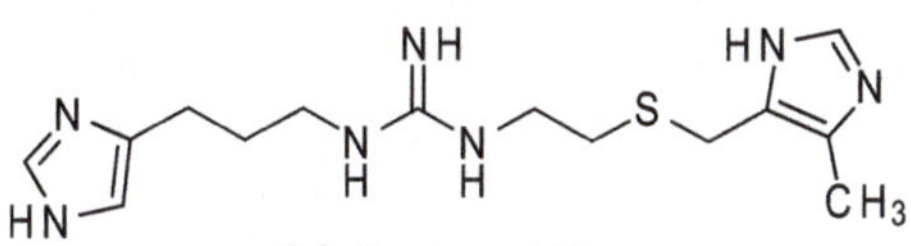

31, **Impromidine**
pKi = 7.6

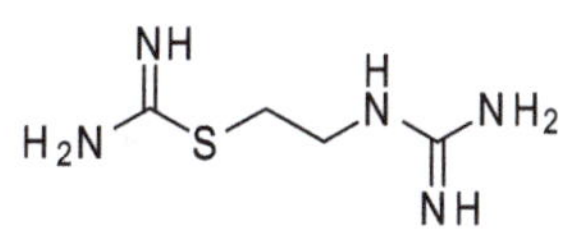

32, **VUF-8430**
pKi = 7.5

Figure 2.7 Impromidine and VUF-8430.

2.2.5. Acylguanidines

N^G-acylated imidazolylpropylguanidines were originally developed as H_2R agonists with strongly reduced basicity in order to improve pharmacokinetic properties (Xie et al., 2006a; Xie et al., 2006b; Ghorai et al., 2008). However, the parent compound, imidazolylpropylguanidine (SK&F-91486, **33**), as well as N^G-alkanoylated analogues such as UR-PI288 (**34**), UR-PI294 (**35**), UR-PI295 (**36**) and UR-PI287 (**37**) were potent hH_4R agonists with pEC_{50} values of 8.1 to 8.4 and with 25- to 300-fold selectivity over the hH_2R (Figure 2.8)(Igel et al., 2009b). At the hH_2R, the potency increased with the chain length, and at the hH_3R the compounds behaved as partial agonists with affinities in the same range as at the hH_4R. The tritiated form of compound **35** ([^{3}H]UR-PI294) proved to be a useful high-affinity radioligand for the study of both, the hH_4R and the hH_3R (Igel et al., 2009c).

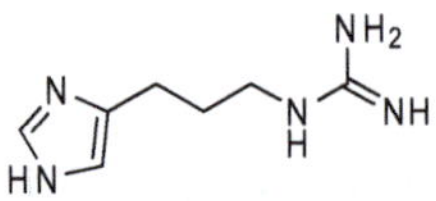

33, SK&F-91486
pEC_{50} = 8.1

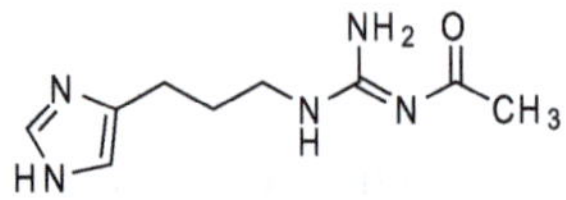

34, UR-PI288
pEC_{50} = 8.3

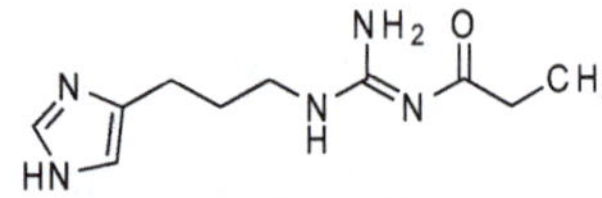

35, UR-PI294
pEC_{50} = 8.5

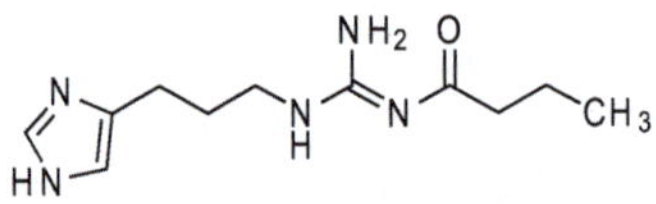

36, UR-PI295
pEC_{50} = 8.6

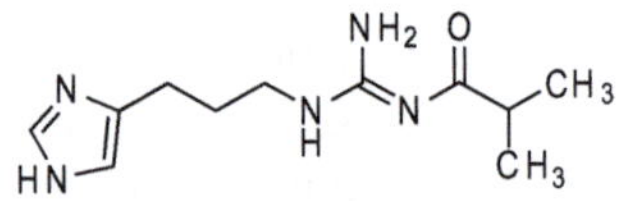

37, UR-PI287
pEC_{50} = 8.4

Figure 2.8 Compounds **33** – **37**.

Introduction of a 3-phenylpropanoyl moiety, irrespective of branching (compounds **38** – **40**, Figure 2.9), slightly reduced intrinsic activity and potency at the hH$_4$R (Ghorai et al., 2008; Igel et al., 2009a; Ghorai et al., 2010). The diphenylpropanoylguanidine **40**, which is reminiscent of diarylalkylguanidine-type H$_2$R agonists (Buschauer et al., 1992), showed about ten-fold selectivity for the hH$_4$R compared to the hH$_2$R and was a neutral antagonist at the hH$_3$R. Bioisosteric replacement of the imidazole ring by an aminothiazole moiety resulted in highly selective agonists for the hH$_2$R (Birnkammer et al., 2012; Kraus et al., 2009). The lack of selectivity over the hH$_3$R is still the major drawback of the most potent acylguanidine-type hH$_4$R agonists. Nevertheless, due to greater than 1000- and 100-fold selectivity over the hH$_1$R and hH$_2$R, respectively, N^G-alkanoyl-substituted imidazolylpropylguanidines including [^{3}H]-labeled version of **35** are applicable to pharmacological experiments on the H$_4$R in native or recombinant systems devoid of the hH$_3$R.

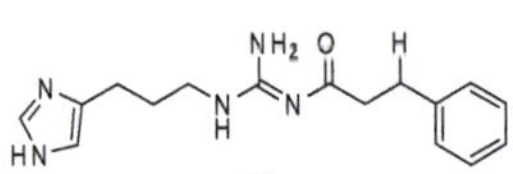

38
-log(EC_{50}) = 7.8

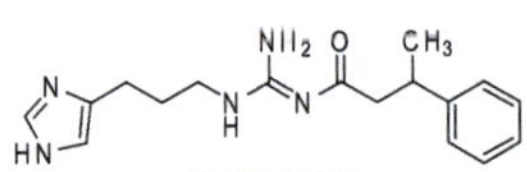

39, UR-AK24
-log(EC_{50}) = 7.8

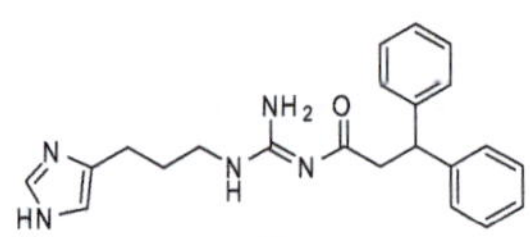

40
-log(EC_{50}) = 8.1

Figure 2.9 Compounds **38** - **40**.

As an example of the putative binding mode of N^G-acylated imidazolylpropylguanidines, UR-PI294 (**35**) in complex with the hH$_4$R is shown in Figure 2.8. The imidazole ring is docked in a similar position as in the case of histamine (Figure 2.3), i.e., bidentate hydrogen bonds of the nitrogen atoms with T178$^{5.42}$ and E182$^{5.46}$ are formed. The acylated guanidino group mimics the amino group of histamine. Protonation causes a salt bridge with D94$^{3.32}$, but also the neutral base may interact with D94$^{3.32}$ via charge-assisted hydrogen bond(s). The binding site readily enables to accommodate the trimethylene spacer between Y95[3.33], F169 (ECL2), Y319[6.51] and F344[7.39]. The terminal propionyl group may interact with an affinity-conferring pocket mainly consisting of hydrophobic amino acids (Y72[2.61], W90[3.28] and W348[7.43]). Unfortunately, the identity of hH_3R and hH_4R residues in these three positions does not provide patterns of hH_3R- and hH_4R-selective interactions.

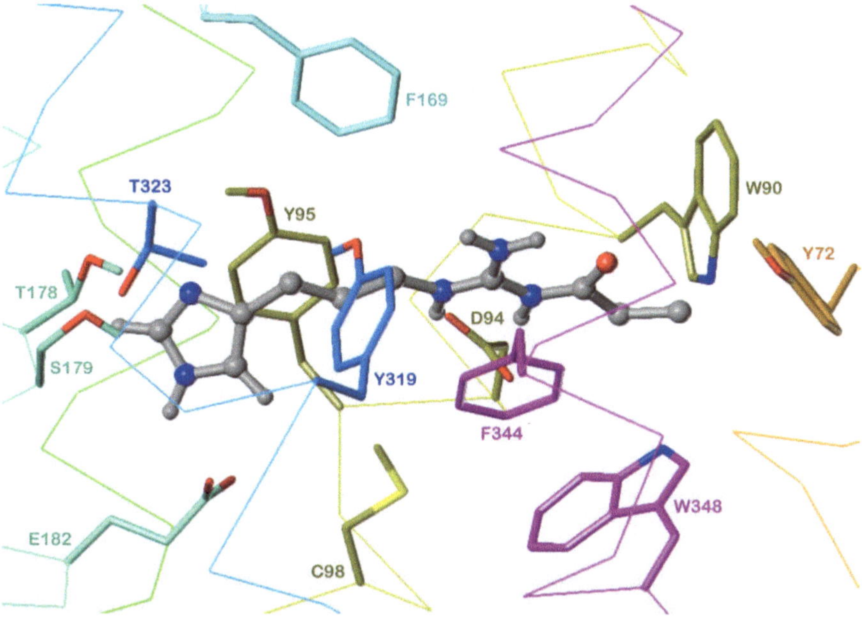

Figure 2.10 Putative binding mode of UR-PI294 (**35**) at the hH_4R. The backbone (Cα trace, lines), C and some essential H atoms of side chains are drawn in spectral colours: TM2, orange; TM3, yellow; TM4, green; TM5, green-blue; TM6, blue; TM7, magenta; nitrogen, blue; oxygen, red; sulfur, yellow. UR-PI294 (C and essential H atoms, grey) is drawn as ball-and-stick model. For source of the model, software and hardware, see Figure 2.3.

2.2.6. Cyanoguanidines

As outlined above, the major drawback of N^G-acylated imidazolylpropylguanidine-type hH_4R agonists is their hH_2R and hH_3R (partial) agonism. With small alkanoyl residues (**33** - **37**) the selectivity for the hH_4R versus the hH_2R was substantially higher, however, there was still some agonistic activity at the hH_2R. Obviously, a second basic group, in addition to the imidazole ring, is essential for H_2R but not H_4R agonism. Differentiation between agonism and antagonism at the H_2R is achieved through replacement of the strongly basic guanidine in impromidine (**31**) by a non-basic cyanoguanidine group. Whereas agonism turns into antagonism at the H_2R (Durant et al., 1985), a cyanoguanidine as well as a thiourea moiety as in compounds **25** - **28** is compatible with agonism at the H_4R. Therefore, the acylguanidine moiety, which is charged under physiological conditions, has been systematically replaced by cyanoguanidine to improve hH_4R selectivity (Igel et al., 2009a).

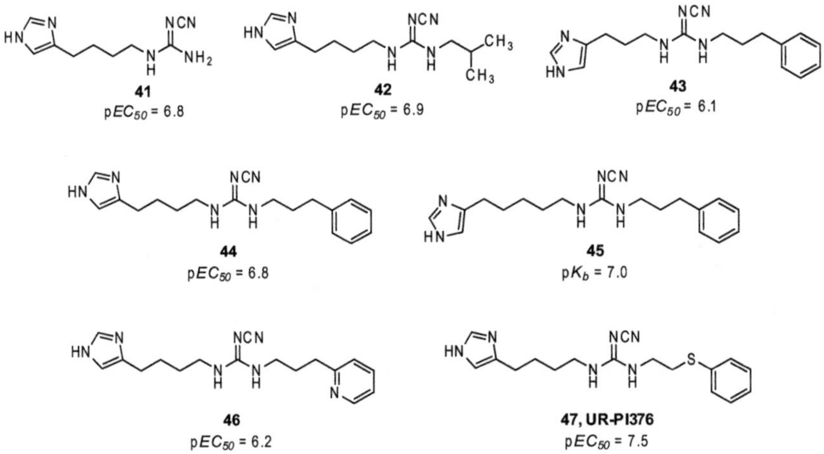

Figure 2.11 Cyanoguanidines **41** -**47**.

Small N^G-substituents were found to improve hH_4R selectivity over the hH_2R in the acylguanidine series (**33-37**) (Igel et al., 2009b). This was not the case for cyanoguanidines (Igel et al., 2009a). The unsubstituted parent compound **41** was a rather potent hH_4R agonist devoid of selectivity compared to the hH_3R. Smaller alkyl groups at the cyanoguanidine group reduced potency at both subtypes. With an isobutyl substituent (**42**), hH_4R agonistic potency was similar to that of compound **41**, whereas at the hH_3R weak inverse agonism became apparent.

Therefore, a N^G-substituent of appropriate size seems to be required for hH_4R vs. hH_3R selectivity. The cyanoguanidine **43** (Igel et al., 2009a), a moderate hH_4R partial agonist (50-fold less potent than its N^G-acylated guanidine analogue **46**), was only a weak hH_2R partial agonist and a moderate hH_3R inverse agonist. The lower homologue of **43**, bearing an imidazolylethyl substituent at the cyanoguanidine moiety, was almost inactive at the hH_4R, whereas the extension of the chain length (**44**) substantially improved hH_4R agonist potency (5-fold) and intrinsic activity (α, 0.9). Upon further elongation of the chain (**45**), hH_4R affinity was retained but the quality of action changed from agonism to inverse agonism. In contrast to the N^G-acylated imidazolylpropylguanidines (e.g. **40**), a bulky diphenylpropyl or phenylbutyl N^G-substituent was unfavourable (decreased hH_4R affinity, inverse agonism or neutral antagonism, respectively). Thus, acylguanidine- and cyanoguanidine-type hH_4R ligands probably prefer different binding modes. The pyridyl analogues of compound **44** (e.g., the 2-pyridyl derivative **46**) had lower hH_4R potency and intrinsic activity. At the hH_3R, all these aralkyl-substituted cyanoguanidines were moderate inverse agonists.

In terms of both hH_4R agonistic potency and selectivity, the replacement of the benzylic methylene group by a sulfur atom (UR-PI376, **47**) turned out to be a key modification. UR-PI376 was the most potent and selective hH_4R agonist in the cyanoguanidine series (Igel et al., 2009a). The selectivity for the hH_4R over the hH_3R was about 30-fold, and the affinities for the hH_1R and the hH_2R were negligible. Compared to the acylguanidine **38**, the potencies at the hH_2R and the hH_3R were reduced by factors of greater than 100 and approximately 400, respectively, whereas hH_4R agonistic potency was retained. In contrast to other selective hH_4R agonists such as 5(4)-methylhistamine (**13**) or VUF-8430 (**32**), UR-PI376 (**47**) did not activate human histamine receptor subtypes other than hH_4R in GTPase assays. Recently, compound **47** was successfully used to study the activation (chemotaxis and calcium response) of human eosinophils via the H_4R (Reher et al., 2012). However, preliminary investigations revealed that compound **47** – like many other hH_4R agonists – is substantially less potent at the mH_4R. Therefore, compound **47** is an interesting tool for pharmacological experiments with human H_4Rs and not suitable for investigations on rodent H_4Rs.

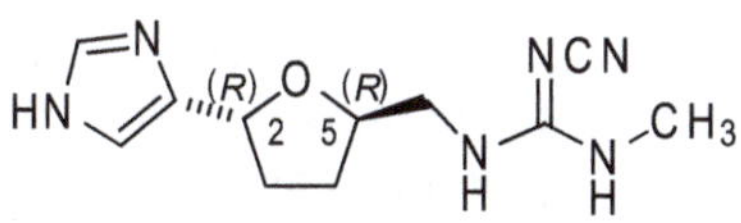

48, OUP-16

$pK_i = 6.9$

Figure 2.12 Chemical structure of OUP-16, a full H_4R agonist.

Also in the imifuramine series, introduction of a non-basic cyanoguanidine moiety instead of the primary amino group substantially reduced hH_3R potency, but increased potency and intrinsic activity of most diastereomers at the hH_4R. In particular, the cyanoguanidine analogue of imifuramine, the (2*R*,5*R*)-configured isomer OUP-16 (**48**, Figure 2.12) (Hashimoto et al., 2003), proved to be a potent full hH_4R agonist with 40-fold selectivity over the hH_3R.

Many hH_4R ligands have been identified among imidazole derivatives originally developed as agonists or antagonists of the earlier discovered histamine receptor subtypes H_1R, H_2R and H_3R. Since the imidazolyl moiety is a structural determinant of the agonistic activity of histamine, most of these compounds are full or partial agonists, too. The key issue is to achieve selectivity over the hH_3R and, to a certain extent, also over the hH_2R. In some structural classes (e.g., acylguanidines and cyanoguanidines), receptor subtype selectivity and quality of action are very sensitive toward minor structural changes. In the light of published data for imidazole-type ligands, the most promising agonists for investigations on the hH_4R are the cyanoguanidines UR-PI376 (**47**) and OUP-16 (**48**), due to moderate to high hH_4R potency and selectivity. UR-PI376 showed no agonistic activity at other human histamine receptor subtypes in GTPase assays.

The value of imidazole derivatives as pharmacological tools for analysis of the (patho)physiological role of the H_4R in translational animal models may be limited. Depending on the individual compound and the pharmacological model, pronounced discrepancies regarding quality of action, affinities, potencies and subtype selectivities at histamine receptor species orthologues were reported (Strasser et al., 2013). In particular, numerous hH_4R agonists turned out to possess considerably lower affinity and potency at rodent H_4Rs. Further investigations of the molecular determinants of receptor subtype and orthologue selectivity are necessary in order to develop novel imidazole-type H_4R ligands which are better suited as pharmacological tools for *in vitro* and *in vivo* studies.

2.3. Benzimidazole Derivatives

The benzimidazole system is a crucial structural feature of a wide range of bioactive compounds, including, e. g., antiparasitics, anticonvulsants, analgesics, antihistaminics, antiulcers, antihypertensives, antiviral, anticancers, proton pump inhibitors, antifungals and anti-inflammatatory agents (McKellar & Scott, 1990; Rossignol & Maisonneuve, 1984; Patil et al., 2008; Dubey & Sanyal, 2010; Boiani & González, 2005; Narasimhan et al., 2012).

2.3.1. Benzimidazoles Related to JNJ-7777120

The first potent H_4R antagonist containing the benzimidazole moiety was deduced from JNJ-7777120 (**11**) by exchanging the indole moiety resulting in VUF-6002 (**49**), also known as JNJ-10191584 (K_i, 27 nM) (Figure 2.13). Both JNJ-7777120 and VUF-6002 are useful tools for the characterization of the H_4R *in vitro* and *in vivo* because of their similar activity at human, rat and mouse H_4R (Jablonowski et al., 2003; Ling et al., 2004; Thurmond, 2004; Varga et al., 2005; Dunford et al., 2006; Dunford et al., 2007). JNJ-7777120 and VUF-6002 exhibit good anti-inflammatory and antinociceptive effects in paw edema and hyperalgesia models. However, another related benzimidazole urea derivative, VUF-6007, does not show any such activity (Coruzzi et al., 2007). Although the overall oral bioavailability of JNJ-7777120 and VUF-6002 in rats is similar (22%

and 27%, respectively), the half-life ($t_{1/2}$) of VUF-6002 is only 1 hr versus that of 3 hr for JNJ-7777120 (Thurmond et al., 2004; Thurmond et al., 2008). Thus, VUF-6002 has even poorer pharmacokinetic properties than JNJ-7777120.

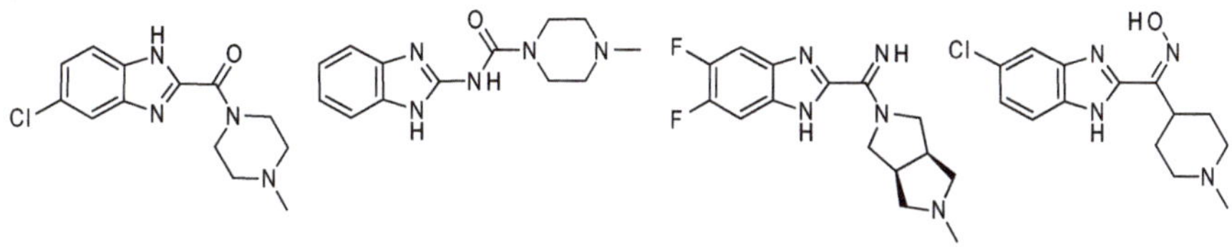

49, VUF-6002 pK_i = 7.6 | **50, VUF-6007** pK_i = 4.7 | **51** pK_i = 8.0 | **52** pK_i = 7.8

Figure 2.13 Benzimidazole analogs of JNJ-7777120.

Pfizer Ltd. succeeded in optimization of VUF-6002 by introducing an amidine instead of an amide moiety and bioisosteric replacement of the *N*-methylpiperazine group with 2-methyloctahydropyrrolo[3,4-c]pyrrole moiety. The amidine **51** is a potent antagonist with improved metabolic stability in human and rat liver microsomes. The amidine function is more polar, resulting in lower oral bioavailability compared to the amide, but due to a larger volume of distribution compound **51** has as longer half-life ($t_{1/2}$= 12 h). Unfortunately, the Pfizer Ltd. compound produced serious adverse effects: dose-dependent lymphoid depletion from spleen, thymus and gut associated lymphoid tissues, decreased reticulocyte counts and decreased erythropoesis. These toxic side effects were not related to the H_4R antagonism but linked to the structure of the compound (Mowbray et al., 2011; Lane et al., 2012). In another approach J&J used VUF-6002 as a starting point and identified compound **52**, the (*Z*)-configured oxime derivative of the corresponding (benzimidazol-2-yl)(1-methylpiperidin-4-yl)ketone, as a ligand for the H_4R of human, mouse, rat, guinea pig, and monkey with excellent selectivity over other histamine receptors as well as many other off-targets (Yu et al., 2010). Exchanging the amide by an oxime leads to agonists with comparable potency at human and rodent H_4R. The combination of high affinity and selectivity make these compounds excellent new tools for further exploring and understanding H_4R pharmacology (Thurmond & Wadsworth, 2009).

In conclusion, benzimidazoles derived from JNJ-7777120 are highly selective H_4R ligands, but further lead optimization is needed to eliminate unfavourable pharmacokinetic and toxic effects. The discovery of potent and selective oxime-type agonists with benzimidazole and indole core structure may pave a way to overcome the drawbacks of imidazole derivatives that are more closely related to histamine, i.e., insufficient selectivity with respect to H_xR subtypes and H_4R species orthologues.

2.3.2. 2-Phenylbenzimidazoles

2-Arylbenzimidazoles having a phenyl group in 2-position were published as high-affinity hH_4R ligands by Lee-Dutra et al. (2006). The initial lead, compound **53** (K_i, 124 nM), was identified as a hit from high-throughput screening of a corporate compound collection. Compound **53** displays similar structural features as VUF-6002 (**49**), both with a benzimidazole and a piperazine group, which is protonated under physiological conditions. However, the distance between these two moieties is significantly longer. In this series a distance of 8.3 Å from the aryl ring to the distal nitrogen atom of the terminal piperazine appeared to be optimal. Modification of the substituents on the aryl ring and on the benzimidazole nucleus led to significantly increased affinities (**54** and **55**).

The terminal basic moiety has major influence on the affinity: Replacement of the *N*-methylpiperazine by *N*-methylhomopiperazine led to compound **59** with a K_i value of 1 nM, being 100-fold lower than that of the initial hit compound **53** (Lee-Dutra et al., 2006). Structure-activity relationships around the H_4R ligand **54** demonstrated that five-membered and open-chain isosteres of the piperazine moiety are unfavourable for binding affinity in this series of benzimidazoles (**56** – **58**, Figure 2.14), whereas increasing the ring size to an *N*-methylhomopiperazine results in a modest increase in activity (**59**). However, the terminal amino moiety has also major influence on the functional activity; incorporation of the constrained histamine analogue spinaceamine and of histamine itself resulted in the antagonist **60** and the very potent agonist **61**, respectively. The conformational restriction of the diamine fragment by a bicyclic system turns a potent agonist into an antagonist. In the context of section 2.2, compound **61** may be regarded as a H_4R ligand combining two putative pharmacophoric entities, a 2-arylbenzimidazole and an imidazolylethyl moiety. Intriguingly, **61** has sub-nanomolar hH_4R affinity and is ranked among the most potent hH_4R agonists described so far (Savall, 2010). Modification of the imidazolylethyl moiety reduced potency and intrinsic activity. Moreover, **61** is reported to possess more than 600- and 1,700-fold selectivity for the hH_4R over the hH_2R (pK_i, 6.9) and the hH_3R (pK_i, 6.4), respectively. The activity at the hH_1R was weak (pK_i < 5). Like many other H_4R agonists, **61** was considerably (greater than 50-fold) less potent at the mH_4R (pEC_{50}, 7.4, α, 0.8) (Figure 2.14; Savall et al., 2010).

J&J disclosed several 2-arylbenzimidazole derivatives in at least four patent applications. Among different linkers between the 2-phenylbenzimidazole moiety and the protonated amine function, propoxy/propylamine or butoxy/butylamine chains were optimal in the *para*- and *meta*-positions, whereas the *ortho*-analogs were inactive.

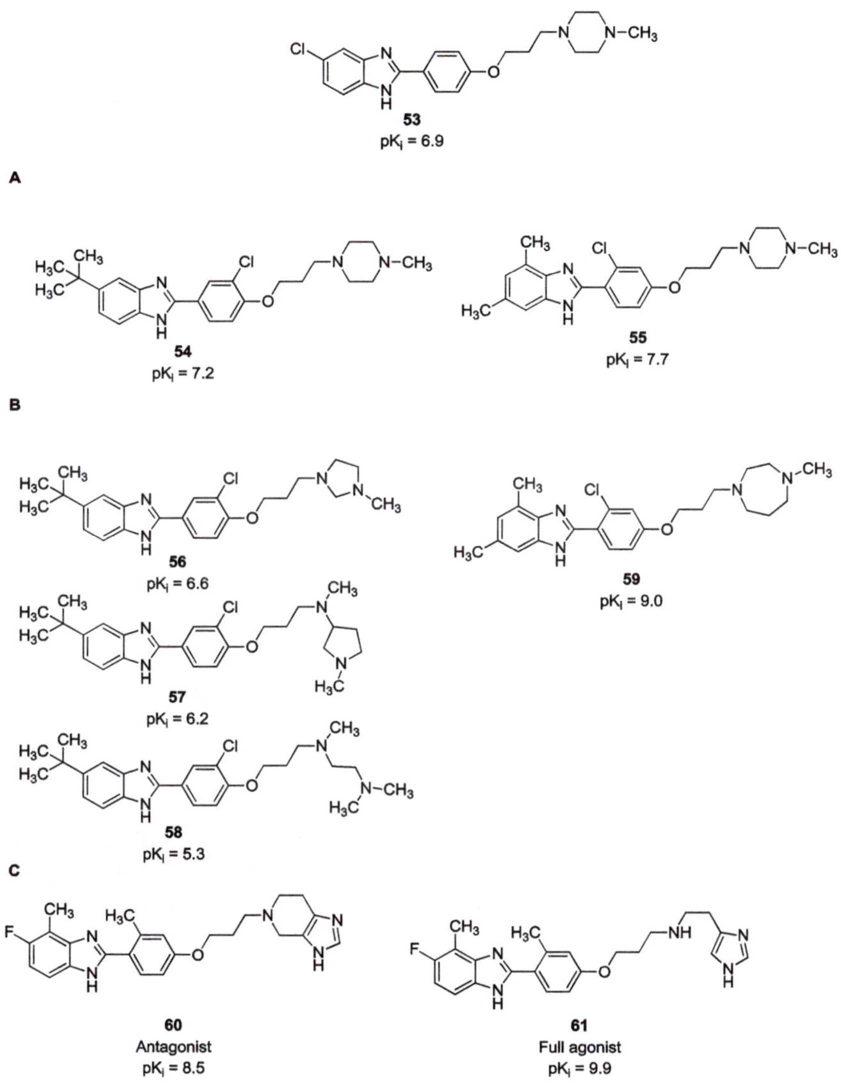

Figure 2.14 Structural variations on 2-arylbenzimidazoles have major influence on H_4R affinity and activity. **A**: Influence of the different substituents on the aryl- and benzimidazole-ring. **B**: Influence of the terminal amino moiety. **C**: Influence of conformational constraint on activity.

2.3.3. Heteroaryl Substituted Benzimidazoles and Related Compounds

The phenyl ring of the class of 2-arylbenzimidazoles was also replaced by heteroaryl rings, e.g., pyrazine or pyrimidine. Replacement of the phenyl ring with pyrimidine afforded a series of H$_4$R antagonists, e.g. compound **62** (Figure 2.15), which proved to have anti-inflammatory effects in *in vitro* models (Thurmond and Wadsworth, 2009). The benzimidazolylpyrazine derivatives (e.g. **65**) showed much lower potency compared with the pyrimidine analogue (**64**). (Arienti et al., 2005; Edwards et al., 2007a; Kindrachuk & Venable, 2011; Daley-Yates et al., 2012).

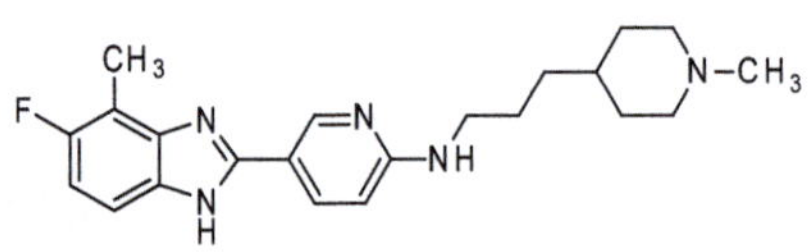

62
pK$_i$ = 7.5

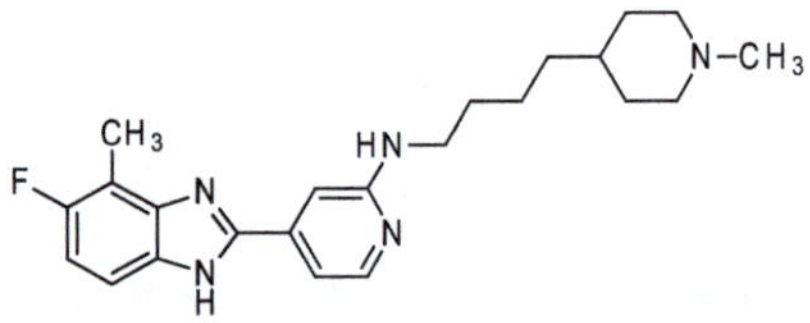

63
pK$_i$ = 8.7

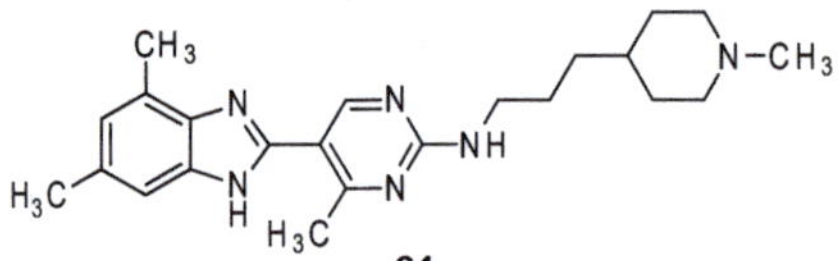

64
pK$_i$ = 8.0

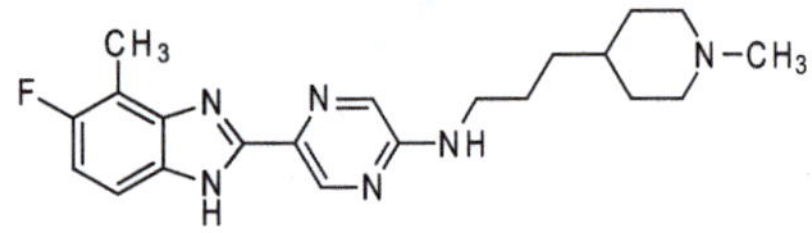

65
pK$_i$ = 6.6

Figure 2.15 Benzimidazol-2-yl-pyridine, -pyrazine and -pyrimidine derivatives.

With the preparation of bicyclic heteroaryl-substituted imidazole compounds, J&J enlarged their library of benzimidazoles and achieved H$_4$R affinities in the subnanomolar range (**66**, Figure 2.16). Pharmacokinetic parameters of these flexible compounds have not been reported yet.

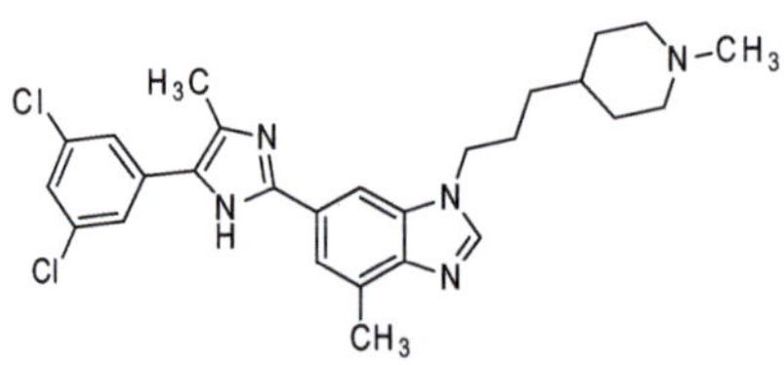

66
pK$_i$ = 9.5

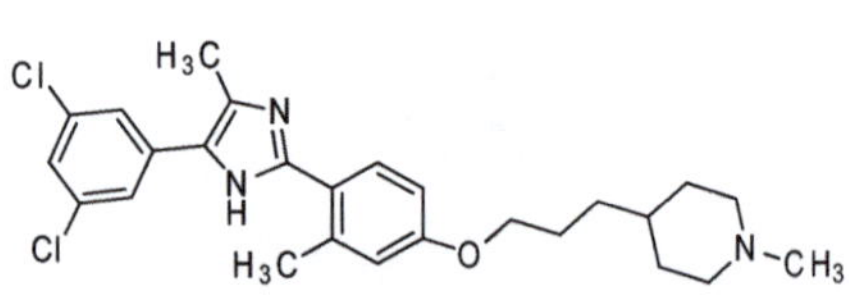

67
pK$_i$ = 7.3

Figure 2.16 Imidazoles **66** and **67**.

Compound **66** is an example of an imidazole-containing compound structurally related to the benzimidazol-2-yl-pyridine class. However, in this case the benzimidazole replaces the pyridine moiety, and the "original" benzimidazole is substituted by a phenylimidazolyl substructure, leading to a slight decrease in H_4R affinity (Buzard et al., 2005).

2.3.4. Benzo-annulated Azole Derivatives

Bioprojet recently published a series of novel benzo-annulated azoles as potent H_4R ligands. The patent application implicates not only benzimidazoles but also benzothiazoles with K_i values below 30 nM. Examples of active compounds (**68**-**70**) are shown in Figure 2.17.

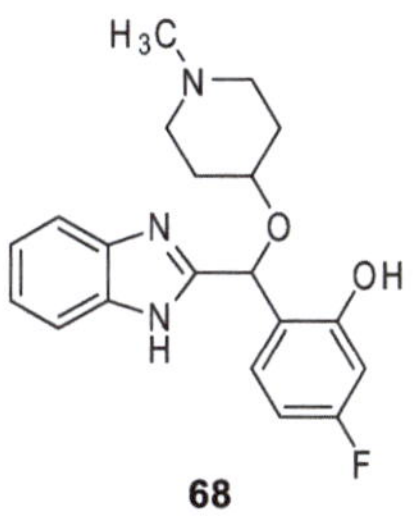

68 $pK_i > 7.5$

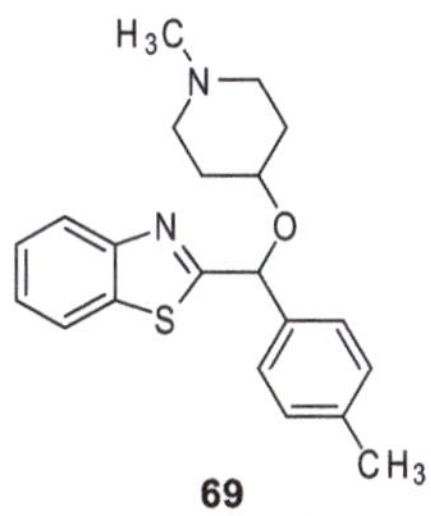

69 $pK_i > 7.5$

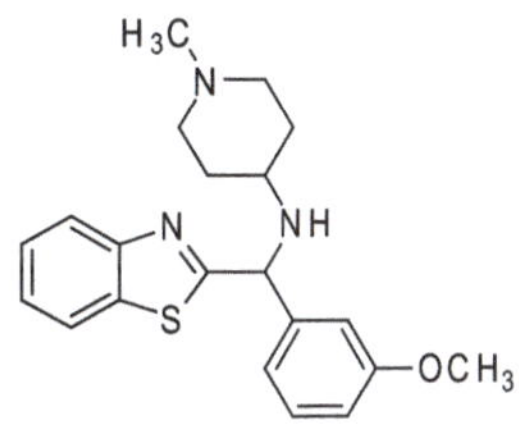

70 $pK_i > 7.5$

Figure 2.17 Benzo-annulated azoles as H_4R ligands.

In the aforementioned benzimidazoles from J&J, the benzimidazole, the aryl group and the protonated piperazine function are in a row whereas Bioprojet showed that the H_4R also tolerates a dendritic orientation around a tertiary carbon atom. The distance between the aryl group and the distal nitrogen is comparable to that in the linear 2-arylbenzimidazoles (**53** - **61**). Figure 2.17 shows only an excerpt from the prepared and pharmacologically characterized compounds. The phenyl ring can be substituted with various moieties. The structural constitution of this series of benzo-annulated azoles shows a high similarity to the established hH_1R agonist histaprodifen (Figure 2.1). As K_i values are not included in the patent application, it is not possible to estimate the contribution of the particular substituents on H_4R affinity (Berrebi-Bertrand et al., 2012).

2.4. Quinoxalines and Quinazolines

Kalypsys Inc. fused an imidazole ring with a quinoxaline resulting in a new class of H_4R ligands. In this series of H_4R antagonists, the protonated piperazine is in 4-position of the imidazo[1,2-a]quinoxaline. Interestingly the substituents on the aryl moiety, R^1 and R^2, have large influence on H_4R and H_1R antagonism (Table 1, Borchardt et al., 2010).

Table 2.1
H_1R and H_4R antagonism influenced by substituents.

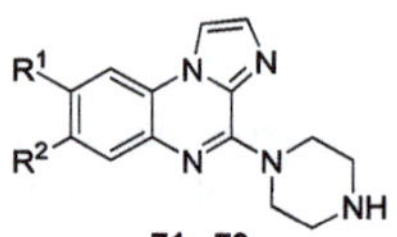

71 - 73

Compound	R^1	R^2	H_1 Antagonist[a]	H_4 Antagonist[a]
71	H	CF_3	-	-
72	F	F	-	+
73	Cl	F	+	+

[a] + $EC_{50} \leq 10$ µM; - $EC_{50} \geq 10$ µM

The first quinoxalines with H_4R affinity were already published in 2005 by J&J (Edwards & Venable, 2005). Smits et al. (2008) used an *in silico* flexible alignment model of a known H_4R agonist (VUF-6884) and an antagonist (JNJ-7777120) and designed a series of potent quinoxaline derivatives (e.g. **74** and **75**, Figure 2.18). Compounds **74** and **75** were also evaluated in vivo and displayed significant anti-inflammatory activity in the carrageenan-induced paw edema model in rats (Smits et al., 2008). After proving the quinazoline heterocycle to be a suitable scaffold for H_4R affinity, sulfonamide moieties were introduced. In the course of this work, several compounds were discovered with excellent affinity for the H_4R in the low nanomolar concentration range. Combinations with other pharmacophoric moieties, aiming at hybrid compounds or compounds addressing multiple targets seem rational. *In vivo*, compound **76**, administered in rat, showed significant reduction of inflammation caused by injection of carrageenan in the paw. Further modifications of the quinazoline heterocycle with substituent elongation were also performed by Yuuki & Tsutomu (2011). Replacement of the quinazoline by a benzoxazine moiety led to potent H_4R ligands with affinities in the nanomolar concentration range (Yuuki & Tsutomu, 2011).

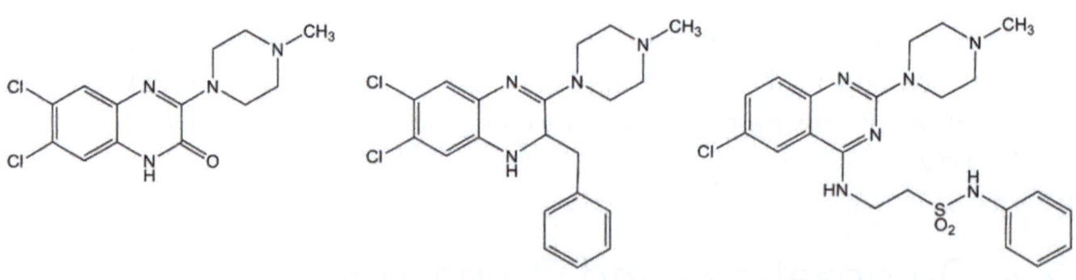

74 $pK_i = 8.3$ | **75** $pK_i = 7.4$ | **76** pKi = 8.3

Figure 2.18 Quinoxaline, quinazoline and benzoxazine based H_4R antagonists.

2.5. Six-membered Nitrogen Containing Heterocycles

Derivatives of six-membered nitrogen-containing heterocycles have become an important class of potent and selective H_4R antagonists/inverse agonists. Among the described structures, compounds with a central pyrimidine scaffold are the most important ones. The pyrimidine scaffold confirms the general pharmacophoric pattern of H_4R ligands (Figure 2.19). As described in section 2.1, the compounds contain a central core (e.g. pyrimidine), a basic group (a saturated nitrogen heterocycle), and a lipophilic moiety (various substitutents are well tolerated e.g. alkyl, aryl, cycloalkyl, cyclo(hetero)aryl or substituted amine).

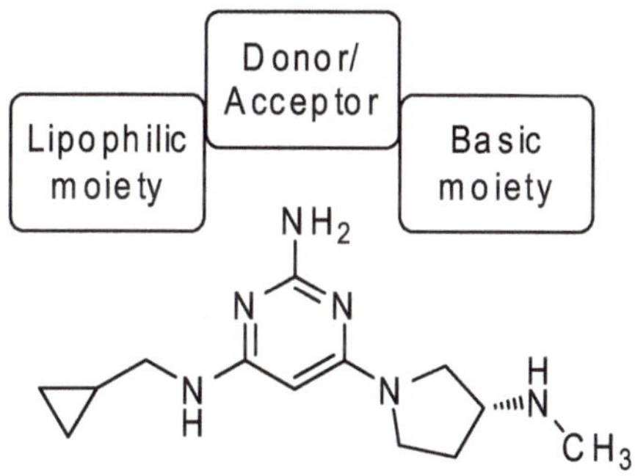

77, ZPL-38937887

Figure 2.19 Structural blueprint for H_4R ligands exemplified by the clinical candidate ZPL-38937887.

So far, only compounds with a central pyrimidine ring (non-fused or annulated) have entered into clinical trials. Promising results from the clinical studies (phase I) for the first clinical H_4R antagonist, UR-63325 (undisclosed structure, but claimed in patent applications from Palau Pharma on mono or fused pyrimidines), were reported [http://www.palaupharma.com]. For another compound, ZPL-38937887 (formerly PF-03893787; **77**; Figure 2.19), clinical phase I studies revealed excellent safety and pharmacokinetic profile [http://www.ziarcopharma.com].

2.5.1. Mono-, di- and triaminopyridines

Three kinds of pyridine derivatives were described as H_4R antagonists: mono-, di- and triaminopyridines.

J&J disclosed a series of *aminopyridines* in a patent application, where all ligands contain *N*-methylpiperazine linked to a heterocycle and a phenyl, biphenyl or heteroaryl substituent (mostly a substituted thiophene), (Gaul et al., 2009). The position of the basic amine relative to the nitrogen in the pyridine ring was also varied (compound **78-81**, Figure 2.20). Compounds with a 2-aryl-4-(4-methylpiperazin-1-yl)pyridine core, as exemplified by compound **82** (Figure 2.20), were most promising.

The potencies mostly depended on the kind of substituent attached to the (hetero)aryl moiety. Thienyl and biphenyl derivatives seemed to be preferred by the H_4R. The most potent compounds in this series are shown in Figure 2.20 (**82** - **84**).

Diaminopyridine ligands were covered by a patent application from J&J (Cai et al., 2009). For example, compound **85** (Figure 2.20) was reported to have a K_i value of 0.9 nM (in radioligand binding assay, hH_4R). UCB Pharma has also designed diaminopyridines, mostly with a 3-methylaminopyrrolidine moiety as a basic core (Hannah et al., 2008). Representative of these structures is **86** (Figure 2.20, K_i = 200-450 nM, EC_{50} = 75-250 nM).

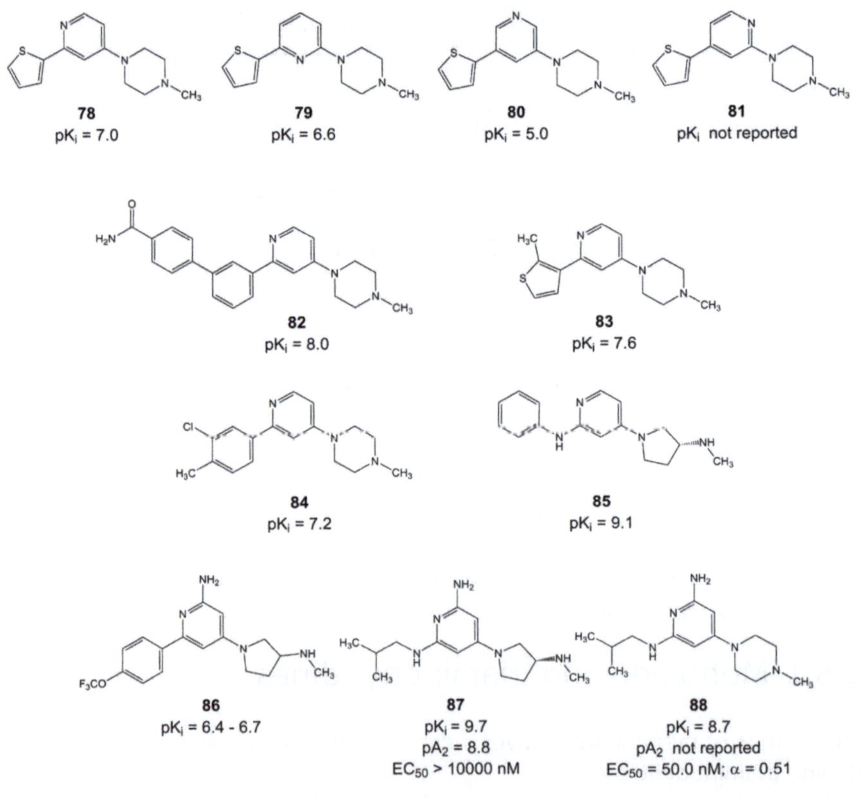

Figure 2.20 Structures of selected aminopyridines.

A structure-activity relationships (SAR) study of *triaminopyridines* has been published (Meduna et al., 2011). The described compounds contain *N*-methylpiperazine or (*R*)-3-methylaminopyrrolidine on the pyridine core, a free amine group and a differently substituted secondary or a tertiary amine. Compounds with substituted secondary amines and a 3-methylaminopyrrolidine ring showed higher antagonistic activities, whereas analogues bearing a 4-methylpiperazin-1-yl substituent behaved as partial agonists (see compounds **87** vs. **88**).

2.5.2. Pyridazine Derivatives

Pyridazines are rarely described as H_4R ligands (Cai et al., 2009). Only a few of the reported diaminopyridazines showed high affinity with hH_4R K_i values <20 nM. The affinity strongly depended on the kind of basic amine (most potent were compounds with an (*R*)-3-methylaminopyrrolidine moiety) and the substituent on the secondary amine (bulky substituents were well tolerated). Representative structures are compounds **89** and **90** (Figure 2.21).

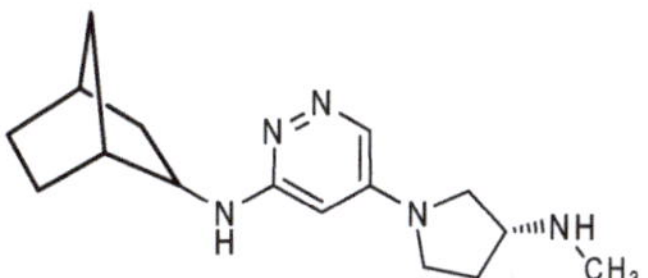

89
$pK_i = 9.0$

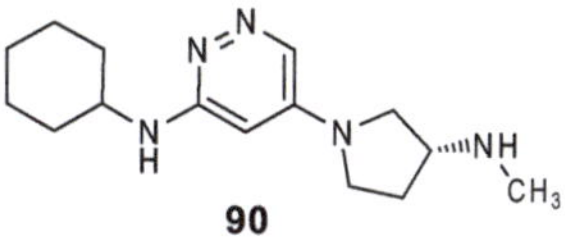

90
$pK_i = 8.3$

Figure 2.21 Structures of selected pyridazines.

2.5.3. Pyrimidine Derivatives

Bayer Healthcare AG disclosed pyrimidine derivatives in two patent applications (Sato et al., 2005a, 2005b). More recently, other groups have also dealt with the class of substituted pyrimidines. Meanwhile, some promising H_4R antagonists have been developed. Pyrimidine derivatives can be divided into non-fused pyrimidines and fused pyrimidines (bicyclic, tricyclic). As the discussion of all these structural variations would be beyond the scope of this review, in the following we focus on non-fused pyrimidines, which can be subdivided into three groups: aminopyrimidines, diaminopyrimidines (2,4-disubstituted and 4,6-disubstituted) and triaminopyrimidines. In most cases, the pyrimidine ring is substituted at 2-, 4- and/or 6-position. Four-fold substitution is rarely described. An additional substituent in 5-position on diamino derivatives may be tolerated.

2.5.3.1. 2,4-Diaminopyrimidines

Diaminopyrimidines devoid of an additional substituent at the heterocycle were, besides related pyridine and pyrazine derivatives, included in a patent application by J&J (Cai et al., 2009). They described a series of compounds with a branched or unbranched alkylamino, a cycloalkylamino or (rare among the most potent H_4R antagonists) an arylalkylamino substituent in 2- position of the pyrimidine ring (e.g. **91-94**, Figure 2.22).

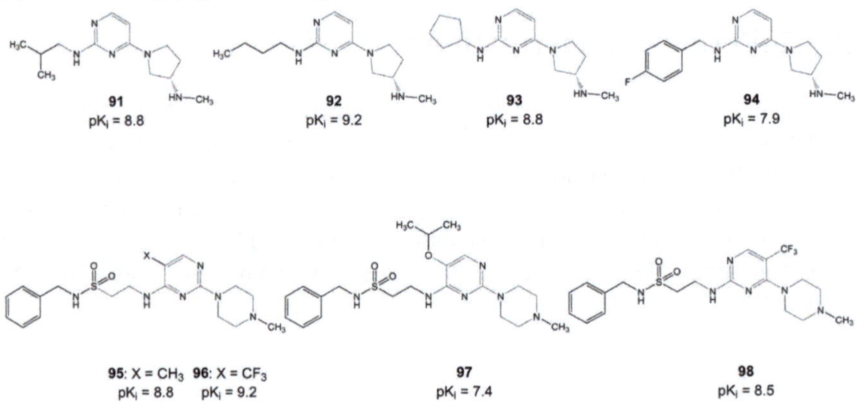

Figure 2.22 Structures of selected 2,4-diaminopyrimidines.

Shionogi & Co. introduced analogues of 2,4-diaminopyrimidines with an additional substituent in 5-position of the pyrimidine moiety (Masuda et al., 2011). The potency depended on the kind of substitutent at this position e.g. **95 – 98** (Figure 2.22).

Compounds bearing a (substituted) aryl moiety in the 6-position, disclosed in patent applications from Bayer Healthcare AG (Sato et al., 2005a, 2005b), were among the first diaminopyrimidine-type H_4R antagonists. The substituent on the aryl ring (mainly phenyl) was in the *meta-* or *para-*positions, and potency depended on the kind of substituent (**99 – 103**, Figure 2.23). Compounds were tested in calcium mobilization assay on CHO cells expressing the H_4R (no exact data; only grouped affinity categories) and also in three other tests (binding assay, migration assay and zymosan-induced pleurisy in mice).

Abbott investigated diaminopyrimidine derivatives in HTS. Upon optimization, compound **103** (Figure 2.23) was identified as a H_4R antagonist with K_i values of 35.0 nM at the hH_4R and 275.0 nM at the rH_4R (Altenbach et al., 2008a, 2008b). The oral bioavailability (*F*) of **103** was 31% and 41% in rats and mice, respectively. Compound **103** showed analgesic effects in various animal models of pain (zymosan peritonitis model in mice, mouse itch model, carrageenan-induced thermal hyperalgesia model).

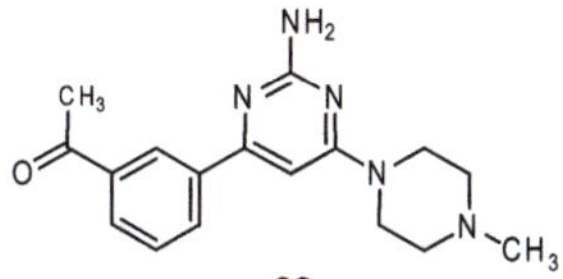

99
$pIC_{50} > 7.7$

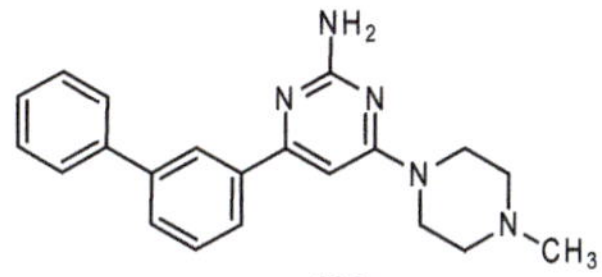

100
$pIC_{50} > 7.0$

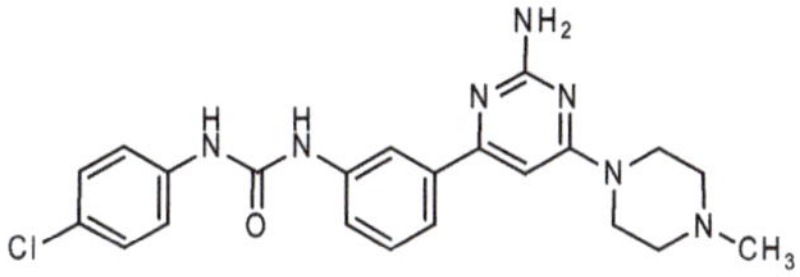

101
$pIC_{50} > 6.3$

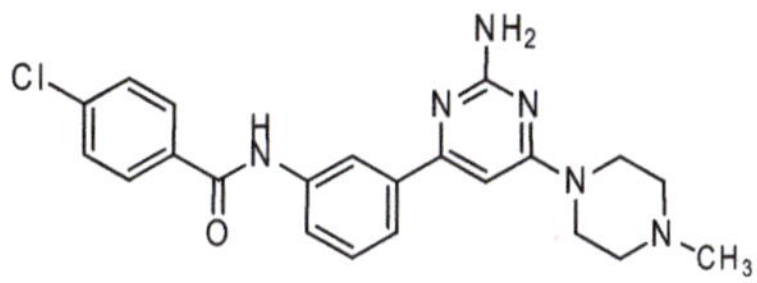

102
$pIC_{50} < 6.3$

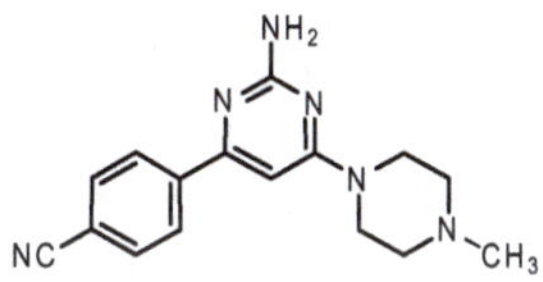

103
$pK_i = 7.5$

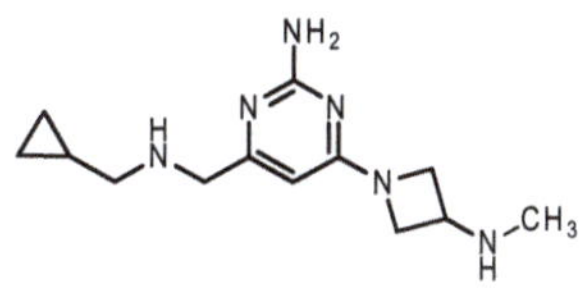

104
> 50 % of binding to hH_4R (1 μM)

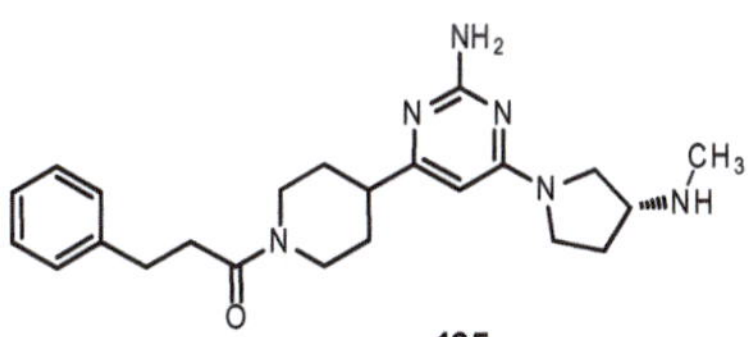

105
> 50 % of binding to hH_4R (1 μM)

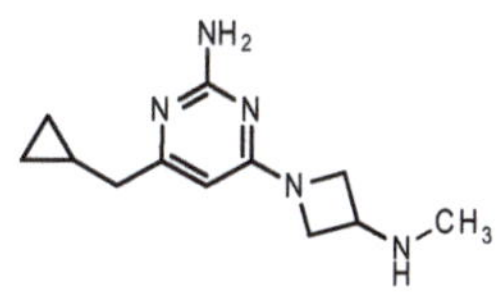

106
> 50 % of binding to hH_4R (1 μM)

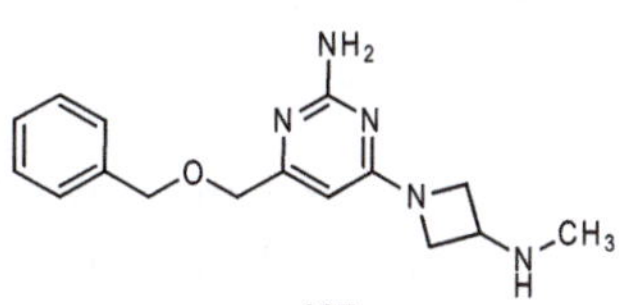

107
> 50 % of binding to hH_4R (1 μM)

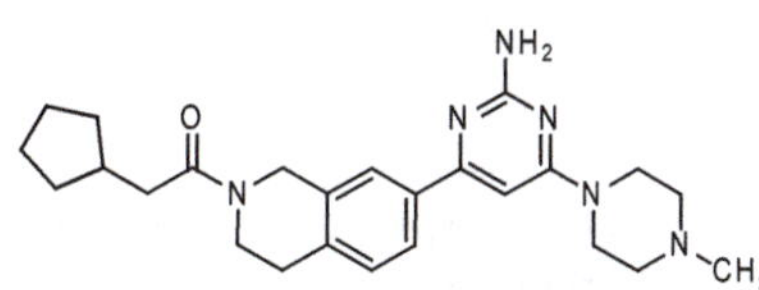

108, INCB38579
$pK_d = 7.8$

Figure 2.23 Structures of selected 2,4-diaminopyrimidines.

Palau Pharma filed several patent applications covering 2,4,6-trisubstituted pyrimidines with branched or unbranched alkyl, cycloalkyl and alkylamino substitutents as exemplified by **104** - **107** (Figure 2.23; Carceller Gonzalez et al., 2009a, 2009b, 2010, 2011). All compounds showed an inhibition of more than 50% of binding to hH_4R at a concentration of 1 μM (no exact data).

6-(1,2,3,4-Tetrahydroisoquinolin-7-yl)pyrimidine-2,4-diamines were described by Incyte Corp. (Zhou et al., 2010; Shin et al., 2012). Compound **108**

(INCB38579; Figure 2.23) is a potent H_4R antagonist with a K_d of 16 nM and at least 80-fold selectivity over the other histamine receptor subtypes. The compound showed inhibitory activity *in vivo* in pain models (carrageenan-induced acute inflammatory pain in rat and formalin-induced pain in rats and mice) and good pharmacokinetic properties in rats and mice. The substance blocked histamine-induced cell shape change, inhibited the migration of human eosinophils as well as the migration of human and mouse dendritic cells and showed antipruritic effects in histamine-induced itch in mice.

2.5.3.2. 4,6-Diaminopyrimidines

Based on high throughput screening (HTS) and further optimization focusing on ligand efficiency (LE; commonly defined as the ratio of free energy of binding to the number of heavy atoms; Masood et al., 2011), Pfizer Ltd. identified compound **109** (Figure 2.24) with an LE value of 0.48, a K_i of 82.0 nM at the hH_4R and selectivity over the other histamine receptor subtypes (hH_3R K_i, 646 nM; hH_2R K_i, 3.6 μM; hH_1R K_i > 10 μM). Furthermore, compound **109** exhibited satisfying metabolic stability in the human liver microsomal assay (HLM: 16 μL min^{-1} mg^{-1}) and moderate stability in the rat liver microsomal assay (RLM: 147 μL/min/mg). Further optimization of this compound resulted in PF03893787 (**77**, cf. section **6.5.4**).

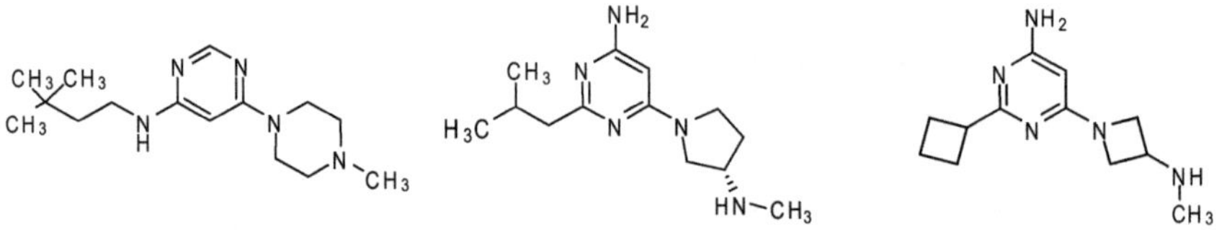

109
pK_i = 7.1

110
> 50% of binding to hH_4R (1 μM)

111
> 50% of binding to hH_4R (1 μM)

Figure 2.24 Selected structures of 4,6-diaminopyrimidines.

Pfizer Ltd. also prepared a series of diaminopyrimidines with an alkyl (branched or unbranched) or a cycloalkyl substituent in 6-position of the pyrimidine scaffold (e.g. **110**, **111**; Figure 2.24) (Carceller Gonzalez et al., 2009c). Compounds tested in binding assay revealed more than 50% of inhibition at 1 μM.

2.5.4. 2,4,6-Triaminopyrimidines

Palau Pharma (Carceller Gonzalez et al., 2007) and Pfizer Ltd. (Bell et al., 2007) published the triaminopyrimidines in early patent applications. Palau Pharma attached the substituted phenyl groups *via* an amine linker to the pyrimidine moiety, e.g., **112** (Figure 2.25). Selected compounds tested at 1 μM in the binding assay ([³H]histamine) at human H_4R displayed at least 50% inhibition (no detailed data, Carceller Gonzalez et al., 2007).

Triaminopyrimidines claimed by Pfizer Ltd. were very closely related to those from Palau Pharma (Bell et al., 2007). The main difference concerns aliphatic substituents instead of aromatic ones e.g. **113**, **77** (ZPL-38937887 formerly PF03893787, Figure 2.25). Some of these compounds were reported to bind to the H_4R in the nanomolar concentration range, to exhibit high selectivity and to possess favorable pharmacokinetic properties (Mowbray et al., 2011). The hH_4R antagonist **77** was extensively investigated in various *in vitro* and *in vivo* assays (e.g. binding and functional assays using recombinant H_4R from various species; assessment of selectivity, genetic toxicity risk, cardiovascular risk, general toxicity). Aiming at a new drug for the treatment of asthma, the compound has completed phase I of clinical studies with single ascending dose and 14 days multiple ascending dose in healthy volunteers (A study to access,

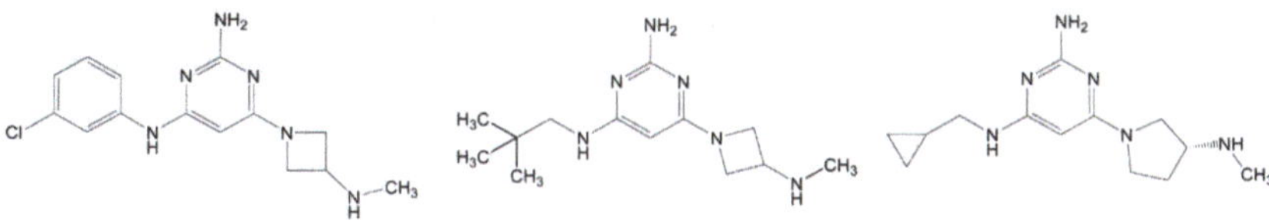

112 >50% of binding to hH_4R (1 μM); **113** pK_i = 8.6; **77, ZPL-38937887** pK_i = 8.8

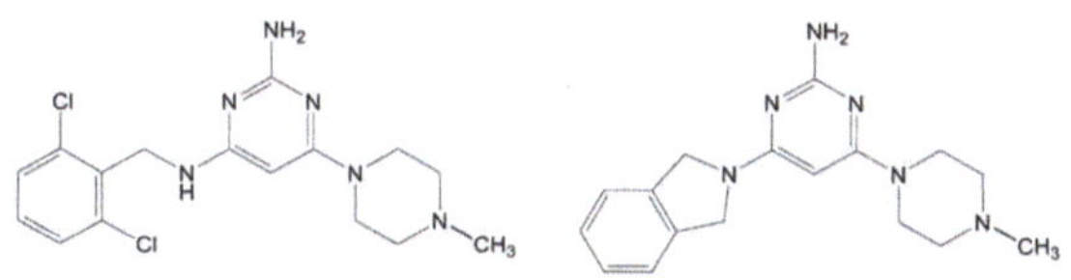

114, ST-1006 pK_i = 7.9; **115, ST-1012** pK_i = 6.6

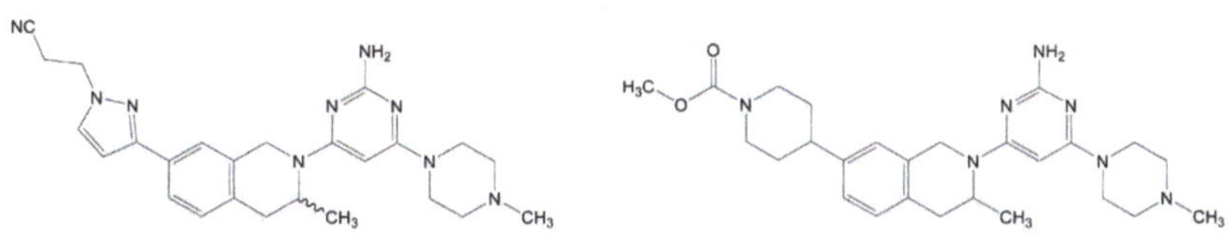

116, enantiomer 1 pIC_{50} = 7.6; **117**, enantiomer 2 pIC_{50} = 6.5; **118** pIC_{50} = 7.9

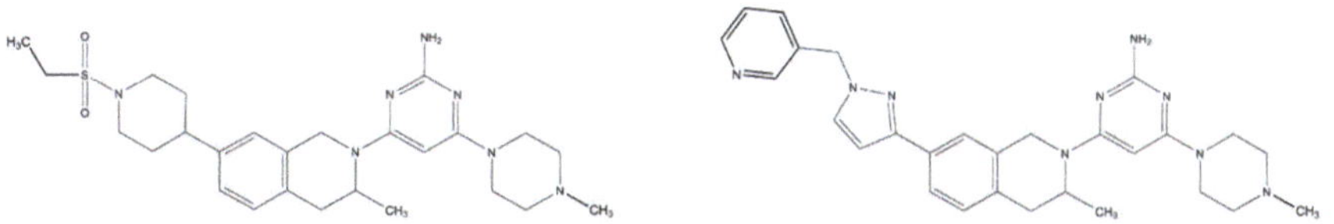

119 pIC_{50} = 7.8; **120** pIC_{50} = 8.1

Figure 2.25 Structures of selected triaminopyrimidines.

2010; A phase 1 study, 2010). Excellent safety and pharmacokinetic profile of this compound at once daily low dose dosing was reported. Now ZPL-38937887 (formerly PF-03893787) is a lead compound of Ziarco Pharma (http://www.ziarcopharma.com)

Structures related to those from Palau Pharma were also found by Sander et al. (2009, 2010) by ligand-based virtual screening. Introduced modifications in a ranked list of compounds led to the identification of the triaminopyrimidine scaffold as a lead structure. Further structural changes and structure-activity relationship studies revealed that different substitution patterns in the benzyl moiety of a molecule cause differences in intrinsic activities and potencies. Derivatives with *ortho*- and *para*-substituents behaved as partial agonists (e.g. **114**, ST-1006; Figure 2.25) whereas *meta*-substituted and rigidified analogues showed inverse agonism (e.g. **115**, ST-1012; Figure 2.25). Modification of a pseudo ionic lock may be responsible for the differences in activity (Tanrikulu et al., 2009; Werner et al., 2010)

An interesting series of triaminopyrimidines from Incyte corp. was filed for patent (Zhang et al., 2010). The common features of this group of compounds were a 2-amino group, a 4-methylpiperazin-1-yl residue and a substituted tetrahydroisoquinolin-2-yl moiety at the 6-position. Methyl substitution in 3-position of the tetrahydroisoquinoline ring introduced a chiral centre. The difference in activity between enantiomers, the eudismic ratio, was up to 14 (see compounds **116** and **117**, Figure 2.25), however, the configuration of the enantiomers was not described. The nature of the substituents at the 7 position of tetrahydroisoquinoline had a crucial influence on the activity of the evaluated compounds. Examples of the most potent H_4R antagonists (with *IC50* values below 20 nM) in this series are **118** – **120** (Figure 2.25).

2.5.5. Triazine Derivatives

J&J described 2-amino-1,3,5-triazine derivatives as H_4R antagonists (Gaul et al., 2009). Some compounds from this series were particularly potent, e.g. **121** – **123** with K_i values below 20 nM. As reported by Łażewska & Kieć-Kononowicz (2012), the introduction of an amino group in 6-position results in a decrease in affinity by more than one order of magnitude (see **124**, Figure 2.26).

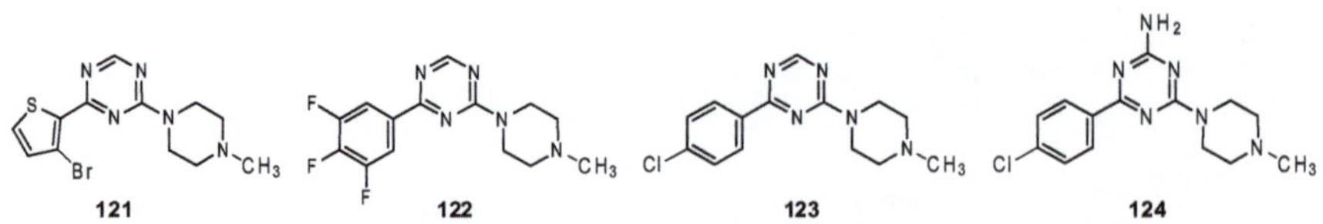

121 pK_i = 8.7 | **122** pK_i = 8.2 | **123** pK_i = 8.0 | **124** pK_i = 6.7

Figure 2.26 Structures of selected triazines.

Recent years have brought progress in the search for potent and selective histamine H_4R antagonists/inverse agonists. Six-membered *N*-heterocycles have become important scaffolds and, among all structures described in the literature (articles and patent applications), pyrimidine derivatives hold most promise. Analysis of the SAR of pyrimidine-type H_4R antagonists confirmed the previously suggested general construction pattern for *N*-heterocycles, which can be extended to (benz)imidazole-type H_4R ligands (Łażewska & Kieć-Kononowicz, 2012, Figure 2.27). Crucial pharmacophoric elements are: a heterocyclic centre (mostly pyrimidine), a basic moiety, and a lipophilic substituent (diverse substitutents are well tolerated e.g. alkyl, aryl, cycloalkyl, cyclo(hetero)aryl or substituted amine etc). Many basic saturated nitrogen heterocycles (mono- or fused, e.g. fused pyrrolidinopiperazine) have been explored; compounds with 3-(methylamino)azetidine, (3*R*)-3-(methylamino) pyrrolidine and 4-methypiperazine substructures proved to be most potent. In most cases a second basic moiety (e.g. $-NH_2$) increased H_4R affinity.

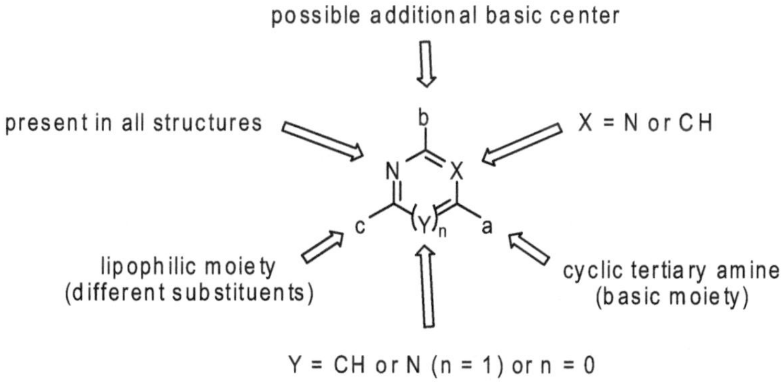

Figure 2.27 General structural features of H_4R antagonists with six-membered *N*-heterocycle (n = 1) or (benz)imidazole (n = 0) core.

2.6. Conclusion

Since the discovery and cloning of the hH_4R at the turn of the millennium, numerous selective H_4R agonists and antagonists have been reported. Among the potent hH_4R antagonists, (benz)imidazoles and six-membered *N*-heterocycles are the major and most promising scaffolds. The number of reported H_4R modulators in clinical trials is reasonably low; so far only compounds with a pyrimidine moiety (non-fused or fused) have entered into clinical trials. Promising results from phase I studies with the first clinical candidate, UR-63325 (structure undisclosed but included in patent applications from Palau Pharma) were reported [http://www.palaupharma.com]. Another compound, ZPL-38937887 (formerly PF-03893787; **77**; Figure 2.19) has completed clinical studies (phase 1) with excellent safety and pharmacokinetic profile [http://www.ziarcopharma.com]. The structure of the phase II candidate JNJ-39758979 on itch and asthma is also undisclosed. Yet another investigational antagonist of the H_4R is UR-65318 from Palau Pharma (structure undisclosed). Current preclinical data on the potential therapeutic value in the treatment of dermatological diseases suggest this compound as a promising candidate for further development. INCB38579 (**108**; Figure 2.23) from Incyte Corp., also revealed encouraging preclinical data and can be used as another tool for future pharmacological characterization of the H_4R. (Shin et al., 2012).

In light of published data for (benz)imidazole-type ligands, the most promising agonists for investigations on the hH_4R are the cyanoguanidines UR-PI376 (**47**) and OUP-16 (**48**), due to moderate to high hH_4R potency and selectivity. However, the value of imidazole derivatives as pharmacological tools for analysis of the (patho)physiological role of the H_4R in translational animal models may be limited. The discovery of potent and selective oxime-type agonists with benzimidazole and indole core structure may pave a way to overcome the H_4R limitations of those imidazoles that are more closely related to histamine. The fact that the disclosed H_4R ligands are derived from only a few scaffolds underlines the urgent need for novel scaffolds that can be used in the development of potent H_4R modulators.

Acknowledgements

The work has been supported by the Else Kröner-Fresenius Foundation (EKFS), Research Training Group Translational Research Innovation – Pharma (TRIP), by the Fraunhofer-Institut für Molekularbiologie und Angewandte Ökologie IME, Projektgruppe Translationale Medizin und Pharmakologie TMP (Fh-TMP), by the Graduate Training Programme (Graduiertenkolleg) GRK 760, "Medicinal Chemistry: Molecular Recognition - Ligand-Receptor Interactions" of the Deutsche Forschungsgemeinschaft and the COST Actions BM0806 (H_4R network), BM1007 and CM1103 of the European Union.

References

Altenbach, R.J., Adair, R.M., Bettencourt B.M et al. (2008a). Structure-activity studies on a series of 2-aminopyrimidine-containing histamine H_4 receptor ligands. *J Med Chem, 51*(20), 6571-6580.

Altenbach, R.J., Lui, H., Drizin, I. et al. (2008b). A method for pain treatment. (Abbott Lab). WO2008060766.

A study to assess the effect of PF-03893787 on lung function following an allergen challenge in asthmatic subjects. (COMPLETED). NCT00856687; 21 January 2010. Available from: www.clinicaltrials.gov [Last accessed 10 December 2012].

A phase 1 study to evaluate the safety and tolerability of different doses of PF-03893787 in healthy adult volunteers. (COMPLETED). NCT00992342; 17 March 2010. Available from: www.clinicaltrials.gov [Last accessed 10 December 2012].

Arienti, K. L., Breitenbucher, J. G., Buzard, D. J. et al. (2005). Benzoimidazole compounds. (Janssen Pharmaceutica). US20050070550.

Bakker, R. A., Schoonus, S. B., Smit, M. J. et al. (2001). Histamine H_1 receptor activation of nuclear factor-kappa B: roles for G beta gamma- and G alpha(q/11)-subunits in constitutive and agonist-mediated signaling. *Mol Pharmacol, 60,* 1133–1142.

Ballesteros, J. A., Weinstein, H. (1995). Integrated methods for the construction of three-dimensional models and computational probing of structure-function relations in G protein-coupled receptors. *Methods Neurosci, 25*, 366-428.

Bell, X.S., Lane, C.A.L., Mowbrary, C.E. et al. (2007). Pyrimidine derivatives. (Pfizer Limited). WO2007072163.

Berrebi-Bertrand, I., Billot, X., Calmels, T. et al. (2012). Benzazoles derivatives as histamine H_4 receptor ligands. Bioprojet. EP2447263A1.

Birnkammer, T., Spickenreither, A., Brunskole, I. et al. (2012). The bivalent ligand approach leads to highly potent and selective acylguanidine-type histamine H_2 receptor agonists. *J Med Chem, 55*(3), 1147-1160.

Black, J. W., Duncan, W. A., Durant, C. J. et al. (1972). Definition and antagonism of histamine H_2 receptors. *Nature, 236*(5347), 385-390.

Boiani, M., González, M. (2005). Imidazole and benzimidazole derivatives as chemotherapeutic agents. *Mini-Rev Med Chem, 5*, 409–424.

Buschauer, A. (1989). Synthesis and in vitro pharmacology of arpromidine and related phenyl(pyridylalkyl)guanidines, a potential new class of positive inotropic drugs. *J Med Chem, 32*(8), 1963-1970.

Buschauer, A., Friese-Kimmel, A., Baumann, G. et al. (1992). Synthesis and histamine H_2 agonistic activity of arpromidine analogues: replacement of the pheniramine-like moiety by non-heterocyclic groups. *Eur J Med Chem, 27*(4), 321-330.

Buzard, D. J., Edwards, J. P., Kindrachuk, D. E. et al. (2005). Imidazole compounds. (Janssen Pharmaceutica). US20050261309A1.

Cai, H., Chavez, F., Dunford, P.J. et al (2009). Diaminopyridine, pyrimidine, and pyridazine modulators of the histamine H_4 receptor. (Janssen Pharmaceutica). WO2009152325.

Carceller Gonzalez, E., Virgili Bernado, M., Soliva Soliva, R. et al. (2011). Aminoalkylpyrimidine derivatives as histamine H_4 receptor antagonists. (Palau Pharma). WO2011076878.

Carceller Gonzalez, E., Medina Fuentes, E.M., Morato Garcia, A. et al. (2010). Piperidin-pyrimidine derivatives as antagonists of histamine H_4 receptor. (Palau Pharma). WO2010094721.

Carceller Gonzalez, E., Medina Fuentes, E.M., Morti Via, J. et al. (2009a). 2-Aminopyrimidine derivatives as histamine H_4 antagonists. (Palau Pharma) WO2009068512.

Carceller Gonzalez, E., Vigili Bernadò, M., Marti Via, J. et al. (2009b). 2-Aminopyrimidine derivatives as histamine H_4 antagonists. (Palau Pharma) WO2009077608.

Carceller Gonzalez, E., Medina Fuentes, E.M., Soliva Solivia, A. et al. (2009c). 4-Aminopyrimidine derivatives. (Palau Pharma) WO2009080721.

Carceller Gonzalez, E., Salas Solana, J., Soliva Soliva, R. et al.(2007). 2-Aminopyrimidine derivatives as modulators of the histamine activity. (Palau Pharma) WO2007031529.

Cesco-Cancia, S., Grimm, J.S., Neelakandla, S.M. et al. (2012). Process for the preparation of benzoimidazol-2-yl pyrimidine derivatives. (Janssen Pharmaceutica). US20120184740.

Chazot, P. L. (2009). Advances in histamine pharmacology reveal new drug targets. *Br J Pharmacol, 157*(1), 1-3.

Cherezov, V., Rosenbaum, D. M., Hanson, M. A. et al. (2007). High-resolution crystal structure of an engineered human β_2-adrenergic G protein-coupled receptor. *Science, 318*(5854), 1258-1265.

Coruzzi, G., Adami, M., Guaita, E. et al. (2007). Antiinflammatory and antinociceptive effects of the selective histamine H_4 receptor antagonists JNJ7777120 and VUF6002 in a rat model of carrageenan-induced acute inflammation. *Eur J Pharmacol, 563*, 240–244.

Daley-Yates, P., Ambery, C., Sweeney, L. et al. (2012). The efficacy and tolerability of two novel H_1/H_3 receptor antagonists in seasonal allergic rhinitis. *Int Arch Allergy Immunol, 158*, 84–98.

Deml, K.-F., Beermann, S., Neumann, D. et al. (2009). Interactions of histamine H_1 receptor agonists and antagonists with the human histamine H_4-receptor. *Mol Pharmacol, 76*(5), 1019-1030.

Desai, P., Thurmond, R. L. (2011). Histamine H_4 receptor activation enhances LPS-induced IL-6 production in mast cells via ERK and PI3K activation. *Eur J Immunol, 41*(6), 1764-1773.

Dove, S., Elz, S., Seifert, R. et al. (2004). Structure-activity relationsships of histamine H_2 receptor ligands. *Mini-Rev Med Chem, 4*, 941-954.

Dubey, A. K., Sanyal, P. K. (2010). Benzimidazoles in a wormy world. *Online Veterinary Journal 2010, 5, 63.*, 63.

Dunford, P. J., O'Donnell, N., Riley, J. P. et al. (2006). The histamine H_4 receptor mediates allergic airway inflammation by regulating the activation of CD4+ T cells. *J Immunol,176*, 7062–7070.

Dunford, P. J., Williams, K. N., Desai, P. J. et al. (2007). Histamine H_4 receptor antagonists are superior to traditional antihistamines in the attenuation of experimental pruritus. *J Allergy Clin Immunol, 119*, 176–183.

Durant, G. J., Ganellin, C. R., Hills, D. W. et al. (1985). The histamine H_2 receptor agonist impromidine: synthesis and structure-activity considerations. *J Med Chem, 28*(10), 1414-1422.

Durant, G. J., Ganellin, C. R., Parsons, M. E. (1975). Chemical differentiation of histamine H_1- and H_2-receptor agonists. *J Med Chem, 18*(9), 905-909.

Edwards, J. P., Kindrachuk, D. E., Venable, J. D. (2007a). Benzoimidazol-2-yl-pyridines as modulators of the histamine H_4 receptor. (Janssen Pharmaceutica). WO2007117400.

Edwards, J.P., Kindrachuduk, D.E., Venable, J.D. (2007b). Benzoimidazol-2-yl pyrimidines and pyrazines as modulators of the histamine H_4 receptor. (Janssen Pharmaceutica). WO2007117399.

Edwards, J. P., Venable, J. D. (2005). Quinoxazoline compounds. (Janssen Pharmaceutica). WO 2005033088.

Gaul, M.D., Zhao, B.-P., Zhu, X. (2009). Substituted nitrogen-containing heteroaryl derivatives useful as modulators of the histamine H_4 receptor. (Janssen Pharmaceutica). WO2009035671.

Ghorai, P., Kraus, A., Birnkammer, T. et al. (2010). Chiral N^G-acylated hetarylpropylguanidine-type histamine H_2 receptor agonists do not show significant stereoselectivity. *Bioorg Med Chem Lett, 20*(10), 3173-3176.

Ghorai, P., Kraus, A., Keller, M. et al. (2008). Acylguanidines as bioisosteres of guanidines: N^G-acylated imidazolylpropylguanidines, a new class of histamine H_2 receptor agonists. *J Med Chem, 51*(22), 7193-7204.

Hannah, D., Watson, R.J., Owen, D. (2008). Novel pyridine derivatives, process for preparing them, pharmaceutical compositions thereof. (UCB Pharma): WO2008122378.

http://www.palaupharma.com [last accessed 14 December 2012]

http://www.ziarcopharma.com [Last accessed 14 December 2012]

Hashimoto, T., Harusawa, S., Araki, L. et al. (2003). A selective human H_4-receptor agonist: (-)-2-cyano-1-methyl-3-{(2*R*,5*R*)-5-[1*H*-imidazol-4(5)-yl] tetrahydrofuran-2-yl}methylguanidine. *J Med Chem, 46*(14), 3162-3165.

Hill, S. J., Ganellin, C. R., Timmerman, H. et al. (1997). International union of pharmacology. XIII. Classification of histamine receptors. *Pharmacol Rev, 49*(3), 253-278.

Igel, P., Dove, S., Buschauer, A. (2010). Histamine H_4 receptor agonists. *Bioorg Med Chem Lett, 20*(24), 7191-7199.

Igel, P., Geyer, R., Strasser, A. et al. (2009a). Synthesis and structure-activity relationships of cyanoguanidine-type and structurally related histamine H_4 receptor agonists. *J Med Chem, 52*(20), 6297-6313.

Igel, P., Schneider, E., Schnell, D. et al. (2009b). N^G-Acylated imidazolylpropylguanidines as potent histamine H_4 receptor agonists: selectivity by variation of the N^G-substituent. *J Med Chem, 52*, 2623-2627.

Igel, P., Schnell, D., Bernhardt, G. et al. (2009c). Tritium-labeled N^1-[3-(1*H*-imidazol-4-yl)propyl]-N^2-propionylguanidine ([^{3}H]UR-PI294), a high-affinity histamine H_3 and H_4 receptor radioligand. *ChemMedChem, 4*(2), 225-231.

Jablonowski, J. A., Grice, C. A., Chai, W. et al. (2003). The first potent and selective non-imidazole human histamine H_4 receptor antagonists. *J Med Chem, 46*, 3957–3960.

Jojart, B., Kiss, R., Viskolcz, B. et al. (2008). Activation mechanism of the human histamine H_4 receptor - An explicit membrane molecular dynamics simulation study. *J Chem Inf Model, 48*(6), 1199-1210.

Jongejan, A., Lim, H. D., Smits, R. A. et al. (2008). Delineation of agonist binding to the human histamine H_4 receptor using mutational analysis, homology modeling, and ab initio calculations. *J Chem Inf Model, 48*(7), 1455-1463.

Kindrachuk, D. E., Venable, J. D. (2011). Bicyclic heteroaryl-substituted imidazoles as modulators of the Histamine H_4 Receptor. (Janssen Pharmaceutica). US8084466.

Kraus, A., Ghorai, P., Birnkammer, T. et al. (2009). N^G-acylated aminothiazolylpropylguanidines as potent and selective histamine H_2 receptor agonists. *ChemMedChem, 4*(2), 232-240.

Lacy, B. E., Talley, N. J., Locke, G. R. et al. (2012). Review article: current treatment options and management of functional dyspepsia. *Aliment Pharmacol Ther, 36,* 3–15.

Lane, C. A., Hay, D., Mowbray, C. E. et al. (2012). Synthesis of novel histamine H_4 receptor antagonists. *Bioorg Med Chem Lett, 22,* 1156–1159.

Łażewska, D. Kieć-Kononowicz, K. (2012). Azines as histamine H_4 receptor antagonists. *Front Biosci, S4,* 967-987.

Lee-Dutra, A., Arienti, K. L., Buzard, D. J. et al. (2006). Identification of 2-arylbenzimidazoles as potent human histamine H_4 receptor ligands. *Bioorg Med Chem Lett-, 16*(23), 6043-6048.

Leurs, R., Smit, M. J., Menge, W. M. et al. (1994). Pharmacological characterization of the human histamine H_2 receptor stably expressed in Chinese hamster ovary cells. *Br J Pharmacol, 112,* 847–854.

Leurs, R., Chazot, P. L., Shenton, F. C. et al. (2009). Molecular and biochemical pharmacology of the histamine H_4 receptor. *Br J Pharmacol, 157*(1), 14-23.

Leurs, R., Vischer, H. F., Wijtmans, M. et al. (2011). En route to new blockbuster anti-histamines: surveying the offspring of the expanding histamine receptor family. *Trends Pharmacol Sci, 32,* 250–257.

Lim, H. D., Adami, M., Guaita, E. et al. (2009). Pharmacological characterization of the new histamine H_4 receptor agonist VUF 8430. *Br J Pharmacol, 157*(1), 34-43.

Lim, H. D., de Graaf, C., Jiang, W. et al. (2010). Molecular determinants of ligand binding to H_4R species variants. *Mol Pharmacol, 77*(5), 734-743.

Lim, H. D., Jongejan, A., Bakker, R. A. et al. (2008). Phenylalanine 169 in the second extracellular loop of the human histamine H_4 receptor is responsible for the difference in agonist binding between human and mouse H_4 receptors. *J Pharmacol Exp Ther, 327*(1), 88-96.

Lim, H. D., Smits, R. A., Bakker, R. A., et al. (2006). Discovery of *S*-(2-guanidylethyl)-isothiourea (VUF 8430) as a potent nonimidazole histamine H_4 receptor agonist. *J Med Chem, 49*(23), 6650-6651.

Lim, H. D., van Rijn, R. M., Ling, P. et al. (2005). Evaluation of histamine H_1-, H_2-, and H_3-receptor ligands at the human histamine H_4 receptor: Identification of 4-methylhistamine as the first potent and selective H_4 receptor agonist. *J Pharmacol Exp Ther, 314*(3), 1310-1321.

Lin, H., Sassano, M. F., Roth, B. L. et al. (2013). A pharmacological organization of G protein-coupled receptors. *Nat Meth, 10,* 140-146-

Ling, P., Ngo, K., Nguyen, S. et al. (2004). Histamine H_4 receptor mediates eosinophil chemotaxis with cell shape change and adhesion molecule upregulation. *Br J Pharmacol, 142,* 161–171.

Lippert, U., Artuc, M., Grutzkau, A. et al. (2003). Human skin mast cells express H_2 and H_4, but not H_3 receptors. *J Investig Dermatol, 123*(1), 116-123.

Liu, C., Ma, X., Jiang, X. et al. (2001). Cloning and pharmacological characterization of a fourth histamine receptor (H_4) expressed in bone marrow. *Mol Pharmacol, 59,* 420–426.

Liu, C., Wilson, S. J., Kuei, C. et al. (2001b). Comparison of human, mouse, rat, and guinea pig histamine H_4 receptors reveals substantial pharmacological species variation. *J Pharmacol Exp Ther, 299*(1), 121-130.

Masuda, T., Tchibana, Y., Miyagawa, M. et al. (2011). Pyrimidine derivatives and pharmaceutical composition containing same. (Shionogi &Co). WO2011078143.

Masood, M.A., Selby, M.D., Bell, A.S. et al. (2011) Discovery of a series of potent and selective human H_4 antagonists using ligand efficiency and libraries to explore structure-activity relationship (SAR). *Bioorg Med Chem Lett, 21* (21), 6591-6595.

McKellar, Q. A., Scott, E. W. (1990). The benzimidazole anthelmintic agents--a review. *J Vet Pharmacol Ther, 13,* 223–247.

Meduna, S.P., Savall, B.M., Cai, H. et al. (2011). Triamino pyrimidines and pyridines as histamine H_4 receptor modulators. *Bioorg Med Chem Lett, 21*, 3113-3116.

Morse, K. L., Behan, J., Laz, T. M. et al. (2001). Cloning and characterization of a novel human histamine receptor. *J Pharmacol Exp Ther, 296*, 1058–1066.

Mowbray, C. E., Bell, A. S., Clarke, N. P. et al. (2011). Challenges of drug discovery in novel target space. The discovery and evaluation of PF-3893787: A novel histamine H_4 receptor antagonist. *Bioorg Med Chem Lett, 21*, 6596–6602.

Narasimhan, B., Sharma, D., Kumar, P. (2012). Benzimidazole: a medicinally important heterocyclic moiety. *Med Chem Res, 21*, 269–283.

Nguyen, T., Shapiro, D. A., George, S. R. et al. (2001). Discovery of a novel member of the histamine receptor family. *J Mol Pharmacol, 59*(3), 427-433.

Oda, T., Morikawa, N., Saito, Y. et al. (2000). Molecular cloning and characterization of a novel type of histamine receptor preferentially expressed in leukocytes. *J Biol Chem, 275*, 36781–36786.

Parsons, M. E., Ganellin, C. R. (2006). Histamine and its receptors. *Br J Pharmacol, 147 Suppl 1*, 127–135.

Patil, A., Ganguly, S., Surana, S. (2008). A systematic review of benzimidazole derivatives. *Rasayan J Chem*, 447–460.

Pertz, H. H., Elz, S., Schunack, W. (2004). Structure-activity relationships of histamine H_1 receptor agonists. *Mini-Rev Med Chem, 4*(9), 935-940.

Reher, T. M., Buschauer, A., Neumann, D. et al. (2012). Incomplete activation of human eosinophils via the histamine H_4 receptor: evidence for ligand-specific receptor conformations. *Biochem Pharmacol, 84*(2), 192–203.

Rossignol, J. F., Maisonneuve, H. (1984). Benzimidazoles in the treatment of trichuriasis: a review. *Ann Trop Med Parasitol, 78*, 135–144.

Sander, K., Kottke, T., Proschak, E. et al. (2010) Lead identification and optimization of diaminopyrimidines as histamine H_4 receptor ligands. *Inflamm Res, 59, Suppl. 2*, S249-S251.

Sander, K., Kottke, T., Stark, H. (2008). Histamine H_3 receptor antagonists go to clinics. *Biol Pharm Bull, 31*, 2163-2181.

Sander K., Kottke, T., Tarnikulu, Y. et al. (2009). 2,4-Diaminopyrimidines as histamine H_4 receptor ligands--Scaffold optimization and pharmacological characterization. *Bioorg Med Chem, 17* (20), 7186-7196.

Sato, H., Fukushima, K., Shimazaki, M. et al. (2005A). 2-Aminopyrimidine derivatives. (Bayer Healthcare AG). WO2005014556.

Sato, H., Tanaka, K., Shimazaki, M. et al (2005b). 2-Aminopyrimidine derivatives. (Bayer Healthcare AG). WO2005054239.

Savall, B. M., Edwards, J. P., Venable, J. D. et al. (2010). Agonist/antagonist modulation in a series of 2-aryl benzimidazole H_4 receptor ligands. *Bioorg Med Chem Lett, 20*, 3367–3371.

Schneider, E. H., Schnell, D., Strasser, A. et al. (2010). Impact of the DRY motif and the missing „ionic lock" on constitutive activity and G-protein coupling of the human histamine H_4 receptor. *J Pharmacol Exp Ther, 333*(2), 382-392.

Schnell, D., Brunskole, I., Ladova, K. et al. (2011a). Expression and functional properties of canine, rat, and murine histamine H_4 receptors in Sf9 insect cells. *Naunyn Schmiedebergs Arch Pharmacol, 383*(5), 457-470.

Schnell, D., Brunskole, I., Ladova, K. et al. (2011b). Expression and functional properties of canine, rat, and murine histamine H_4 receptors in Sf9 insect cells. *Naunyn Schmiedebergs Arch Pharmacol, 383*(5), 457-470.

Seifert, R., Strasser, A., Schneider, E. H. et al. (2013). Molecular and cellular analysis of human histamine receptor subtypes. *Trends Pharmacol Sci, 34*(1), 33-58.

Shin, N., Coates, E., Murgolo, N. J. et al. (2002). Molecular modeling and site-specific mutagenesis of the histamine-binding site of the histamine H_4 receptor. *Mol Pharmacol, 62*(1), 38-47.

Shin, N., Covington, M., Bian, D. et al. (2012) INCB38579, a novel and potent histamine H_4 receptor small molecule antagonist with anti-inflammatory pain and anti-pruritic functions. *Eur J Pharmacol, 675* (1-3), 47-56.

Smits, R. A., Lim, H. D., Hanzer, A. et al. (2008). Fragment based design of new H_4 receptor–ligands with anti-inflammatory properties in vivo. *J Med Chem, 51*, 2457–2467.

Sterk, G. J., van der Goot, H., Timmerman, H. (1986). The influence of guanidino and isothiourea groups in histaminergic compounds on H_2-activity. *Agents Actions, 18*(1/2), 137-140.

Strasser, A., Wittmann, H.-J., Buschauer, A., et al. (2013). Species-dependent activities of GPCR ligands: lessons from histamine receptor orthologs. *Trends Pharmacol Sci, 34*(1), 13-32.

Tanrikulu, Y., Proschak, E., Werner, T. et al. (2009). Homology model adjustment and ligand screening with a pseudoreceptor of the human histamine H_4 receptor. *ChemMedChem, 4,* 820-847.

Thurmond, R. L., Gelfand, E. W., Dunford, P. J. (2008). The role of histamine H_1 and H_4 receptors in allergic inflammation: the search for new antihistamines. *Nat Rev Drug Discov, 7,* 41-53.

Thurmond, R. L., Desai, P. J., Dunford, P. J. et al. (2004). A potent and selective histamine H_4 receptor antagonist with anti-inflammatory properties. *J Pharmacol Exp Ther, 309*(1), 404-413.

Thurmond, R. L. Wadsworth, S. A. (2009). Use of histamine H_4 antagonist for the treatment of post-operative adhesion. (Janssen Pharmaceutica). WO2009152287.

Tiligada, E., Zampeli, E., Sander, K. et al. (2009). Histamine H_3 and H_4 receptors as novel drug targets. *Expert Opin Investig Drugs, 18*(10), 1519-1531.

Varga, C., Horvath, K., Berko, A. et al. (2005). Inhibitory effects of histamine H_4 receptor antagonists on experimental colitis in the rat. *Eur J Pharmacol, 522,* 130–138.

Vollinga, R. C., Menge, W. M. P. B., Leurs, R. et al. (1995). New analogs of burimamide as potent and selective histamine H_3 receptor antagonists: The effect of chain length variation of the alkyl spacer and modifications of the *N*-thiourea substituent. *J Med Chem, 38*(12), 2244-2250.

Walter, M., Stark, H. (2012). Histamine receptor subtypes: A century of rational drug design. *Front Biosci (Schol Ed), S4,* 461-488.

Werner T., Sander, K., Tanrikulu, Y. et al. (2010) In silico characterization of different ligand binding modes in the human histamine H_4 receptor and their impact on receptor activation. *ChemBioChem, 11,* 1850-1855.

Xie, S.-X., Ghorai, P., Ye, Q.-Z. et al. (2006a). Probing ligand-specific histamine H_1- and H_2-receptor conformations with N^G-acylated imidazolylpropylguanidines. *J Pharmacol Exp Ther, 317*(1), 139-146.

Xie, S.-X., Kraus, A., Ghorai, P. et al. (2006b). N^1-(3-cyclohexylbutanoyl)-N^2-[3-(1*H*-imidazol-4-yl)propyl]guanidine (UR-AK57), a potent partial agonist for the human histamine H_1- and H_2-receptors. *J Pharmacol Exp Ther, 317*(3), 1262-1268.

Yanai K, Rogala B, Chugh K et al. (2012). Safety considerations in the management of allergic diseases: focus on antihistamines. *Curr Med Res Opin* 28(4):623–642.

Yu, F., Wolin, R. L., Wei, J. et al. (2010). Pharmacological characterization of oxime agonists of the histamine H_4 receptor. *J Receptor Ligand Channel Res, 3*, 37-49.

Yuuki, T., Tsutomu, M. (2011). Aromatic fused heterocyclic derivative and pharmaceutical composition containing same. (Shionogi & Co). WO2011024987.

Zhang, C., Qian, D.-Q., Zhuo, J. et al. (2010). 4,6-Disubstituted 2-amino-pyrimidines as histamine H_4 receptor modulators. (Incyte Corporation). WO2010075270.

Zhou, J., Maduskuie, T.P., Qian, D. et al. (2010). Substituted pyrimidine derivatives as antagonists of the histamine H_4 receptor. (Incyte Corporation). WO2010108059.

Zhu, Y., Michalovich, D., Wu, H. et al. (2001). Cloning, expression, and pharmacological characterization of a novel human histamine receptor. *Mol Pharmacol, 59*, 434–441.

Chapter 3

Histamine Metabolism

H.G. Schwelberger[1], F. Ahrens[2], W.A. Fogel[3], F. Sánchez-Jiménez[4]

[1]Molecular Biology Laboratory, Department of Visceral, Transplantation and Thoracic Surgery, Medical University Innsbruck, Austria, e-mail: hubert.schwelberger@i-med.ac.at

[2]Department of Veterinary Science, Institute of Animal Physiology, Ludwig-Maximilians University Munich, Germany

[3]Department of Hormone Biochemistry, Medical University of Lodz, Poland

[4]Department of Molecular Biology and Biochemistry, University of Malaga, Spain

Abstract

Histamine is formed by decarboxylation of the amino acid L-histidine, a process catalyzed by histidine decarboxylase (HDC) and can be inactivated either by methylation of the imidazole ring, catalyzed by histamine N-methyltransferase (HMT) or by oxidative deamination of the primary amino group, catalyzed by diamine oxidase (DAO). This chapter describes the enzymatic reactions and the properties of the enzymes involved, including their structures, their cellular localization, their genes, expression and regulation, and the determination of their enzymatic activities. It also addresses cellular histamine transport, storage and release. Further, it discusses alterations in histamine metabolism associated with human diseases and how this might affect histamine receptor signaling.

3.1. Introduction

Histamine [2-(1*H*-Imidazol-4-yl)ethanamine] is an important mediator of many biological processes including inflammation, gastric acid secretion, neuromodulation, and regulation of immune function (Falus et al., 2004). Due to its potent pharmacological activity even at very low concentrations, the synthesis, transport, storage, release, and degradation of histamine have to be carefully regulated to avoid undesirable reactions. Histamine may also be generated by microbiological action in the course of food processing and spoilage and may therefore be present in substantial amounts in many fermented foodstuffs and beverages (Sarkadi, 2004).

Histamine is formed by decarboxylation of the amino acid L-histidine in a reaction catalyzed by the enzyme histidine decarboxylase (HDC) (Darvas and Falus, 2004). The major routes of histamine inactivation in mammals are methylation of the imidazole ring, catalyzed by histamine N-methyltransferase (HMT) (Schwelberger, 2004a), and oxidative deamination of the primary amino group, catalyzed by diamine oxidase (DAO) (Schwelberger, 2004b) (Figure 3.1 and Figure 3.2). The primary histamine inactivation products are inactive at the histamine receptors and are further metabolized for transport and secretion.

Although the principles of histamine formation and degradation have been known for several decades, the details of the individual reactions and the enzymes involved have only been investigated recently. Cloning of genes and cDNAs encoding histamine metabolizing enzymes were crucial for clarifying the identity of the proteins involved in this pathway and for studying their regulation, expression and processing. The determination of the three-dimensional structures of the proteins facilitated the understanding of the reaction mechanisms. Despite these advances, several key issues remain to be addressed in the future including the regulation of tissue-specific gene

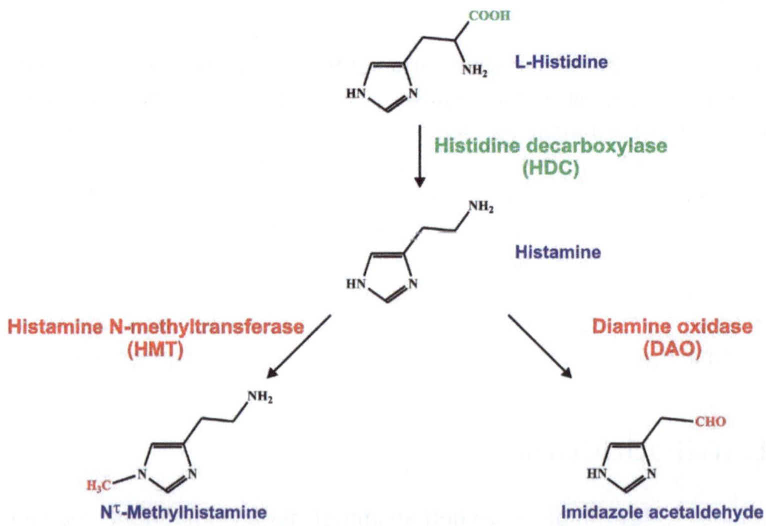

Figure 3.1 Formation and inactivation of histamine. Histamine is formed by decarboxylation of L-histidine, catalyzed by histidine decarboxylase (HDC), and can be inactivated either by methylation of the imidazole ring, catalyzed by histamine N-methyltransferase (HMT), or by oxidative deamination of the primary amino group, catalyzed by diamine oxidase (DAO).

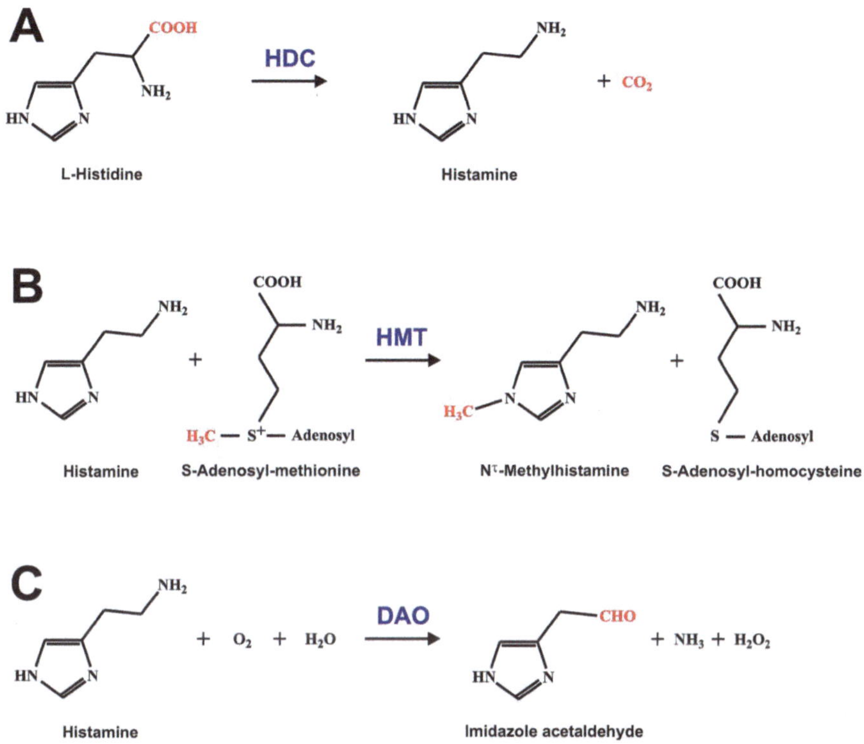

Figure 3.2 Details of the reactions leading to the formation and inactivation of histamine. The reactions catalyzed by (A) histidine decarboxylase (HDC), (B) histamine N-methyltransferase (HMT), and (C) diamine oxidase (DAO) are depicted with all co-substrates and co-products.

expression and both the cellular and subcellular localization, especially of the human enzymes. In the sections below we summarize the current knowledge of the proteins involved in histamine metabolism and transport and discuss their potential involvement in human diseases.

3.2. Histamine Formation

3.2.1. Histidine Decarboxylase Reaction Mechanism

Histamine is synthetized by decarboxylation of the amino acid L-histidine. In Gram positive bacteria, the reaction catalyst is a pyruvoyl-containing enzyme (Snell and Guirard, 1986; Gallager et al., 1989); however, in Gram negative bacteria and Metazoa, including humans, the histidine decarboxylase (HDC) reaction (EC 4.1.1.22) is assisted by the cofactor pyridoxal 5'-phosphate (PLP), a vitamin B6 derivative acting as a universal ligand to the alpha-amino group of α-amino acids in different catalytic processes (racemases, amino transferases, α and β-decarboxylases, synthases, transferases, and lyases) (Hayashi, 1995; Hayashi et al., 1990 and 2010; Mehta and Christen, 2000). The reaction mechanism of the PLP-dependent enzyme proceeds as shown in Figure 3.3. It has been extensively described in previous references (Olmo et al., 2002; Rodriguez-Caso et al., 2003; Moya-García et al., 2008; Pino-Ángeles et al., 2010). Briefly, in the free form of the enzyme, the carbonyl group of PLP forms a Schiff's base (internal aldimine) with the ε-amino group of a specific lysine group of the polypeptide. When the substrate L-histidine enters the catalytic center, its α-amino group acts as a nucleophilic group towards the Schiff's base. Therefore, the L-histidine α-amino group displaces the lysine group of the protein forming the external aldimine (transaldimination). Then a

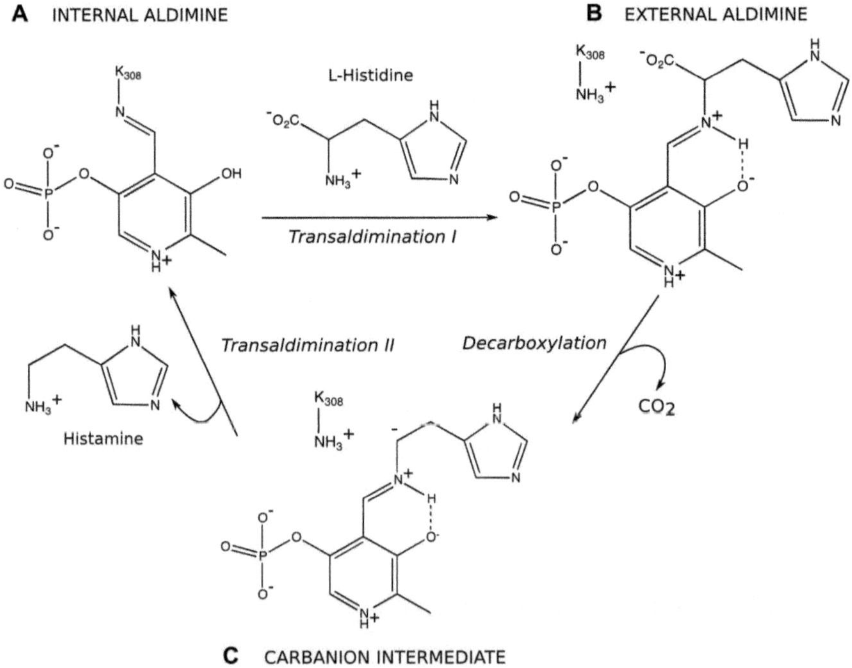

Figure 3.3 Reaction mechanism of mammalian HDC. Reaction mechanism of mammalian HDC as experimentally deduced from rat HDC (Olmo et al., 2002). (A) Major enolimine tautomeric form of the PLP-enzyme complex (internal aldimine). (B) Major ketoenamine form of the PLP-substrate intermediate (external aldimine). (C) instable carbanionic intermediate produced during decarboxylation.

carbanionic intermediate is generated by deprotononation which is most likely stabilized by the extended π system of the PLP pyridine ring (Eliot and Kirsch, 2004; Hayashi, 1995; John, 1995). This intermediate is named as the quinonoid intermediate due to its quinone-like structure. These steps are common to all PLP dependent enzymes. The way to proceed after the quinonoid formation determines the C_α that will be broken, that is to say the type of catalysis carried out by the PLP-dependent enzyme. In the case of HDC, it is the C_α-COO-bond (decarboxylation reaction). In rat HDC, it is the limitant step of the reaction as demonstrated by Olmo et al. (2002). In fact, this step is extremely slow leading to very low catalytic constants (K_{cat} in the range of 0.1 s^{-1}) for the mouse and rat enzymes. However, a recent publication on the catalytic properties of the human enzyme describes it as two orders of magnitude more efficient one (similar affinity as the rodent enzyme but higher K_{cat}) (Komori et al., 2012). Once L-histidine is decarboxylated, a reverse transaldimination occurs, so the polypeptidic ε-amine group (K^{308} in rat HDC and K^{305} in human HDC) recovers its bond to PLP, and histamine is released.

The mentioned PLP-adducts can present enol-keto tautomeric forms. Thus enolimine forms are more stable in a hydrophobic environment and ketoenamine forms are the major forms in a polar environment. These characteristics are useful to study PLP-dependent reactions, as the cofactor derivatives change their absorption properties (300-500 nm) depending on their bonds and even the major tautomeric forms occurring during a PLP-dependent reaction, giving valuable information on both the chemical properties and dynamics of the intermediates along the catalytic process. This strategy has been used to characterize many PLP-dependent enzymes from many different sources (Hayashi, 1995), including mammalian HDCs (Olmo et al., 2002; Rodríguez-Caso et al., 2003a), as well as HDC inhibitors affecting PLP-derivative status (Bertoldi et al. 2002; Rodríguez-Caso et al., 2003a and b).

At present only a few HDC inhibitors have been described, none of which are currently used for human therapies. Some substrate analogs are used as inhibitors for research purposes, such as histidine-methylester (HME), a competitive inhibitor able to form an external aldimine that cannot be further solved, and α-fluoromethyl histidine (FMH) that acts as a suicide substrate (Rodriguez-Caso et al., 2003a). These inhibitors are common for both PLP-dependent enzymes, the bacterial and the mammalian HDCs (Kubota et al., 1984; Hayashi et al., 1986). More recently, Wu and Gehring described new mammalian HDC inhibitors that mimic the transition state of the reaction, i.e. a pyridoxyl-histidine methyl ester conjugate (Wu and Gehring, 2008). Its development was facilitated by the information provided by the 3D model of rat HDC (Moya-García et al., 2005). In addition, it has been found that tea polyphenols such as epigallocatechin 3-gallate (EGCG) are able to inhibit both HDC and its homologous DDC (Bertoldi et al., 2002; Rodríguez-Caso et al., 2003b). A molecular dynamics study predicts that the polyphenol could bind to the catalytic center entry occluding the substrate entrance with the EGCG-HDC catalytic site interaction being stronger for mammalian DDC than for HDC (Ruiz-Pérez et al., 2012).

3.2.2. Mammalian Histidine Decarboxylase Protein

The active mammalian enzyme is a homodimer of approximately 110 kDa. The exact number of amino acids in the mature monomer is still unknown. The primary translation product, a 74 kDa polypeptide is composed of 656 amino acid residues in the case of rat HDC (UniProtKB code P16453, DCHS_RAT) and 662 in the human enzyme (UniProtKB code P19113, DCHS_HUMAN). The precursor needs to be processed by proteolytic removal of a carboxy-terminal fragment of about 150 amino acids to become a fully active enzyme. Activity-polypeptide length dependence is not a distinguishing feature, as different carboxy-terminal truncated versions of rat HDC exhibit different activities (Fleming et al., 2004a). Most of the kinetic studies on mammalian HDC have been carried out using as control monomeric HDC fragments comprising the first 480-516 amino acids (Olmo et al., 1996 and 2002; Rodríguez-Caso et al., 2003; Fleming et al., 2004b; Moya-García et al., 2012; Komori et al., 2012).

The fully active 54 kDa form was reported to be associated with the lumenal side of the endoplasmic reticulum (ER) (Furuta et al., 2006). It was postulated that it would be a SRP-independent ER import mechanism assisted by the carboxy-terminus of the full-length translated polypeptide (Tanaka, 2000). It was demonstrated that the C-terminal 10-kDa portion of the full-length polypeptide contains a signal necessary for HDC to be targeted to the ER membrane (Suzuki et al., 1998). After carboxy-terminus-assisted trafficking, this terminal portion is removed. The homology of this carboxy-terminal fragment with any other polypeptide is unclear. In addition to PEST regions it is predicted to contain a high percentage of random coil structure, which suggest a short half-life (Viguera et al., 1994; Olmo et al., 1999).

Primary structures of mammalian HDCs were known since 1989 (Joseph et al., 1990; Yamamoto et al., 1990). In the first 480 amino acids of its primary sequence, the mammalian enzyme is by more than 50% identical to aromatic L-amino acid decarboxylase or dopa decarboxylase (DDC, EC 4.1.1.28, UniProtKB code P20711, DDC_HUMAN in the case of the human enzyme). It was deduced that both enzymes in Metazoa, HDC and DDC, evolved from a common ancestor (Mehta and Christen, 2000; Sandmeier et al., 1994). There are several facts reinforcing this idea. Primary sequence similarity is higher in the common fragment between both mammalian enzymes than between bacterial PLP-dependent HDC and mammalian HDC (Moya-García et al., 2006). DDC is able to convert histamine but at higher concentrations than other DDC substrates such as serotonin and DOPA. More recently, Komori et al. (2012) reported that an $S^{354}G$ mutation in the active site of human HDC reduced the enzyme affinity for L-histidine but confered the ability to bind L-DOPA.

Mammalian HDC has evaded detailed characterization for a long time. Several features contributed to the lack of structural and functional information on this enzyme. HDC is expressed in considerable quantities only in a small set of mammalian cells (mast cells, histaminergic neurons, enterochromaffin-like cells), all of which are located dispersed among other histamine non-producing cells. Even in these cell types, HDC is a minor element in the proteome. In addition, the enzyme is extremely unstable, with a half-life on the order of a few hours *in vivo* (Dartch et al., 1998 and 1999). *In vitro*, HDC in diluted extracts tends to form inactive aggregates. The primary translation product contains at least one PEST sequence (a Pro, Glu, Ser, Thr enriched fragment acting as degradation promoter motif) in the amino-terminus of the monomer and at least another one in the C-terminal portion of the protein (the one that is lost during the enzyme maturation). At least 3 different proteolytic systems have been demonstrated to accept mammalian HDC as a substrate, including the 26S proteasome (ATP and ubiquitin-dependent process) (Viguera et al., 1994; Olmo et al., 1999 and 2000), calpain (Ca^{2+}-dependent process) (Rodríguez-Agudo et al., 2000) and caspase 9 (Furuta et al., 2007).

The first 3D structure of mammalian HDC (rat HDC) could not be obtained by X-ray chrystallography, but was determined in silico by applying comparative modeling and molecular dynamics methods (Rodríguez-Caso et al. 2003a; Moya-García et al., 2005) using the X-ray solved pig DDC as the template (Burkhard et al., 2002). These in silico approaches allowed us to determine the topology of the quaternary structure of the mature enzyme (Moya-García et al., 2012), to locate its catalytic site and to analyze the mechanism of decarboxylation (Moya-García et al., 2008). All the predicted features were validated by experimental methods (Olmo et al., 2002; Fleming et al., 2004b; Moya-García et al., 2012). More recently, Komori et al. (2012b) could solve the crystal structure of a stabilized active form of human HDC in the presence of the substrate analog HME (PDB ID: 4E1O). The crystallized form was stabilized by two C/S substitutions without any significant alteration in the K_{cat} of the mutant enzyme with respect to the wild-type enzyme.

Three different domains can be distinguished in the structure of the mature enzyme (Fig. 3.4). The amino-terminus is mainly involved in the intertwining among both monomers of the protein. A PEST sequence is located in this region that could be involved in the degradation of the active enzyme (Moya-García et al., 2005). The central portion of the mature polypeptide contains the PLP-binding domain that is characterized by 7 beta sheets, which is a common feature among all group II L-amino acid decarboxylases (Pino-Ángeles et al., 2006). Another important fragment in the protein is the flexible loop (residues 331-349 in rat HDC), which is essential for the catalytic mechanism. A conformational change

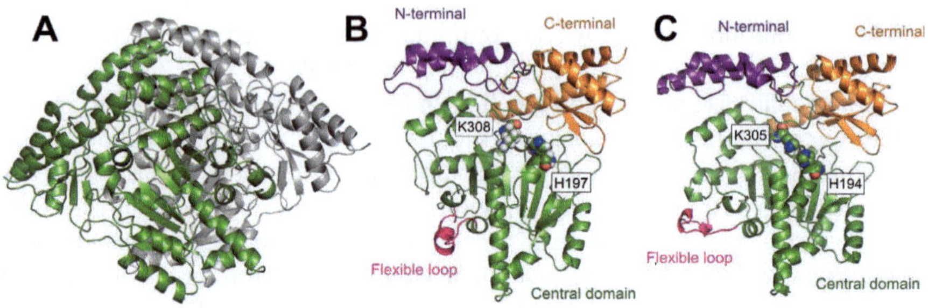

Figure 3.4 Structure of mammalian histidine decarboxylase (HDC). (A) Homodimeric structure of active mammalian HDC as modeled by Moya-García et al., 2005; one monomer is depicted in grey and the other one in green. (B) Structure of the rat HDC monomer as modeled by Moya-García et al., 2005. (C) Structure of the human HDC monomer as deduced for the crystal structure solved by Komori et al., 2012. In panels B and C, the amino-terminal domains (corresponding to fragment 4-71 in rat HDC and its counterpart in human HDC) are coloured in purple, the PLP-binding domains in green (fragment 72-371 in rat HDC and its counterpart in human HDC), and the carboxy-terminal fragments in orange. The Lys residue involved in transaldimination (K^{308} in rat HDC and K^{305} in human HDC) and the His residue involved in the PLP-pyridine ring stabilization (H^{197} in rat HDC and H^{194} in human HDC) are represented in coloured Van der Waals' spheres to locate the core of one of the two catalytic sites of the active enzyme. The flexible loop (fragment 331-344 in rat HDC and its counterpart in human HDC) is depicted in magenta and shown in its closed (panel B) and open conformations (panel C).

of the loop during substrate binding occludes the catalytic site entry, stabilizing the external aldimine in the catalytic site. In turn, the closed conformation of the loop on the catalytic site entry prevents it from proteolytic cleavage (Pino-Ángeles et al., 2010). Conformational changes of the loop are coordinated with global changes in the enzyme conformation during reaction that can be observed by chromatographic and electrophoretic analyses (Rodríguez-Caso et al., 2003; Fleming et al., 2004b). The major part of the carboxy-terminal domain of the active enzyme does not take part either in the dimerization surface or in the catalytic site (Moya-García et al., 2008; Moya-García et al., 2011). Figure 3.5 depicts the residues that are located in the catalytic center in both the rat HDC model and the human HDC crystal after a molecular dynamics refinement of the latter. It is composed of residues from both monomers and both catalytic centers are close in the quaternary structure. For example, the distance between the lysine residues of both catalytic centers (K^{308} in rat HDC and K^{305} in human HDC) is shorter than 3 nm. An ionic patch in the dimerization surface of each monomer has been detected that contains residues (at least both D^{315} in rat HDC) essential to maintain the enzymatic activity located very close to the catalytic sites (Moya-García et al., 2011) (Fig. 3.5). All of this knowledge will be helpful to find new HDC intervention strategies.

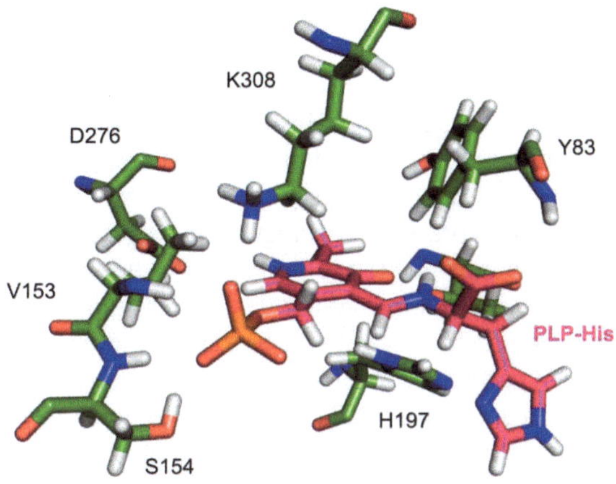

Figure 3.5 External aldimine in rat HDC. The figure shows the position of the PLP-histidine adduct (external aldimine) with respect to residues of the PLP-binding site as predicted by Moya-García et al., 2008 using quantum mechanics/molecular mechanics methods. The carbon skeleton of the adduct is represented in magenta and the skeletons of the polypeptide residues are in green. Nitrogen and oxygen atoms are depicted in blue and red, respectively, in both the external aldimine and the protein residues. The depicted residues have been located close to the external aldimine in both rat and human HDCs (Moya-García et al., 2008; Kumori et al., 2012).

3.2.3. Mammalian Histidine Decarboxylase Expression

As mentioned before, HDC is only expressed in a limited number of mammalian cell types. Human HDC corresponds to the Ensembl code ENSG00000140287 (www.ensembl.org). It is located on chromosome 15: 50,534,144-50,558,259, reverse strand. The encoding region is composed of 12 exons, however, multiple alternatively spliced forms have been detected as described in Ensembl and by Yatsunami et al. (1994). Of note, a truncated form of the human protein has been associated with Tourette syndrome in a family with several members suffering this rare disease (Ercan-Sencicek et al., 2010).

Methylation/demethylation of CpG islands in the mammalian HDC gene seems to be involved in regulation of mammalian HDC expression (Kuramasu et al., 1998). Other regulatory motifs have been detected in the human HDC promoter including TATA, GC, CACC, Sp1 and GATA boxes (Yatsunami et al., 1994). Nakagawa et al. (1997) validated the existence of two positive and one negative regulatory *cis*-elements (-855 to -841 and -532 to -497 relative to the transcription start site) and a positive one (-829 to -821), as well as the binding of a *trans*-element to a putative c-Myb binding motif TAACTG in basophilic cells but not in histamine non-producing cells, suggesting a role of c-Myb in the tissue-specific expression of HDC.

In addition to mast cells and other immune cells, gastric enterochromaffin-like cells (ECLC) are an important source of histamine (Hocker, 2004). The expression of HDC in human ECLC and in cultured histamine-producing gastric cancer cells has been studied mainly by T. Wang's group. Working on these models, they demonstrated that HDC expression is upregulated by gastrin, involving at least 3 gastrin-responsive elements located in the coding sequence and able to bind 3 different trans-acting elements. In contrast, gastric HDC expression is repressed by Kruppel-like factor 4 by competing with the Sp1 site complex, as well as interfering with the downstream gastrin responsive elements (Ai et al., 2004). Moreover, two nuclear factors, Ying-yang1 and SREBP-1a form a complex to inhibit HDC gene expression through the upstream GC box (Ai et al., 2006). GATA proteins have also been proposed as negative regulators of HDC expression in gastric epithelium (Walson et al., 2002). Of interest is the promotion of gastric HDC expression by Helicobacter pyroli that involves activation of the MEK1-2/ERK1-2 cascade through cAMP-dependent stimulation of Rap1 and B-Raf (Wessler et al., 2002).

3.3. Histamine Transport, Storage and Release

Histamine is soluble in aqueous solutions, as it is mainly protonated to a singly charged cation under physiological pH conditions. Histamine has two basic centers: the aliphatic amino group (pKa = 9.7) and one of the two nitrogens of the imidazole ring that does not readily bind a proton (pKa = 5.9; measured at 30°C; Levy, 1935). The positive charge of histamine limits permeability through membranes by diffusion. As histamine is stored in specific granules inside cells and local clearance of released (interstitial) histamine is attributed in part to the action of cytosolic histamine N-methyltransferase, there is an obvious requirement for transmembrane transport of histamine for uptake, storage and release. Although histamine transport in cells has been investigated for a long time, major advances have been made after cloning transport proteins and expression of these proteins in cell lines or Xenopus laevis oocytes to study transport and electro-physiological properties.

This section comprises the description of different carrier proteins relevant for histamine transport, their properties and transport mechanisms. Histamine transport is proposed to be conducted by organic cation transporter (OCT), plasma membrane monoamine transporter (PMAT), vesicular monamine transporter (VMAT), and endo-/exocytotic uptake/release in/from cellular vesicles or granules (Fig. 3.6 and Fig. 3.7). As the distribution of these transporters differs between tissues and organs, different functions and interactions make it necessary to look at the most important tissues/organs where histamine transport plays a crucial role.

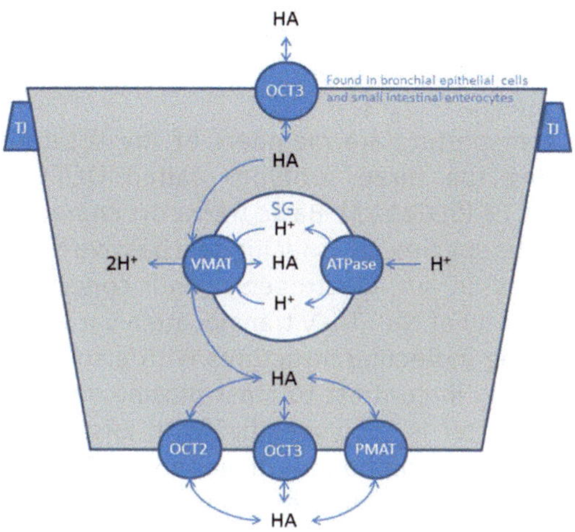

Figure 3.6 Histamine transport through plasma and vesicle membranes. Histamine can be passively transported through the plasma membrane by organic cation transporter (OCT) 2 and 3, and by the plasma membrane monoamine transporter (PMAT). If cells are polarised such as epithelial cells, transport proteins are commonly detected at the basolateral side while OCT3 has been found at the apical membrane in bronchial epithelial cells and small intestinal enterocytes. Transport of histamine from the cytoplasm into vesicles is mediated by the vesicular monamine transporter (VMAT) 2, a secondary active antiport using an electrochemical gradient of protons established by a H^+-ATPase. Abbreviations: HA = histamine; SG = secretory granule; TJ = tight junction.

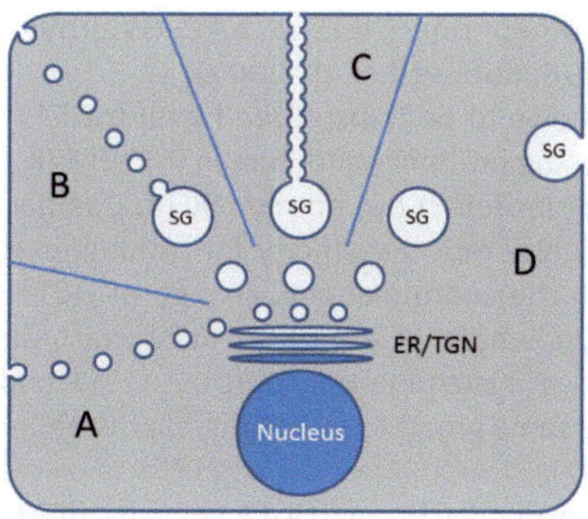

Figure 3.7 Histamine secretion pathways. (A) Histamine secretion may occur constitutively through small vesicular carriers after integration of VMAT2 in those vesicles and subsequent uptake of histamine from the cytoplasm. (B) One of the regulated exocytosis pathways of histamine is "piecemeal degranulation" detected in mast cells and basophils. It is mediated by vesicular transport of granule content from secretory granules to the plasma membrane. (C) This "piecemeal degranulation" is increased if vesicles are forming channels between the granule and the plasma membrane. (D) Regulated exocytosis of histamine by fusion of granules with the cell surface. Abbreviations: ER/TGN = endoplasmic reticulum/trans-Golgi network; SG = secretory granule.

3.3.1. Histamine Transport by Organic Cation Transporter (OCT)

Organic cation transporters are members of the solute carrier family 22 (SLC22) containing the three subtypes called OCT1 (SLC22A1), OCT2 (SLC22A2), and OCT3 (SLC22A3). Basic transport characteristics of OCT1-3 are similar in various species. OCTs facilitate a passive translocation through plasma membranes in an electrogenic and concentration dependent manner independently of Na^+. They translocate a variety of organic cations with widely differing molecular structures with a small relative molecular mass (below 500 Da). In contrast to most plasma membrane transporters, this class of transporter is "polyspecific" with large variations in affinity and turnover of different compounds. In general, OCTs are characterized by low affinity and high capacity in their transport properties (Koepsell et al. 2007).

OCTs participate in the translocation of endogenous amines and xenobiotics in the liver, kidney, intestine, and other organs critical for elimination. In addition to this general function, OCT3 has also been proposed to represent the Uptake2 in the nervous system that constitutes a low affinity neurotransmitter removal system, facilitating the termination of the neurotransmitter action in combination with Uptake1, representing specific transporters with high affinity of the SLC6 monoamine transporter family including transporters for serotonin (SERT; SLC6A4), dopamine (DAT; SLC6A3), and norepinephrine (NET; SLC6A2) (Whitby et al., 1961; Iversen, 1965; Wu et al., 1998; Koepsell et al., 2007).

Histamine was shown to be transported by rat (r)OCT2 (K_M = 540 μM, V_{max} = 8.5 ± 0.6 nmol min^{-1} mg $protein^{-1}$) and human (h)OCT3 (K_M = 180 μM, V_{max} = 5.6 ± 0.7 nmol min^{-1} mg $protein^{-1}$) but not by rOCT1. Compared with OCT2 and 3, OCT1 has only a slightly reduced affinity for histamine. However, histamine is not compatible with the conformational change of OCT1 occurring during the transport cycle (Gründemann et al., 1999). In contrast, another group showed comparable kinetics of histamine transport by rOCT1 and rOCT2 (Arndt et al., 2001). However, the same group found negligible histamine transport by hOCT1 (Koepsell et al., 2007).

Histamine-transporting OCT2 and OCT3 show a different tissue distribution and membrane localisation. Human OCT2 is detectable in epithelial cells and neurons and is most strongly expressed in the kidney but also in a variety of other organs like the small intestine, lung, skin, and brain. In polarised cells such as epithelial cells, OCT2 is normally located at the basolateral plasma membrane. In contrast, the tissue expression pattern of OCT3 is very broad and includes not only epithelial cells and neurons but also muscle and glial cells. In man, the strongest expression has been found in skeletal muscle, liver, placenta and heart. The OCT3 protein was found to be localised at the basolateral plasma membrane of hepatocytes and of renal proximal tubular epithelial cells. In contrast, the OCT3 protein was shown to be localised at the luminal membranes of bronchial epithelial cells and of enterocytes in the small intestine (Koepsell et al., 2007).

Polymorphisms in the SLC22A2 and SLC22A3 genes, encoding OCT2 and OCT3, respectively, have been identified. For human (h)OCT2, all variants were functionally active and showed no or no dramatic functional difference compared to wild-type hOCT2 (Koepsell et al., 2007). For hOCT3, two variants exhibited a significant reduction in catecholamine uptake while another variant showed a significantly increased uptake of histamine (Chen et al., 2010).

3.3.2. Histamine Transport by Plasma Membrane Monoamine Transporter (PMAT)

The plasma membrane monoamine transporter belongs to the equilibrative nucleoside transporter family (ENT, SLC29) and is therein named ENT4 (SLC29A4) (Engel et al., 2004). Together with OCT3 (SLC22A3), PMAT represents the most prominent Uptake2 transporter for endogenous monoamines. Similar to OCT3, basic transport characteristics of PMAT are sodium independency, low substrate affinity but high transport capacity and PMAT facilitates a passive translocation through plasma membranes in an electrogenic and concentration dependent manner (Engel and Wang, 2005).

When comparing hOCT3 with hPMAT, hOCT3 displayed higher transport activities for histamine, epinephrine, and norepinephrine while PMAT had higher transport rates for dopamine and serotonin. Kinetic studies in stably transfected hPMAT expressing HEK293 cells resulted in a calculation of $K_M = 4379 \pm 679$ µM and $V_{max} = 42.4 \pm 4.1$ nmol min^{-1} mg protein^{-1} for histamine transport in these cells compared with $K_M = 641 \pm 24$ µM and $V_{max} = 34.6 \pm 0.1$ nmol min^{-1} mg protein^{-1} for hOCT3 measured in parallel. So PMAT transports histamine at a high capacity (comparable to OCT3) whereas its affinity for histamine is much lower than that for other biogenic amines such as dopamine and serotonin (Duan and Wang, 2010).

Similar to hOCT3 mRNA, hPMAT mRNA is detectable in many tissues. However, expression of hPMAT is much higher in most brain regions, representing the predominant Uptake2 system in the CNS, while that of hOCT3 dominates in tissues outside the CNS like in the adrenal gland and in skeletal muscles (Duan and Wang, 2010). In contrast to cortisol, corticosterone can easily cross the blood-brain barrier (Karssen et al., 2001) and can potently inhibit transport by OCT3 but not PMAT (Engel et al., 2004), which could explain the tissue distributions of these two transporters very well.

3.3.3. Histamine Transport by Vesicular Monoamine Transporter (VMAT)

Vesicular monoamine transporters (VMATs) belong to the solute carrier family 18 (SLC18), containing three subtypes called VMAT1 (SLC18A1) and VMAT2 (SLC18A2), both transporting biogenic amines, and VAChT (SLC18A3), an acetylcholine transporter. This class of transporter is integrated in the lipid bilayer membrane of secretory vesicles in neuronal and endocrine cells. Positively charged amines are exchanged by VMATs against two protons from the vesicle lumen. The necessary driving force is an electrochemical gradient across the vesicular membrane established by proton pumping into the vesicle via a H^+-ATPase located at the same membrane (Eiden et al., 2004).

Biogenic amines are synthesized in the cytoplasm of neuronal and endocrine cells and concentrated into the vesicles, which then discharge them into the extracellular space by exocytosis. Following the rules of the mass action law, high affinity transport by VMATs reduces the concentration of the amines in the cytoplasm to sustain the building reaction. Therefore, the amines are accumulated in the vesicles to a final substrate concentration of up to 500 mM, exceeding the cytosolic concentration by 10,000-fold (Parsons, 2000).

The two vesicular monoamine transporters differ in both their functional properties and their distribution. VMAT2 has a 2- to 3-fold higher affinity for most monoamine substrates than VMAT1. As histamine lacks the hydroxyl groups, which are generally considered necessary for recognition as a substrate by VMATs, it shows different transport kinetics. Histamine was shown to be transported by VMAT2 with a relatively low affinity (K_M = 24 μM) compared with other amines (Merickel and Edwards, 1995). The affinity of VMAT1 for histamine is more than 30-fold lower precluding the transport of histamine by VMAT1 (Peter et al., 1994). Consequently, the presence of histidine decarboxylase (HDC) in neuronal and endocrine cells is associated with the presence of VMAT2 in these cells (Weihe and Eiden, 2000).

3.3.4. Histamine Transport in Different Tissues and Cells

Histamine transport is important in processes of storage (basophils, mast cells) and elimination (kidney, gut, nerve synapse). Within the last years, interesting results yielded new insights about the function and regulation of histamine transporting carriers, suggesting that these transporters are not static, rather are actively involved in the regulation of intra- and extracellular histamine concentration.

Murine immature basophils are capable of taking up histamine and are characterised by low granulation and interleukin (IL)-3 inducible histamine and pro-Th2 cytokine synthesis. Stimulated in that way, newly generated histamine is not stored in granules in these cells but is immediately released into the extracellular space from these cells. OCT3 was identified to be responsible for the release and the uptake of histamine. In the latter case, histamine inhibited its own synthesis and that of the pro-Th2 cytokines. Inhibition of OCT3, which is possible by H_3/H_4 histamine receptor ligands, increased the histamine content and decreased pro-Th2 cytokines. Consequently, OCT3 modulates the intracellular histamine level in these basophils and thereby controls histamine and pro-Th2 cytokine synthesis (Schneider et al., 2005).

Enterocytes of the porcine proximal colon are able to secrete histamine into the gut lumen. This secretion was abolished by application of 1-methyl-4-phenylpyridinium (MPP), a permanently charged prototypical organic cation and high-affinity substrate for OCTs. In addition, enterocytes contain both histamine degrading enzymes HMT and DAO and thereby represent a system for bioelimination of systemic histamine, which was unrecognized for a long time. Both enzymes are localized intracellularly with HMT being a cytosolic protein and DAO being a secretory protein stored in secretory vesicles. Therefore, transport of histamine into the enterocyte is necessary to get in contact with these enzymes. This process was demonstrated by increased unmetabolised histamine fluxes across the epithelium subsequent to inhibition of the inactivating enzymes. In this context it was shown that the HMT product N^{τ}-methylhistamine appeared exclusively at the basolateral side of the cells (Aschenbach et al., 2009). This data provided additional evidence that transport of histamine and its metabolites and catabolism are closely interrelated as shown earlier in HEK293 cells (Ogasawara et al., 2006).

Similar to other neurotransmitters, histamine must be cleared from the synaptic cleft after release from histaminergic neurons. So far proof for an Uptake1 system (see above) in neurons is missing, despite speculations that such a plasma membrane transporter for histamine should exist (Hoffman et al., 1998). However, it was shown that glial cells, i.e. astrocytes, can take up histamine and thus represent the main histamine inactivation site (Rafalowska et al., 1987). Histamine can be taken up in rat cortical astrocytes by a high affinity system, which is time-, concentration-, temperature-, and sodium-dependent. However, the transport system remains unidentified and participation of OCTs and specific transporters for other monoamines (i.e. SLC6A2-4) have been excluded (Perdan-Pirkmajer et al., 2012).

3.3.5. Histamine Storage

Despite the fact that HDC is an inducible enzyme that can be expressed in many cell types, histamine is mainly present in mast cells and their circulating counterparts, i.e. basophils. Histamine is stored in secretory granules (SG) that most likely emanate from vesicles of the trans-Golgi network (Dvorak, 1998). Numerous other components and immune mediators are stored together with histamine in these SGs. The presence of lysosomal hydrolases such as β-hexosaminidase and cathepsin-D indicates that these organelles are not just secretory vesicles but also have a lysosomal function. Therefore, they are often called secretory lysosomes to stress the close connection between lysosomal (endosomal) and secretory organelles even if it is unusual that both functions are active in the same compartment (Lorentz et al., 2012).

The contents of the SGs are defined by endoplasmic reticulum export activities, most likely including carrier proteins like VMAT2 for histamine uptake into the SGs, and transport and processing through the Golgi network (Glick and Nakano, 2009; Lorentz et al., 2012). Electron micrographs provided a description of granule patterns. In human mast cells, four basic granule patters have been identified, including scrolls, crystals, particles, and a mixed pattern. Mixed granule refers to a majority of the granules in individual mast cells that display mixtures of the other three basic patterns. Basophil granules are larger and less numerous than their counterparts in mast cells. They are filled with dense particles that vary in the density of packing within the granules (Dvorak, 1998).

The amount of histamine that can be released from secretory granules depends on the stimulus and is proportional to the number of SGs and the storage density of histamine in those granules, referred to as quantal release (Travis et al., 2000). It was assumed that it depends on the activity of HDC and the presence of VMAT2 in the cells. However, studies on enterochromaffin-like (ECL) cells, which stimulate gastric acid secretion by parietal cells through histamine release in a paracrine manner, showed that HDC expression is not influenced by fasting although acid secretion is at a basal level in that nutritional state and that mucosal histamine is significantly decreased. Instead the availability of the HDC substrate L-histidine was decreased by increased expression of the histidine catabolyzing enzymes histidase and urocanase and by decreased expression of the cellular L-histidine uptake transporter SN2. Moreover, expression and protein levels of VMAT2 were reduced explaining the low concentration of histamine in ECL cells during fasting (Lambrecht et al., 2007). The vesicular storage of histamine is not only dependent on VMAT2 but is also promoted by its association with the negatively charged heparin matrix in SGs. This became clear from studies showing that the storage of histamine and serotonin is regulated differentially in VMAT2$^{+/-}$ mast cells where only 50% of VMAT2 is present compared to wild-type mast cells. Serotonin release from VMAT2$^{+/-}$ mast cell granules was 43% of that from wild-type cells whereas histamine release from those cells was 68% (Travis et al., 2000).

3.3.6. Histamine Release

Histamine release from cells can occur in three different ways. The most important mechanism of histamine release from mast cells and basophils is through immunological stimulation. If these cells are sensitized by IgE antibodies attached to receptors on their membrane surfaces they degranulate when exposed to the appropriate antigen due to cross-linking of the antibodies and the IgE receptors. It is a regulated rapid release reaction involving simultaneous fusion of a large number of granules with the cell surface for bulk release of histamine and other mediators (Fig. 3.7D; Dvorak, 1998; Lorentz et al., 2012). Upon cross-linking of the high-affinity receptor for IgE, degranulation is induced by the activation of phospholipase Cγ and protein kinase C and the increased mobilization of calcium (Gilfillan and Tkaczyk, 2006). This can occur in either a sequential or in a multi-vesicular manner. In sequential exocytosis, vesicles fuse with the plasma membrane followed by the fusion of adjoining vesicles with the first vesicle. In multi-vesicular exocytosis, vesicles fuse with each other before interacting with the plasma membrane (Alvarez de Toledo and Fernandez, 1990). This process involves several membrane fusion proteins referred to as SNAREs. They are able to break up the membranes of the granules and the plasma membrane and fuse these membranes with each other. Most of these proteins contain sequences that can be modified by phosphorylation, lipid composition of the membrane, or nitrosylation. These modifications are involved in the regulation of SNARE complex formation which is linked to cell signalling (Lorentz et al., 2012).

A second release mode is called "piecemeal degranulation", which describes the release of granule contents in the absence of typical granule extrusion from basophils and mast cells. This process can be triggered by cytokines and is mediated by vesicular transport due to fusion of a vesicle with a histamine containing granule, exchange of contents, and subsequent vesicular traffic to the plasma membrane where the contents is released (Fig. 3.7B). This form of mediator release can be boosted if a multitude of vesicles are forming channels between granules and the plasma membrane, from which the entire granule contents is released rather than vesicle-sized parts of granules (Fig. 3.7C; Dvorak, 1998).

A third release mode is a constitutive release of histamine from cells. This is made possible either by passive transport of histamine operated by the above mentioned transport proteins located at the plasma membrane (Fig. 3.6) or by secretory vesicles from the trans-Golgi network that are routed by the cytoskeleton directly to the plasma membrane (Fig. 3.7A). Release of histamine due to OCT3 transport has been shown in immature murine basophils (Schneider et al., 2005). Constitutive release conducted by secretory vesicles has been described for cytokines in innate immune cells (Lacy and Stow,

2011). However, proof for histamine release in that way is lacking. Although many publications discussed a constitutive release of histamine providing a regulatory feedback mechanism in an autocrine manner, comparatively little is known about the mechanisms involved. The situation is complicated by the fact that the terms "constitutive release" and "piecemeal degranulation" are often used interchangeably.

3.4. Histamine Inactivation

In mammals there are two alternative routes for inactivation of histamine: methylation of the imidazole ring at the tele position by histamine N-methyltransferase (HMT) and oxidative deamination of the primary amino group by diamine oxidase (DAO) (Maslinski and Fogel, 1991). Both reactions convert histamine into products that are virtually inactive at histamine receptors. The primary inactivation products are further converted for transport and secretion.

3.4.1. Inactivation of Histamine by Histamine N-methyltransferase

HMT (EC 2.1.1.8) inactivates histamine by transfer of a methyl group from S-adenosyl-L-methionine (SAM) to the N_{e2} nitrogen of the imidazole ring forming N^{τ}-methylhistamine (Fig. 3.2B). HMT is a highly specific enzyme that does not show significant mehylation of other substrates. K_M values for histamine and SAM of 6-13 μM and 6-10 μM, respectively, have been reported for HMT enzymes purified from different mammalian tissues. The enzyme is inhibited by the reaction products N^{τ}-methylhistamine and S-adenosyl-L-homocystein with K_i values in the low micromolar range as well as by SH-group reagents and by antimalarial drugs (Schwelberger, 2004a). The activity of HMT is usually determined by methylation of histamine with S-adenosyl-L-[methyl-^{14}C]methionine followed by extraction of the radioactively labelled N^{τ}-methylhistamine and quantitation by liquid scintillation analysis (Küfner et al., 2001).

Native HMT proteins were purified to homogeneity from different mammalian tissues and recombinant enzymes were obtained by expression of the cDNAs encoding rat and human HMT in bacteria. Characterization of the purified proteins showed that HMT is a soluble monomeric protein of ca. 33 kDa that does not possess any post-translational modifications and does not require any cofactors for enzymatic activity (Schwelberger, 2004a). Determination of the three-dimensional structure of recombinant human HMT by X-ray crystallography revealed that HMT has a two-domain structure (Fig. 3.8A; Horton et al., 2001). The larger N-terminal domain is a classical

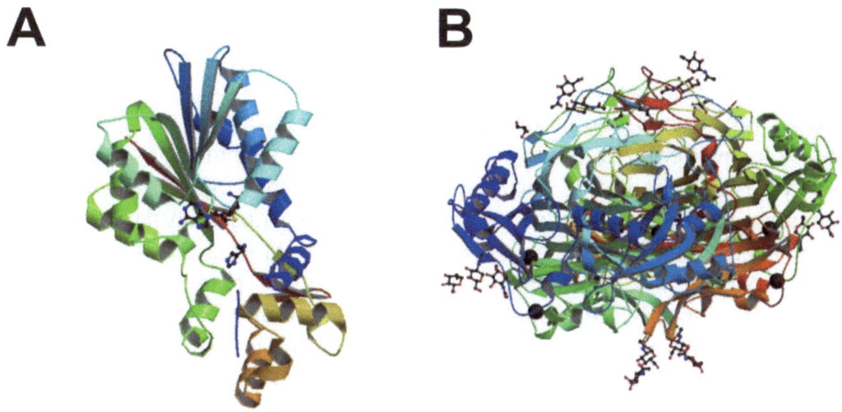

Figure 3.8 Structures of the human HMT and DAO proteins. (A) Structure of monomeric human HMT complexed with histamine and S-adenosyl-L-homocysteine (PDB ID: 1JQD; Horton et al., 2001). (B) Structure of homodimeric human DAO containing 2 Cu^{++} and 4 Ca^{++} ions and 12 N-acetyl-D-glucosamine residues (PDB ID: 3HI7; McGrath et al., 2009). Images were downloaded from the Protein Data Bank website (www.rcsb.org) using standard cartoon style and secondary structure coloring options.

methyltransferase fold with a SAM binding domain whereas the smaller C-terminal domain partly resembles a motif in the bacterial chemotaxis receptor methyltransferase and contributes only one residue to histamine binding. Analysis of ternary complexes of HMT with histamine and S-adenosyl-L-homocysteine revealed that binding of substrate and co-substrate places the target N_{e2} atom of histamine close to the donor methyl group. The bound histamine is buried in an almost-closed acidic pocket at the interface of the two HMT domains. Histamine binding involves polar and hydrophobic interactions with several amino acid residues including C^{196}, which explains the inhibitory effect of SH-group reagents.

Cloning and characterization of HMT cDNA and genomic sequences from different species revealed that HMT primary structures are highly conserved in mammals (Takemura et al., 1992; Girard et al., 1994; Yamauchi et al., 1994; Aksoy et al., 1996; Wang et al., 2001; Schwelberger and Drasche, 2002). All amino acid residues that have been shown to interact with histamine and S-adenosyl-L-homocysteine in human HMT (Horton et al., 2001) are absolutely conserved in all species analyzed. Human HMT has 292 amino acid residues and is encoded by a single gene designated HNMT, which spans ca. 35 kbp on chromosome 2q22.1 and has 6 exons (Aksoy et al., 1996). The promoter region of the HMT genes has not been characterized except for the notion that canonical TATA, CAAT, or initiator sequences are not present near the transcription start site which is similar to other cytosolic methyltransferase genes. Very little information is currently available on the regulation of expression of mammalian HMT genes.

While HMT occurs ubiquitously in vertebrate species, it does not appear to be present in invertebrates, plants or microorganisms. In mammals, HMT enzymatic activity has been found in many tissues (Maslinski and Fogel, 1991) and HMT expression has been confirmed in many human tissues by analysis of its mRNA (Girard et al., 1994; Preuss et al., 1998). In man, highest levels of expression were found in kidney and liver, considerable expression in spleen, prostate, ovary, intestine, and spinal cord, along with a small amount of detectable HMT mRNA in most other tissues analyzed. An immunohistochemical study in the guinea pig showed specific cytoplasmic staining of the HMT protein in many tissues and cell types and the tissue-specific staining pattern corresponded well with the enzymatic activity in the respective tissue (Tahara et al., 2000). Recently, we were able to produce highly specific and sensitive monoclonal antibodies for human HMT, which will facilitate the cellular and subcellular localization of the enzyme in human tissues.

Biochemical, structural and immunohistochemical data indicate that HMT is a cytosolic protein, which is responsible for the intracellular inactivation of histamine that is either synthesized in the cell or taken up from the extracellular space, likely after binding to one of its receptors present on the cell surface. The ubiquitous expression of HMT suggests that there is a widespread requirement for this catabolic pathway and that many cells have to deal with histamine. It can be speculated that most cells that express HMT also receive histamine as a signal, therefore analysis of HMT expressing cells for the presence of histamine receptors may yield interesting results.

Alterations of histamine metabolism have been implicated in various diseases including asthma, inflammatory bowel disease, and neurological disorders, with the diminished HMT activity proposed to play a critical role. HMT expression and activity show considerable inter-individual variation and it is difficult to assess what is normal and what is pathological. Recently, a $C^{314}T$ polymorphism of the human HNMT gene has been described that leads to the expression of a HMT protein with a $T^{105}I$ amino acid substitution, which has a reduced enzymatic activity and stability compared to the wild-type enzyme (Preuss et al., 1998). In a study with 192 asthmatic patients, a significantly increased frequency of the I^{105} allele was found in patients compared to controls (Yan et al., 2000). Other studies with a similar number of patients failed to show an increased frequency of the I^{105} allele in patients with schizophrenia and in patients with inflammatory bowel disease (Yan et al., 2000; García-Martin et al., 2006). Certainly more trials are necessary to assess the relative risk of this and other polymorphisms of the human HNMT gene for the development of different histamine-related disorders.

3.4.2. Inactivation of Histamine by Diamine Oxidase

DAO (EC 1.4.3.22) uses molecular oxygen to oxidatively deaminate histamine to imidazole acetaldehyde, ammonia and hydrogen peroxide (Fig. 3.2C). The original name of this enzyme was histaminase until it was recognized that histamine is not the only substrate (Buffoni, 1966). Whereas DAO enzymes from different mammalian tissues show significant oxidation of many other diamine substrates with primary amino groups such as aliphatic diamines and polyamines, they commonly have a much higher affinity for histamine with K_M values of ca. 20 μM (Schwelberger and Bodner, 1997). DAO does not convert monoamines and is inhibited by copper-chelating agents such as cyanide and by carbonyl group reagents such as aminoguanidine and semicarbazide (Schwelberger, 2004b). The activity of DAO is usually determined radiometrically by conversion of [1,4-^{14}C]putrescine to 4-amino-[1,4-^{14}C]butyraldehyde, which in alkaline solution spontaneously forms the cyclic compound Δ_1-[1,4-^{14}C]pyrroline that can be extracted into an organic solvent for quantitation (Schwelberger et al., 1995).

Based on the characterization of DAO proteins from different mammalian tissues and on the cloning of the corresponding cDNAs and genes it is now clear that DAO is a member of the class of copper-containing amine oxidases (AOC) that have in their active-sites a Cu^{2+} ion bound by three conserved histidine residues and the cofactor 2,4,5-trihydroxyphenylalanine quinone (TPQ), formed post-translationally from a conserved tyrosine residue (Klinman, 1996). Mammalian DAO is a homodimeric glycoprotein of ca. 200 kDa with subunits being linked by disulfide bonds formed by conserved cysteine residues. DAO proteins from different species and tissues show considerable molecular weight variation due to heterogenous glycosylation (Schwelberger, 2004b). Determination of the structure of recombinant human DAO expressed in insect cells by X-ray crystallography confirmed the typical folding pattern found in other AOC family members (Fig. 3.8B; McGrath et al., 2009). This was crucial for understanding the catalytic mechanism and the restrictions in substrate binding and allowed the assignment of residues important for catalytic function, including the TPQ precursor tyrosine, the three histidine residues binding the Cu^{2+} ion, and an aspartic acid residue functioning as a catalytic base. Furthermore, it identified the disulfide forming cysteine residues, the glycoside carrying asparagine residues, and the presence of 4 Ca^{2+} ions.

Characterization of cDNAs and genes encoding DAO from various tissues of different species revealed a high degree of sequence conservation (Barbry et al., 1990; Lingueglia et al., 1993; Chassande et al., 1994; Schwelberger, 1999; Hütter and Schwelberger, 2000; Schwelberger et al., 2000). Amino acid residues with important catalytic or structural function (McGrath et al., 2009) were found to be absolutely conserved in all DAO proteins. Human DAO is formed from a 751 amino acid precursor polypeptide by cleavage of a 19 amino acid N-terminal signal peptide, followed by dimerisation and attachment of asparagine-linked oligosaccharide chains, which occurs during transport via the endoplasmatic reticulum and the Golgi network. Human DAO is encoded by a single gene designated ABP1 (from its original characterization as amiloride binding protein-1), which has 5 exons and spans ca. 10 kbp on chromosome 7q36.1 (Chassande et al., 1994). The promoters of the DAO genes lack TATA and CAAT boxes or consensus initiator sequences. The proximal promoter region of the human DAO gene contains several consensus sequences for the binding of nuclear factors, including a PU-1 site, two Sp-1 sites, a cAMP response element, several AP-2 half-sites, and many AP-1 and glucocorticoid response element sequences. Close to the transcription start site there is an extensive palindromic sequence (E-PAL) that has been shown to confer epithelial-specific gene expression (Chassande et al., 1994). Nevertheless, our understanding of the regulation of DAO expression is still very limited.

In contrast to HMT, expression of DAO is restricted to certain tissues and to specific cells (Maslinski and Fogel, 1991; Schwelberger, 2004b). High levels of DAO are invariably present in the intestine, the placenta, and pregnancy plasma of all species studied to date. High amounts of DAO are also present in the kidneys of many species including man, pig, cow, horse, dog, and cat, whereas DAO is at the limit of detection in kidney tissue from rodents and rabbits. On the other hand, DAO is not detectable in the liver of most species except for rabbit and guinea pig. In tissues containing appreciable amounts of DAO, the cellular and subcellular localization of the enzyme has been studied by immunohistochemistry using specific antibodies. DAO was found to be present in the decidual cells of the maternal part of human placenta (Weisburger et al., 1978) and, using confocal laser scanning fluorescence microscopy, DAO was localized in vesicular structures associated with the basolateral plasma membrane in proximal tubular epithelial cells of porcine kidney and in differentiated epithelial cells of porcine small intestine (Schwelberger et al., 1998). Very recently, we were able to confirm the cellular and subcellular localization of DAO in human tissues using highly specific and sensitive monoclonal antibodies (Schwelberger et al., 2013).

DAO has also been detected in body fluids including pregnancy plasma, lymph, and seminal plasma (Maslinski and Fogel, 1991). Apart from pregnancy, DAO is usually not detectable in blood plasma but tissue-bound DAO can be released into the circulation following administration of the glycosaminoglycan heparin. The major sites of release appear to be intestine and kidney, the tissues containing the highest amounts of DAO (Biebl et al., 2002). The current model of DAO action proposes that DAO is produced and stored in epithelial cells of certain tissues and is released from these stores triggered by specific signals that are generated when histamine or other substrates need to be degraded (Schwelberger, 2004b). The best characterised DAO release stimulator is heparin which is released together with histamine by activated mast cells. The signal to activate the degradation of histamine is generated when histamine is released thus facilitating a temporal and local control of mediator action.

The intestine is a major site of DAO expression in all species and probably one of the most important functions of DAO is the prevention of resorption of histamine and other diamines present in food (Schwelberger, 2004b; Schwelberger, 2010). The term enteral histaminosis has been coined for a situation where intestinal DAO activity is compromised to such an extent that histamine, present abundantly in fermented food such as cheese and red wine or produced by enteric bacteria, can enter the circulation to exhibit undesirable effects (Sattler et al., 1988; Sattler and Lorenz, 1990). It has also been proposed that certain forms of food intolerance may result when the degradative capacity of DAO is overcharged leading to the resorption of histamine and other biologically active diamines in unmetabolized form (Maintz and Novak, 2007). DAO has also been implicated as an important factor in inflammatory bowel disease such as ulcerative colitis and Crohn´s disease where a decreased DAO activity and an increased histamine concentration was found in the affected region (Schmidt et al., 1990; Raithel et al., 1995). Although it appears that the amount of DAO stored in intestinal epithelial cells is abundant, many commonly used drugs have been described to inhibit DAO activity at least partially in vitro (Sattler et al., 1985). Additionally, the expression levels of DAO in human intestine show considerable inter-individual differences and there might be a DAO-related genetic predisposition for developing the diseases discussed above.

3.4.3. Catabolism of N^{τ}-methylhistamine and Imidazole Acetaldehyde

The catabolic pathways of the primary histamine inactivation products, N^{τ}-methylhistamine and imidazole acetaldehyde, are shown in Figures 3.9 and 3.10, respectively. In mammals, the methylation product of histamine, N^{τ}-methylhistamine, is present in tissues as well as in body fluids such as in the brain, the lung, the liver or gastric mucosa and plasma, cerebrospinal fluid and urine. It is a main product of histamine in the stomach, while extragastric metabolites of histamine are imidazole acetic acid, N^{τ}-methylimidazole acetic acid, acetylhistamine, and histaminol (Code et al., 1976). Interestingly enough, histaminol was recently identified as a component of wine (Bordiga et al., 2010). In the brain, DAO appears not to be expressed and therefore N^{τ}-methylhistamine and N^{τ}-methylimidazole acetic acid are major histamine metabolites. Their level in the CNS was shown to be influenced by age and gender. In older subjects, the concentration of both metabolites in the cerebrospinal fluid was significantly higher and positively correlated with age, females always having higher levels than males (Prell et al., 1991). Clinical studies also revealed N^{τ}-methylhistamine levels to be elevated in the cerebrospinal fluid of

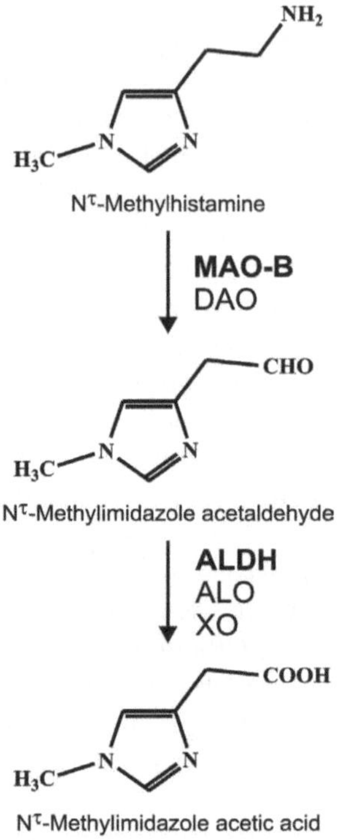

Figure 3.9 Metabolism of Nτ-methylhistamine. MAO-B, monoamine oxidase type B; DAO, diamine oxidase; ALDH, aldehyde dehydrogenase; ALO, aldehyde oxidase; XO, xanthine oxidase.

schizophrenics (Ito, 2004). Biogenic and dietary monoamines in the brain and in the periphery are oxidatively deaminated by monoamine oxidases (MAO; E.C. 1.4.3.4), which in contrast to DAO are members of the flavin-containing amine oxidase family. Two existing isoenzymes, MAO-A and MAO-B, are encoded by two different genes and share 70% amino acid sequence identity. The isoenzymes differ in both their substrate specificity and inhibitor sensitivity. While MAO-A preferentially oxidizes serotonin, noradrenaline and adrenaline and is inhibited by clorgyline, MAO-B preferentially oxidizes phenylethylamine and N-methylhistamine and is irreversibly inhibited by deprenyl. In the brain, MAO-A is predominantly found in catecholaminergic neurons whereas MAO-B is more abundant in serotonergic and histaminergic neurons and glial cells (Shih et al., 1999). MAO-B enzyme activity in the human brain is altered by the single nucleotide polymorphism located in intron 13 of the MAOB gene (Balciuniene et al., 2002). In man, MAO-B of the same amino acid sequence as the brain enzyme is present in platelets (Chen et al., 1993). The platelet MAO-B is expressed in a stable fashion over several decades but then increases in aging people. Patients suffering from pernicious anaemia and the neuro-degenerative disorders Alzheimer's disease, Parkinson's disease and Huntington's chorea were shown to have increased platelet MAO-B, whereas low levels of platelet MAO-B were recorded in alcoholics of type 2, recurring criminals and those expressing antisocial behavior. MAO-B activity in platelets correlates with such personality traits as sensation seeking, impulsiveness and extraversion. Platelet MAO-B has a high degree of heritability and the regulation of MAO-B gene expression seems to explain most of the inter-individual differences in activity (Oreland et al., 2002). It has been suggested that platelet MAO-B is a genetic marker for the size or functional capacity of the central monoamine systems and the serotonergic system in particular. Chronic treatment with different antipsychotic drugs may differentially regulate the gene expression of catabolic enzymes of biogenic amine neurotransmitters, among them MAO-B, resulting in the different clinical efficacy and/or side effects of antipsychotics (Chen et al., 2007). Oxidation of N^{τ}-methylimidazole acetaldehyde, generated in the metabolism of N^{τ}-methylhistamine by oxidative deamination by MAO-B or by MAO-B/DAO outside the central nervous system (Fig. 3.9), is accomplished by cytoplasmic aldehyde dehydrogenase (Gitomer et al., 1983; Henehan et al., 1985).

Likewise, imidazole acetaldehyde, the product of oxidative deamination of histamine by DAO, is mainly oxidized to imidazole acetic acid, with a small portion reduced to imidazolylethanol or histaminol. Aldehyde dehydrogenase (EC 1.2.1.3) was claimed to be the only enzyme in the human liver capable of catalyzing dehydrogenation of aldehydes arising via actions of monoamine, diamine and plasma amine oxidases (Ambroziak et al., 1991). Imidazole acetic acid can be conjugated with phosphoribosyl-pyrophosphate to form imidazole acetic acid ribotide. The reaction is catalyzed by imidazole acetic acid-phosphoribosyl transferase (ImPRT; E.C. 6.3.4.8) and ATP serves as an energy source rather than as a substrate. Ribosylated imidazole acetic acid is present in urine and plasma and in micromolar concentrations also in the central nervous system where it is a product of the L-histidine transamination pathway (Fig. 3.10). There is evidence suggesting that being an endogenous ligand of imidazoline receptors and adrenergic receptors, imidazole acetic acid ribotide can act as a neuroregulator in the brain, i.e. a regulator of general sympathetic drive and in particular of systemic blood pressure, or exert hormone-like activity in the periphery. The conjugate imidazole acetic acid ribotide may be dephosphorylated by phosphatases and 5'-nucleotidases to yield imidazole acetic acid riboside, which is far less active than the parent compound (Crowley, 1964; Friedrich et al., 2007).

The oxidation of aldehydes to the corresponding acids can be catalyzed by aldehyde oxidases (EC 1.2.3.1). These are cytosolic enzymes which belong to the family of molybdo-flavoenzymes, along with xanthine oxidoreductase. Aldehyde oxidases have a broad substrate specificity; in humans, a single aldehyde oxidase

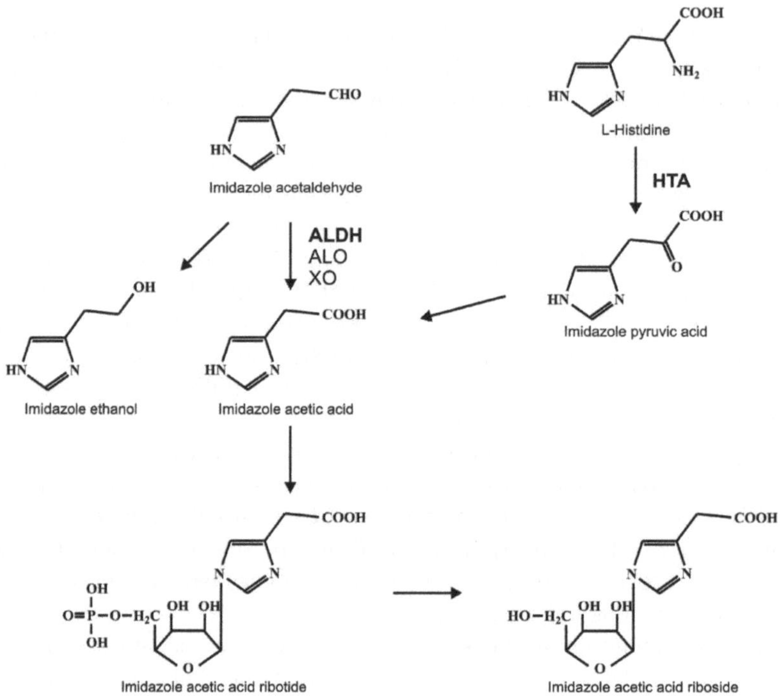

Figure 3.10 Metabolism of imidazole acetaldehyde. ALDH, aldehyde dehydrogenase; ALO, aldehyde oxidase; XO, xanthine oxidase; HTA, histidine transaminase.

gene (AOX1) and two pseudogenes clustering on a short stretch of chromosome 2q have been described. Human AOX1 and its homologous proteins are best known as drug metabolising enzymes, while the physiological substrates and functions remain obscure (Garattini et al., 2009).

Alcohol dehydrogenase, also known as aldehyde reductase or aldo-keto reductase family 1 member A1 (EC 1.1.1.2) is an enzyme involved in the reduction of biogenic and xenobiotic aldehydes to alcohol derivatives and is present in virtually all tissues. The enzyme in humans is encoded by the AKR1A1 gene. Alternative splicing of this gene results in two transcript variants encoding the same protein (Bohren et al., 1989; Fujii et al., 1999). Liver alcohol dehydrogenase (ADH) is metalloenzyme containing Zn^{2+}. ADH catalyzes interconversion of a large variety of saturated and unsaturated aliphatic and aromatic alcohols and the corresponding aldehydes and ketones utilizing NAD(H). Human ADH is heterogeneous, and, at least, five genes encode for polypeptides which, by dimerization, form different isoenzymes (Pietruszko, 1975). Whether one or both of these enzymes contribute to production of alcohol metabolites of histamine/ methylhistamine is unknown as yet.

3.5. Conclusion

Despite the fact that some details of the formation and degradation of histamine still have to be worked out, the main facts about the pathway of histamine metabolism in mammals have been well established. There is extensive information on each of the enzymes involved, including protein sequences, three-dimensional structures and modifications, the reaction mechanisms, the genes encoding the enzymes and their basic regulation of expression. What needs to be addressed in the future is the interplay of the histamine metabolic pathway with histamine transport and histamine signaling in different tissues as well as alterations of the components that lead to disease.

Acknowledgements

The authors would like to thank their coworkers at the respective institutions for their input and contributions. This work was supported by COST Action BM0806.

References

Ai, W., Liu, Y., Langlois, M. et al. (2004). Kruppel-like factor 4 (KLF4) represses histidine decarboxylase gene expression through an upstream Sp1 site and downstream gastrin responsive elements. *J Biol Chem, 279*, 8684-8693.

Ai, W., Liu, Y., Wang, T. C. (2006). Yin yang 1 (YY1) represses histidine decarboxylase gene expression with SREBP-1a in part through an upstream Sp1 site. *Am J Physiol Gastrointest Liver Physiol, 290*, G1096-1104.

Aksoy, S., Raftogianis, R., Weinshilboum, R. (1996). Human histamine N-methyltransferase gene: structural characterization and chromosomal localization. *Biochim Biophys Res Commun, 219*, 548-554.

Alvarez de Toledo, G., Fernandez, J.M. (1990). Compound versus multigranular exocytosis in peritoneal mast cells. *J Gen Physiol, 95*, 397-409.

Ambroziak, W., Pietruszko, R. (1991). Human aldehyde dehydrogenase: Activity with aldehyde metabolites of monoamines, diamines, and polyamines. *J Biol Chem, 266*, 13011-13018.

Arndt, P., Volk, C., Gorboulev, V. et al. (2001). Interaction of cations, anions, and weak base quinine with rat renal cation transporter rOCT2 compared with rOCT1. *Am J Physiol Renal Physiol, 281*, F454-F468.

Aschenbach, J.R., Honscha, K.U., von Vietinghoff, V. et al. (2009). Bioelimination of histamine in epithelia of the porcine proximal colon of pigs. *Inflamm Res, 58*, 269-276.

Balciuniene, J., Emilsson, L., Oreland, L. et al. (2002). Investigation of the functional effect of monoamine oxidase polymorphisms in human brain. *Hum Genet, 110*, 1-7.

Barbry, P., Champe, M., Chassande,O. et al. (1990). Human kidney amiloride-binding protein: cDNA structure and functional expression. *Proc Natl Acad Sci USA, 87*, 7347-7351.

Biebl, M., Klocker, J., Perkmann, R. et al. (2002). Heparin-induced diamine oxidase release into the circulation in pigs. *Inflamm Res, 51* Suppl 1, S93-94.

Bohren, K.M., Bullock, B., Wermuth, B. et al. (1989). The aldo-keto reductase superfamily. cDNAs and deduced amino acid sequences of human aldehyde and aldose reductases. *J Biol Chem, 264*, 9547-9551.

Bordiga, M., Travaglia, F., Locatelli, M. et al. (2010). Histaminol: identification and HPLC analysis of a novel compound in wine. *J Agric Food Chem, 58*, 10202-10208.

Buffoni, F. (1966). Histaminase and related amine oxidases. *Pharmacol Rev, 18*, 1163-1199.

Burkhard, P., Dominici, P., Vorri-Voltattorni, C. et al. (2001). Structural insight into Parkinson's disease treatment from drug-inhibited DOPA decarboxylase. *Nat Struct Biol, 8*, 1847-1862.

Chassande, O., Renard, S., Barbry, P. et al. (1994). The human gene for diamine oxidase, an amiloride binding protein: molecular cloning, sequencing, and characterization of the promoter. *J Biol Chem, 269*, 14484-14489.

Chen, K., Wu, H.F., Shih, J.C. (1993). The deduced amino acid sequences of human platelet and frontal cortex monoamine oxidase B are identical. *J Neurochem, 61*, 187-190.

Chen, M.L., Chen, C.H. (2007). Chronic antipsychotics treatment regulates MAOA, MAOB and COMT gene expression in rat frontal cortex. *J Psychiatr Res, 41*, 57-62.

Chen, L., Pawlikowski, B., Schlessinger, A. et al. (2010). Role of organic cation transporter 3 (SLC22A3) and its missense variants in the pharmacologic action of metformin. *Pharmacogenet Genomics, 20*, 687-699.

Code, C.F., Green, W.E., Kennedy, J.C. et al. (1976). Metabolism of histamine in secreting and isolated canine stomach. *Am J Physiol, 230*, 219-227.

Crowley, G.M. (1964). The enzymatic synthesis of 5´-phosphoribosylimidazole acetic acid. *J Biol Chem, 239*, 2593-2601.

Dartsch, C., Chen, D., Persson, L. (1998). Multiple forms of rat stomach histidine decarboxylase may reflect posttranslational activation of the enzyme. *Regul Pept, 77,* 33-41.

Dartsch, C., Chen, D., Hakanson, R. et al. (1999). Histidine decarboxylase in rat stomach ECL cells: relationship between enzyme activity and different molecular forms. *Regul Pept, 81,* 41-48.

Darvas, Z., Falus, A. (2004). Histidine decarboxylase (HDC) enzyme and gene. In A. Falus (Ed.), *Histamine: Biology and medical aspects* (pp. 31-42). Budapest: SpringMed Publishing.

Duan, H., Wang, J. (2010). Selective transport of monoamine neurotransmitters by human plasma membrane monoamine transporter and organic cation transporter 3. *J Pharmacol Exp Ther, 335,* 743-753.

Dvorak, A.M. (1998). Histamine content and secretion in basophils and mast cells. *Prog Histochem Cytochem, 33,* 169-320.

Eiden, L.E., Schäfer, M.K., Weihe, E. et al. (2004). The vesicular amine transporter family (SLC18): amine/proton antiporters required for vesicular accumulation and regulated exocytotic secretion of monoamines and acetylcholine. *Pflugers Arch, 447,* 636-640.

Engel, K., Zhou, M., Wang, J. (2004). Identification and characterization of a novel monoamine transporter in the human brain. *J Biol Chem, 279,* 50042-50049.

Engel, K., Wang, J. (2005). Interaction of organic cations with a newly identified plasma membrane monoamine transporter. *Mol Pharmacol, 68,* 1397-1407.

Ercan-Sencicek, A.G., Stillman, A.A., Ghosh, A.K. et al. (2010). L-histidine decarboxylase and Tourette's syndrome. *N Engl J Med, 362,* 1901-1908.

Falus, A., Grosman, N., Darvas, Z. (2004). Histamine: Biology and Medical Aspects. Budapest: SpringMed Publishing.

Fleming, J.V., Fajardo, I., Langlois, M.R. et al. (2004a). The carboxyl-terminus of rat L-histidine decarboxylase (HDC) specifically inhibits enzymatic activity and disrupts PLP-dependent interactions with L-histidine substrate analogues. *Biochem J, 381,* 769-778.

Fleming, J.V., Sánchez-Jiménez, F., Moya-García, A.A. et al. (2004b). Mapping of catalytically important residues in the rat L-histidine decarboxylase enzyme using bioinformatic and site directed mutagenesis approaches. *Biochem J, 379*, 253-261.

Friedrich, V.L., Martinelli, G.P., Prell, G.D. et al. (2007). Distribution and cellular localization of imidazoleacetic acid-ribotide, an endogenous ligand at imidazol(in)e and adrenergic receptors, in rat brain. *J Chem Neuroanat, 33*, 53-64.

Fujii, J., Hamaoka, R., Matsumoto, A. et al. (1999). The structural organization of the human aldehyde reductase gene, AKR1A1, and mapping to chromosome 1p33-->p32. *Cytogenet Cell Genet, 84*, 230-232.

Furuta, K., Ichikawa, A., Nakayama, K. et al. (2006). Membrane orientation of the precursor 74-kDa of L-histidine decarboxylase. *Inflamm Res, 55*, 185-191.

Furuta, K., Nakayama, K., Sugimoto, Y. et al. (2007). Activation of histidine decarboxylase through post-translational cleavage by caspase-9 in a mouse mastocytoma P-815. *J Biol Chem, 282*, 13438-13446.

García-Martin, E., Mendoza, J. L., Martínez, C. et al. (2006). Severity of ulcerative colitis is associated with a polymorphism at diamine oxidase gene but not at histamine N-methyltransferase gene. *World J Gastroenterol, 12*, 615-620.

Garattini, E., Fratelli, M., Terao, M. (2009). The mammalian aldehyde oxidase gene family. *Hum Genomics, 4*, 119-130.

Gilfillan, A.M., Tkaczyk, C. (2006). Integrated signalling pathways for mast-cell activation. *Nat Rev Immunol, 6*, 218-230.

Girard, B., Otterness, D.M., Wood, T.C. et al. (1994). Human histamine N-methyltransferase pharmacogenetics: cloning and expression of kidney cDNA. *Mol Pharmacol, 45*, 461-468.

Gitomer, W.L., Tipton, K.F. (1983). The role of cytoplasmic aldehyde dehydrogenase in the metabolism of N-tele-methylhistamine. *Pharmacol Biochem Behav, 18* Suppl 1, 113-116.

Glick, B.S., Nakano, A. (2009). Membrane traffic within the Golgi apparatus. *Annu Rev Cell Dev Biol, 25*, 113-132.

Gründemann, D., Liebich, G., Kiefer, N. et al. (1999). Selective substrates for non-neuronal monoamine transporters. *Mol Pharmacol, 56*, 1-10.

Hayashi, H. (1995). Pyridoxal enzymes: mechanistic diversity and uniforminity. *J Biochem, 118*, 463-473.

Hayashi, H., Tanase, S., Snell, E.E. (1986). Pyridoxal 5'-phosphate-dependent histidine decarboxylase. Inactivation by alpha-fluoromethylhistidine and comparative sequences at the inhibitor- and coenzyme- binding sites. *J Biol Chem, 261*, 11003-11009.

Hayashi, H., Wada, H., Yoshimura, T. et al. (1990). Recent topics in pyridoxal 5'-phosphate enzyme studies. Annu Rev Biochem, 59, 87-110.

Henehan, G.T., Ward, K., Kennedy, N.P. et al. (1985). Subcellular distribution of aldehyde dehydrogenase activities in human liver. *Alcohol, 2*,107-110.

Hocker, M. (2004). Molecular mechanisms of gastrin-dependent gene regulation. *Ann N Y Acad Sci, 1014*, 97-109.

Hoffman, B.J., Hansson, S.R., Mezey, E. et al. (1998). Localization and dynamic regulation of biogenic amine transporters in the mammalian central nervous system. *Front Neuroendocrinol, 19*, 187-231.

Horton, J.R., Sawada, K., Nishibori, M. et al. (2001). Two polymorphic forms of human histamine methyltransferase: structural, thermal, and kinetic comparisons. *Structure, 9*, 837-849.

Hütter, E., Schwelberger, H. G. (2000). Characterization of cDNA clones encoding mouse diamine oxidase. *Inflamm Res, 49* Suppl 1, S57-59.

Ito, C. (2004). The role of the central histaminergic system in schizophrenia. *Drug News Perspect, 17*, 383-387.

Iversen, L.L. (1965). The uptake of catechol amines at high perfusion concentrations in the rat isolated heart: A novel catechol amine uptake process. *Br J Pharmacol Chemother, 25*, 18-33.

Joseph, D.R., Sullivan, P.M., Wang, Y.M. et al. (1990). Characterization and expression of the complementary DNA encoding rat histidine decarboxylase. *Proc Natl Acad Sci USA, 87*, 733–737.

Karssen, A.M., Meijer, O.C., van der Sandt, I.C. et al. (2001). Multidrug resistance P-glycoprotein hampers the access of cortisol but not of corticosterone to mouse and human brain. *Endocrinology, 142*, 2686-2694.

Klinman, J.P. (1996). New quinocofactors in eukaryotes. *J Biol Chem, 271,* 27189-27192.

Koepsell, H., Lips, K., Volk, C. (2007). Polyspecific organic cation transporters: structure, function, physiological roles, and biopharmaceutical implications. *Pharm Res, 24,* 1227-1251.

Komori, H., Nitta, Y., Ueno, H. et al. (2012). Structural study reveals that Ser-354 determines substrate specificity on human histidine decarboxylase. *J Biol Chem, 287,* 29175-29183.

Kubota, H., Hayashi, H., Watanabe, T. et al. (1984). Mechanism of inactivation of mammalian L-histidine decarboxylase by (S)-α-fluoromethyl histidine. *Biochem Pharmacol, 33,* 983-990.

Küfner, M.A., Ulrich, P., Raithel, M. et al. (2001). Determination of histamine degradation capacity in extremely small human colon samples. *Inflamm Res, 50* Suppl 2, S96-97.

Lacy, P., Stow, J.L. (2011). Cytokine release from innate immune cells: association with diverse membrane trafficking pathways. *Blood, 118,* 9-18.

Lambrecht, N.W., Yakubov, I., Sachs, G. (2007). Fasting-induced changes in ECL cell gene expression. *Physiol Genomics, 31,* 183-192.

Levy, M. (1935). A note on the titration constants of imidazole derivatives. *J Biol Chem, 109,* 361-363.

Lingueglia, E., Renard, S., Voilley, N. et al. (1993). Molecular cloning and functional expression of different molecular forms of rat amiloride-binding proteins. *Eur J Biochem, 216,* 679-687.

Lorentz, A., Baumann, A., Vitte, J. et al. (2012). The SNARE machinery in mast cell secretion. *Front Immunol, 3,* 143.

Maintz, L., Novak, N. (2007). Histamine and histamine intolerance. *Am J Clin Nutr, 85,* 1185-1196.

Maslinski, C., Fogel, W.A (1991). Catabolism of histamine. In B. Uvnäs (ed.), *Handbook of experimental pharmacology. Vol 97. Histamine and histamine antagonists* (pp. 165-189). Berlin, Heidelberg: Springer.

McGrath, A.P., Hilmer, K.M., Collyer, C.A. et al. (2009). Structure and inhibition of human diamine oxidase. *Biochemistry, 48,* 9810-9822.

Mehta, P.K., Christen, P. (2000). The molecular evolution of pyridoxal-5'-phosphate-dependent enzymes. *Adv Enzymol Relat Areas Mol Biol, 74,* 129-184.

Merickel, A., Edwards, R.H. (1995). Transport of histamine by vesicular monoamine transporter-2. *Neuropharmacology, 34,* 1543-1547.

Moya-García, A.A., Medina, M.A., Sánchez-Jiménez, F. (2005). Mammalian histidine decarboxylase: from structure to function. *Bioessays, 27,* 57-63.

Moya-García, A.A., Pino-Ángeles, A., Sánchez-Jiménez, F. (2006). New structural insights to help in the search for selective inhibitors of mammalian pyridoxal-5'-phosphate-dependent histidine decarboxylase. *Inflamm Res, 55* Suppl 1, S55-S56.

Moya-García, A.A., Ruíz-Pernía, J., Martí, S. et al. (2008). Analysis of the decarboxylation step in mammalian histidine decarboxylase. A computational study. *J Biol Chem, 283,* 12393-12401.

Moya-García, A.A., Pino-Ángeles, A., Gil-Redondo, R. et al. (2009). Structural features of mammalian histidine decarboxylase reveal the basis for specific inhibition. *Br J Pharmacol, 157,* 4-13.

Nakagawa, S., Okaya, Y., Yatsunami, K. et al. (1997). Identification of multiple regulatory elements of human L-histidine decarboxylase gene. *J Biochem, 121,* 935-940.

Ogasawara, M., Yamauchi, K., Satoh, Y. et al. (2006). Recent advances in molecular pharmacology of the histamine systems: organic cation transporters as a histamine transporter and histamine metabolism. *J Pharmacol Sci, 101,* 24-30.

Olmo, M.T., Rodríguez-Agudo, D., Medina, M.A. et al. (1999). The pest regions containing C-termini of mammalian ornithine decarboxylase and histidine decarboxylase play different roles in protein degradation. *Biochem Biophys Res Commun, 257,* 269-272.

Olmo, M.T., Urdiales, J.L., Pegg, E. et al. (2000). In vitro study of proteolytic degradation of rat histidine decarboxylase. *Eur J Biochem, 267,* 1527-1531.

Olmo, M.T., Sánchez-Jiménez, F., Medina, M.A. et al. (2002). Spectroscopic analysis of recombinant rat histidine decarboxylase. *J Biochem, 132,* 433-439.

Oreland, L., Damberg, M., Hallman, J. et al. (2002). Risk factors for the neurohumoral alterations underlying personality disturbances. *Neurotox Res, 4*, 421-426.

Parsons, S. M. (2000). Transport mechanisms in acetylcholine and monoamine storage. *FASEB J, 14*, 2423-2434.

Perdan-Pirkmajer, K., Pirkmajer, S., Černe, K. et al. (2012). Molecular and kinetic characterization of histamine transport into adult rat cultured astrocytes. *Neurochem Int, 61*, 415-422.

Peter, D., Jimenez, J., Liu, Y. et al. (1994). The chromaffin granule and synaptic vesicle amine transporters differ in substrate recognition and sensitivity to inhibitors. *J Biol Chem, 269*, 7231-7237.

Pietruszko, R. (1975). Mammalian liver alcohol dehydrogenases. *Adv Exp Med Biol, 56*, 1-31.

Pino-Angeles, A., Morreale, A., Negri, A. et al. (2010). Substrate uptake and protein stability relationship in mammalian histidine decarboxylase. *Proteins, 78*, 154-61.

Prell, G.D., Khandelwal, J.K., Burns, R.S. et al. (1991). Influence of age and gender on the levels of histamine metabolites and pros-methylimidazoleacetic acid in human cerebrospinal fluid. *Arch Gerontol Geriatr, 12*, 1-12.

Preuss, C.V., Wood, T.C., Szumlanski, C.L. et al. (1998). Human histamine N-methyltransferase pharmacogenetics: common genetic polymorphisms that alter activity. *Mol Pharmacol, 53*, 708-717.

Rafalowska, U., Waskiewicz, J., Albrecht, J. (1987). Is neurotransmitter histamine predominantly inactivated in astrocytes? *Neurosci Lett, 80*, 106-110.

Raithel, M., Matek, M., Baenkler, H.W. et al. (1995). Mucosal histamine content and histamine secretion in Crohn´s disease, ulcerative colitis and allergic enteropathy. *Int Arch Allergy Immunol, 108*, 127-133.

Raychowdhury, R., Fleming, J.V., McLaughlin, J.T. et al. (2002). Identification and characterization of a third gastrin response element (GAS-RE3) in the human histidine decarboxylase gene promoter. *Biochem Biophys Res Commun, 297*, 1089-1095.

Rodríguez-Agudo, D., Sánchez-Jiménez, F., Medina, M.A. (2000). Rat histidine decarboxylase is a substrate for m-calpain in vitro. *Biochem Biophys Res Commun, 271*, 777-781.

Rodríguez-Caso, C., Rodríguez-Agudo, D., Moya-García, A.A. et al. (2003a). Local changes in the catalytic site of mammalian histidine decarboxylase can affect its global conformation and stability. *Eur J Biochem, 270*, 4376-4387.

Rodríguez-Caso, C., Rodríguez-Agudo, D., Sánchez-Jiménez, F. et al. (2003b). Green tea epigallocatechin-3-gallate is an inhibitor of mammalian histidine decarboxylase. *Cell Mol Life Sci, 60*, 1760-1763.

Sandmeier, E., Hale, T.I., Christen, P. (1994). Multiple evolutionary origin of pyridoxal-5-phosphate dependent amino acid decarboxylases. *Eur J Biochem, 221*, 997-1002.

Sarkadi, L. (2004). Histamine in food. In A. Falus (Ed.), *Histamine: Biology and medical aspects* (pp. 176-185). Budapest: SpringMed Publishing.

Sattler, J., Häfner, D., Klotter, H.J. et al. (1988). Food-induced histaminosis as an epidemiological problem: plasma histamine elevation and haemodynamic alterations after oral histamine administration and blockade of diamine oxidase (DAO). *Agents Actions, 23*, 361-365.

Sattler, J., Hesterberg, R., Lorenz, W. et al. (1985). Inhibition of human and canine diamine oxidase by drugs used in an intensive care unit: relevance for clinical side effects? *Agents Actions, 16*, 91-94.

Sattler, J., Lorenz, W. (1990). Intestinal diamine oxidases and enteral-induced histaminosis: studies on three prognostic variables in an epidemiological model. *J Neural Transm, 32* Suppl, 291-314.

Schmidt, W.U., Sattler, J., Hesterberg, R. et al. (1990). Human intestinal diamine oxidase (DAO) activity in Crohn's disease: a new marker for disease assessment? *Agents Actions, 30*, 267-270.

Schneider, E., Machavoine, F., Pleau, J.M. et al. (2005). Organic cation transporter 3 modulates murine basophil functions by controlling intracellular histamine levels. *J Exp Med, 202*, 387-393.

Schwelberger, H.G. (1999). Molecular cloning of mammalian diamine oxidase genes and cDNAs. *Inflamm Res, 48* Suppl 1, S79-80.

Schwelberger, H.G. (2004a). Histamine N-methyltransferase (HNMT) enzyme and gene. In A. Falus (Ed.), *Histamine: Biology and medical aspects* (pp. 53-59). Budapest: SpringMed Publishing.

Schwelberger, H.G. (2004b). Diamine oxidase (DAO) enzyme and gene. In A. Falus (Ed.), *Histamine: Biology and medical aspects* (pp. 43-52). Budapest: SpringMed Publishing.

Schwelberger, H.G. (2010). Histamine intolerance: a metabolic disease? *Inflamm Res, 59* Suppl 2, S219-221.

Schwelberger, H.G., Bodner, E. (1997). Purification and characterization of diamine oxidase from porcine kidney and intestine. *Biochim Biophys Acta, 1340*, 152-164.

Schwelberger, H.G., Drasche, A. (2002). Structure and expression of porcine histamine N-methyltransferase. *Inflamm Res, 51* Suppl 1, S85-86.

Schwelberger, H.G., Drasche, A., Hütter, E. (2000). Analysis of mammalian diamine oxidase genes. *Inflamm Res, 49* Suppl 1, S51-52.

Schwelberger, H.G., Feurle, J., Houen, G. (2013): New tools for studying old questions: antibodies for human diamine oxidase. *J Neural Transm*, 120, 1019-1026.

Schwelberger, H.G., Hittmair, A., Kohlwein, S.D. (1998). Analysis of tissue and subcellular localization of mammalian diamine oxidase by confocal laser scanning fluorescence microscopy. *Inflamm Res, 47* Suppl 1, S60-61.

Schwelberger, H.G., Klocker, J., Sattler, J. et al. (1995). Determination of the activity of diamine oxidase in extremely small tissue samples. *Inflamm Res, 44* Suppl 1, S94-95.

Shih, J.C., Chen, K., Ridd, M.J. (1999) Monoamine oxidase: from genes to behavior. *Annu Rev Neurosci, 22*, 197-217.

Snell, E.E., Guirard, B.M. (1986). Pyridoxal-dependent histidine decarboxylase from Morganella AM-15. *Methods Enzymol, 122*, 139-143.

Suzuki, S., Tanaka, S., Nemoto, K. et al. (1998). Membrane targeting and binding of the 74-kDa form of mouse L-histidine decarboxylase via its carboxyl-terminal sequence. *FEBS Lett, 437*, 44-48.

Tahara, A., Nishibori, M., Ohtsuka, A. et al. (2000). Immunohistochemical localization of histamine N-methyltransferase in guinea pig tissues. *J Histochem Cytochem, 48,* 943-954.

Takemura, M., Tanaka, T., Taguchi, Y. et al. (1992). Histamine N-methyltransferase from rat kidney: cloning, nucleotide sequence, and expression in Escherichia coli cells. *J Biol Chem, 267,* 15687-15691.

Tanaka, S. (2000). SRP-independent targeting pathway to endoplasmic reticulum: histidine decarboxylase as a model protein. Seikagaku, 72, 121-125.

Travis, E.R., Wang, Y.M., Michael, D.J. et al. (2000). Differential quantal release of histamine and 5-hydroxytryptamine from mast cells of vesicular monoamine transporter 2 knockout mice. *Proc Natl Acad Sci USA, 97,* 162-167.

Weisburger, W.R., Mendelsohn, G., Eggleston, J.C. et al. (1978). Immunohistochemical localization of histaminase (diamine oxidase) in decidual cells of human placenta. *Lab Invest, 38,* 703-706.Viguera, E., Trelles, O., Urdiales, J. L. et al. (1994). L-amino acid decarboxylases producing 1,4-diamines: analogies among differences. *Trends Biochem Sci, 19,* 318-319.

Wang, L., Yan, L., McGuire, C. et al. (2001). Mouse histamine N-methyltransferase: cDNA cloning, expression, gene cloning and chromosomal localization. *Inflamm Res, 50,* 300-308.

Weihe, E., Eiden, L.E. (2000). Chemical neuroanatomy of the vesicular amine transporters. *FASEB J, 14,* 2435-2449.

Wessler, S., Rapp, U.R., Wiedenmann, B. et al. (2002). B-Raf/Rap1 signaling, but not c-Raf-1/Ras, induces the histidine decarboxylase promoter in Helicobacter pylori infection. *FASEB J, 16,* 417-419.

Whitby, L.G., Axelrod, J., Weil-Malherbe, H. (1961). The fate of H3-Norepinephrine in animals. *J Pharmacol Exp Ther, 132,* 194-201.

Wu, X., Kekuda, R., Huang, W. et al. (1998). Identity of the organic cation transporter OCT3 as the extraneuronal monoamine transporter (uptake2) and evidence for the expression of the transporter in the brain. *J Biol Chem, 273,* 32776-32786.

Wu, F., Yu, J., Gehring, H. (2008). Inhibitory and structural studies of novel coenzyme-substrate analogs of human histidine decarboxylase. *FASEB J, 22,* 890-897.

Yamamoto, J., Yatsunami, K., Ohmori, E. et al. (1990). cDNA-derived amino acid sequence of L-histidine decarboxylase from mouse mastocytoma P-815 cells. *FEBS Lett, 276*, 214-218.

Yamauchi, K., Sekizawa, K., Suzuki, H. et al. (1994). Structure and function of human histamine N-methyltransferase: critical enzyme in histamine metabolism in airway. *Am J Physiol, 267*, L342-349.

Yan, L., Galinsky, R.E., Bernstein, J.A. et al. (2000). Histamine N-methyltransferase pharmacogenetics: association of a common functional polymorphism with asthma. *Pharmacogenetics, 10*, 261-266.

Yan, L., Szumlanski, C.L., Rice, S.R. et al. (2000). Histamine N-methyltransferase functional polymorphism: lack of association with schizophrenia. *Am J Med Genet, 96*, 404-406.

Yatsunami, K., Ohtsu, H., Tsuchikawa, M. et al. (1994). Structure of the L-histidine decarboxylase gene. *J Biol Chem, 269*, 1554-1559.

Chapter 4

Histamine Receptors and Inflammatory Cells

Madeleine Ennis[1*], Milan Ciz[2], Karim Dib[1], Sheli Friedman[3], Roopesh Singh Gangwar[3], Bernhard F. Gibbs[4], Francesca Levi-Schaffer[3], Antonin Lojek[2], Helena Migalovich-Sheikhet[3], Liam O'Mahony[5], Tomas Perecko[2], Francisca Sánchez-Jiménez[6], José L. Urdiales[6], Ondrej Vasicek[2]

[1]Centre for Infection and Immunity, Health Sciences Building, School of Medicine, Dentistry and Biomedical Sciences, The Queen's University Belfast, 97 Lisburn Road, Belfast, BT9 7AE, UK

[2]Institute of Biophysics AS CR, v.v.i., Kralovopolska 135, 612 65 Brno, Czech Republic

[3]Institute for Drug Research, School of Pharmacy, Faculty of Medicine, The Hebrew University of Jerusalem, Jerusalem, Israel

[4]Medway School of Pharmacy, University of Kent, Chatham Maritime, ME4 4TB, UK

[5]Molecular Immunology, SIAF, Obere Strasse 22, CH-7270 Davos Platz, Switzerland

[6]Deparment of Molecular Biology and Biochemistry and CIBER de Enfermedades Raras (CIBERER). Faculty of Sciences. University of Málaga. 29071 Málaga, Spain

*Corresponding author:
Professor Madeleine Ennis, Centre for Infection and Immunity, Health Sciences Building, School of Medicine, Dentistry and Biomedical Sciences, The Queen's University Belfast, 97 Lisburn Road, Belfast, BT9 7AE, UK, Email: m.ennis@qub.ac.uk

4.1. Introduction

Up to now, four subtypes of human histamine receptors (HR)s have been identified: H_1R, H_2R, H_3R, and H_4R, all of them being GPCRs with seven transmembrane (TM)-spanning helices, which are expressed on various cell types.

The H_1R is expressed upon various cell types including eosinophils (Simons & Akdis, 2009). This receptor plays a significant role in allergic inflammation and immune modulation, as it increases the release of histamine and other mediators, chemotaxis of eosinophils and neutrophils, and enhances the antigen-presenting capacity and co-stimulatory activity of B cells.

H_2R has a key role in stimulating gastric acid secretion (Code, 2008; Hill *et al.*, 1997) while also posessing immunoregulatory effects (Akdis & Simons, 2006) and a role in allergic inflammatory responses by mediating mucus production in the airway, vasodilation, and bronchial smooth muscle relaxation (Parsons & Ganellin, 2006). Histamine via H_2R differentially reduces fMLP[1]-induced O_2^- generation in neutrophils and eosinophils (Reher *et al.*, 2012). Selective H_2R antagonists (famotidin, tiotidine and zolantadine) reduce histamine-induced effects of induction of intracellular cAMP accumulation in neutrophils and eosinophils (Reher *et al.*, 2012).

Classically H_3R acts as a presynaptic auto-receptor that inhibits the synthesis and release of histamine in the histaminergic neurons in the central nervous system (CNS) (Martinez-Mir *et al.*, 1990). H_3R is also highly expressed in eosinophils, dendritic cells and monocytes (Simons & Akdis, 2009), while low amounts are expressed in the peripheral tissues (Hill *et al.*, 1997). Activation of H_3R inhibits adenylate cyclase, reduces production of cAMP and inhibits Ca^{2+} influx (Gaudy-Marqueste, 2010). Histamine through H_3R may increase pruritus without the involvement of mast cells and also increase nasal congestion (Simons & Akdis, 2009). Selective agonists are in development for the disorders of central nervous system (Simons & Akdis, 2009). Chemotaxis of eosinophils via H_3R was controversial (Raible *et al.*, 1994); however, recent studies (see below) suggest that the newly described receptor, H_4R, mediates this effect.

The H_4R is a relatively novel histamine receptor structurally and pharmacologically related to the H_3R receptor (Liu *et al.*, 2001). H_4R is highly expressed in various cells of the immune system such as bone marrow and peripheral hematopoietic cells, eosinophils, mast cells, neutrophils, dendritic cells, T-cells and basophils (Parsons & Ganellin, 2006). H_4R mediates the chemotaxis of mast cells and eosinophils *in vivo*. Histamine through H_4R can increase calcium flux in human eosinophils (Simons & Akdis, 2009), and together with H_2R it can increase the IL-16 release from the lymphocyte (Parsons & Ganellin, 2006).

This chapter address the role of histamine and in particular that mediated by the H_4R receptor in different immune cells involved in the innate response. The role and function of the cells are described followed by an examination of the effects of histamine on the cells.

1 N-formyl-methionyl-leucyl-phenylalanine

4.2. Mast cells and Basophils

Mast cells and basophils are considered to be major allergic effector cells and are generated from CD34+ stem cells, which are produced in the bone marrow (reviewed in Schroeder, 2011; Metcalfe, 2008; Falcone *et al.*, 2011). Although both cell types share many morphological features and functions, basophils mature and are released into the blood from the bone marrow while mast cells arise from precursors that usually mature within tissues such as the skin, lung and gastrointestinal tract. Mast cells depend on stem cell factor (SCF), a cytokine produced by eosinophils, bone marrow endothelial cells and fibroblasts, for their survival, proliferation and maturation *in vitro* (Piliponsky *et al.*, 2002). Mast cells are relatively large and their cytoplasm is highly granulated (Figure 4.1). These granules contain and release pharmacologically active compounds such as histamine, tryptase, chymase and proteoglycans (Galli *et al.*, 2008). In humans, granules of tissue mast cells may contain either tryptase alone (found in lung alveoli, small intestinal mucosa and the mucosa in allergic eye disease) or tryptase together with mast cell-specific chymase, cathepsin G and carboxypeptidase A (mainly located in normal skin, blood vessels, the submucosa, and synovium) (Hsu & Boyce, 2009). A specific growth factor for basophils has not yet been discovered but they can be grown *in vitro* from CD34+ cells using IL-3, although in mice this cytokine exhibits some degree of redundancy since IL-3 knockout mice still produce some basophils (Lantz *et al.*, 1998; Shen *et al.*, 2008).

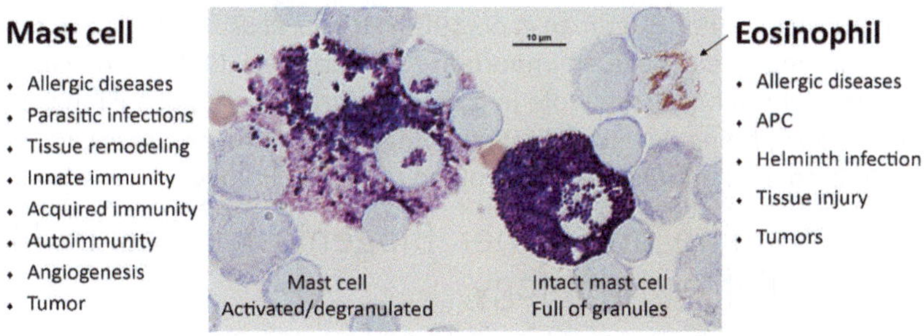

Figure 4.1 Mast cells and eosinophil from the peritoneal lavage of CD48 knockout mice stained with Giemsa. Mast cells, full of granules are stained in violet color and one eosinophil on the top right corner is stained orange-purple. The key functions of the mast cells are written on the left and those of eosinophil on the right.

4.2.1. Function

Basophils migrate from the blood into tissues affected by allergic inflammation (reviewed in Falcone, Knol & Gibbs, 2011). These cells, along with their tissue-fixed mast cell counterparts are major sources of histamine in humans, which is rapidly released upon allergen binding to antigen-specific IgE when bound to high affinity IgE receptors (FcεRI). Allergen-mediated activation of mast cells and basophils also results in the release of a large number of inflammatory mediators that govern the signs and symptoms of allergy. These include the release of eicosanoids (primarily PGD2 and LTC4, where mast cells release both and basophils release only LTC4) and various inflammatory cytokines. In humans, basophils are particularly prominent sources of the Th2-type cytokines IL-4 and IL-13, which cause IgE class switching in B cells leading to polyclonal IgE synthesis (Yanagihara *et al.*, 1998) whereas mast cells generate IL-8 and TNF-α (Gibbs *et al.*, 2001). These cell types, therefore, not only contribute to the symptoms of acute allergic inflammation but also appear to play immunomodulatory roles in supporting pro-allergic immunity (particularly basophils) and inflammation associated with allergic disease. These cells also respond to infections with certain parasites and it is thought that their main biological function may be related to parasite expulsion (Falcone *et al.*, 2001; Knight *et al.*, 2008), although their exact biological roles are still poorly understood.

Mast cells together with dendritic cells are the first immune system cells that respond to and interact with antigens, invading pathogens or environmentally derived toxins. Mast cells can enhance host resistance by promoting clearance of bacteria during several bacterial infections (Kalesnikoff & Galli, 2008; Piliponsky *et al.*, 2010; Thakurdas *et al.*, 2007). FcεRI mediates important interactions between mast cells and bacteria. IgE antibodies against *Staphylococcus aureus* or its toxin have been found in patients suffering from atopic dermatitis (Bunikowski *et al.*, 1999; Friedman *et al.*, 1985; Motala, *et al.*, 1986). These reports suggest that mast cells have an important role in bacterial infection. Mast cells are also thought to have some role in autoimmunity (Eller & Rosenkranz, 2012). New models are being developed for studying mast cell functions *in vivo* (Reber *et al.*, 2012) to help understand their role in pathophysiology.

4.2.2. Histamine and Other Biogenic Amines in Mast Cell Pathophysiology

Mast cells are the major histamine producers in humans (Schneider *et al.*, 2010). As in other important histamine producing cells, such as histaminergic neurons and enterochromaffin-like cells (ECLC), the newly synthesized histamine is stored in specialized vesicles until a stimulus promotes exocytosis (Woska & Gillespie, 2012). The enzyme responsible for histamine synthesis, histidine decarboxylase (HDC, EC 4.4.4.22), is associated with the internal membrane of endoplasmic reticulum (ER) in its active form (Suzuki *et al.*, 1998; Moya-García *et al.*, 2005; Furuta *et al.*, 2006). Thus, the first location of nascent histamine must be the ER lumen. From there, histamine is conducted to specialized granules. This raises the question of why histamine needs to be partitioned from the other intracellular components. Histamine is a mono- or dication depending on the polarity of the environment, due to the presence of an imidazole moiety. Imidazole groups are often involved in acid/base catalytic reactions as well as in electrostatic binding to aromatic moieties of many different macromolecules with specific functional consequences (Pilbak, *et al.*, 2012; Mutti *et al.*, 2011). These chemical characteristics may explain why a cell producing large amounts of histamine has evolved to keep this reactive amine separate from other molecules in the cytoplasm. In addition to 5 membrane receptors (histamine receptors 1-4 and N-methyl aspartate receptor), histamine can bind to several types of ion transporters, members of Cytochrome P450 (Cyp) family, and nucleic acids (Ruiz-Chica *et al.*, 2006; Schneider *et al.*, 2010).

An increase in free, newly synthesized histamine leads to a decrease in the cellular growth rate, as demonstrated in transfected cell cultures unable to store the amine in appropriate vesicles (Abrighach *et al.*, 2010), where the G1/S transition is blocked in HEK cells transfected to express active HDC. This could explain the lack of stable transfected cell models overexpressing active mammalian HDC. These observations on transfected cultured cells are consistent with the fact that histamine is not actively produced by mast cell precursors during the first stages of mouse bone marrow cell differentiation *in vitro*, when more active cell proliferation occurs (García-Faroldi *et al.*, 2009a). This suggests an antiproliferative role for newly synthesized free histamine in mammalian cells.

Mast cell granules are mainly composed of several mast cell specific proteases (e.g. tryptase and chymase), peptidoglycans (e.g. serglycin), and biogenic amines. However, histamine is not the only biogenic amine present in mast cell granules. Some rodent mast cell subsets contain serotonin (the decarboxylation product of the HDC paralogue, the aromatic L-amino acid decarboxylase or dopa decarboxylase (DDC, EC 4.1.1.28) (Moya-García *et al.*, 2005). In addition, polyamines are essential for normal mast cell differentiation (García-Faroldi *et al.*, 2010). The natural polyamines, putrescine, spermidine and spermine are ubiquitous aliphatic amines that have low molecular weight and are highly charged cations under physiological conditions. Putrescine is formed by decarboxylation of ornithine, a reaction catalysed by the enzyme ornithine decarboxylase (ODC, EC 4.1.1.17). Higher polyamines, spermidine and spermine, are synthesized by adding an aminopropyl group from decarboxylated S-adenosyl-methionine to putrescine and spermidine, respectively (Urdiales *et al.*, 2001). Spermidine and spermine are polycations (3 and 4 positive charges at physiological pH) able to interact with anionic macromolecules, such as nucleic acids and proteoglycans (Belting *et al.*, 2003; Poulin *et al.*, 2012; Ruiz-Chica *et al.*, 2001a-b and 2003), and other protein targets for histamine (e.g. NMDA receptor and CytP450) (Colwell & Levine, 1997; LaBella & Brandes, 2000). Taking into account these chemical and docking properties, it is not surprising that polyamines contribute to the normal mast cell granule conformation as is the case for the other biogenic amines, histamine and serotonin (Ringvall *et al.*, 2008; Rönnberg *et al.*, 2012). Thus, mast cells provide an excellent model to study the biochemical and physiological interplay among biogenic amines (Medina *et al.*, 2005; Sánchez-Jiménez *et al.*, 2007).

Higher polyamines are essential to maintain cell proliferation and viability (Kahana, 2009; Pegg, 2009). In contrast, excess of non-compartmentalized histamine can be an anti-proliferative factor (Abrighach *et al.*, 2010). In addition, an excess of free polyamines can be deleterious as a source of reactive oxygen species and toxic aldehydes produced during their degradation (Babbar *et al.*, 2007). In mammalian cells, including mast cells, there are clear insights to indicate opposite activity patterns among the biosynthetic pathways of both histamine and polyamines (García-Faroldi *et al.*, 2009b). Thus, histamine interferes with both ornithine and spermine uptake systems and is a negative non-direct modulator of ODC activity in mammalian cells (Abrighach *et al.*, 2010; Fajardo *et al.*, 2001a-b). These intertwined regulatory mechanisms are probably part of the molecular coordination of cell proliferation/differentiation programs in mammalian cells. In fact, in mouse bone marrow-derived mast cells, ODC is expressed and the highest spermidine and spermine levels are observed during the initial differentiation stages, in contrast to histamine concentrations (García-Faroldi *et al.*, 2009a).

The antiproliferative role of histamine, in agreement with the observation on transfected HDC overexpressing cells and both the HDC expression profile and histamine production during mast cell differentiation, is apparently in contrast with high HDC expression observed in mast-cell derived malignancies, such as mastocytosis. Valent's group reported that human mastocytosis mast cells display increased expression of HDC and aberrant granules and proposed HDC as a marker of disease malignancy (Krauth *et al.*, 2006). More recently, bone marrow cell hybridomas have also been obtained with high HDC expression (Kawahara, 2012). However, neoplastic cells have severe alterations of their signal transduction mechanisms (such as c-Kit in the case of the most aggressive forms of mastocytosis) (García-Montero *et al.*, 2006), so the behaviour of transformed mast cells with respect to intracellular histamine signalling may be altered compared to normal cells. For instance, in chronic myeloid leukaemia, the oncoprotein BCR/ABL induces HDC expression; of course in these cells, functional properties of some important elements for cell life/death equilibrium are altered (Aichberger *et al.*, 2006). Another possibility associating the survival of neoplastic cells with increased histamine production could be that the neoplastic histamine-producing cells would have higher or new capacities to "hidden" histamine with respect to their normal counterparts. This hypothesis should be pursued since transformed mast cells have altered granule morphologies and histamine (as well as polyamines) can bind to nucleic acids and has been located in cancer cell nuclei (Medina *et al.*, 2008; Ruiz-Chica *et al.*, 2006).

To our knowledge, there are no pharmacological methods to prevent histamine synthesis by mast cells. Two different inhibitors have been described as substrate analogues and their mechanisms of action have been characterized: monofluoromethyhistidine (alpha-FMH) and histidine methyl ester (HME) (Olmo *et al.*, 2002; Rodríguez-Caso *et al.*, 2003). However, neither can be used clinically. In contrast, the tea polyphenol epigallocatechin 3-gallate (EGCG) has been described as an inhibitor of HDC and DDC. EGCG can inhibit both paralogue enzymes by altering the environment of the cofactor and most probably occluding the catalytic site entrance for the substrate (Ruiz-Pérez *et al.*, 2012). Consequently, these results could be the beginning for development of new preventive and/or therapeutic agents to be used, at least topically, for inflammatory pathologies where mast cell-derived biogenic amines could be an important part of the problem (e.g. prevention of atopic allergy reactions, atopic dermatitis, parasite infections, etc). The antitumourigenic and anti-inflammatory properties described for EGCG (as an inhibitor of angiogenesis and NF-κB signalling) suggest that EGCG or its derivatives may be promising drugs against mastocytosis (Melgarejo *et al.*, 2007, 2009 and 2010 a and b). Other rare diseases could also benefit from the development of new histamine synthesis modulators (Pino-Ángeles *et al.*, 2012).

An overview of the interactions between histamine, polyamines and various cell types is given in Figure 4.2. Histamine can have autocrine effects on mast cells usually *via* the H_1R and H_4R. Modulation of the activity of these receptors to discover new and more effective anti-inflammatory strategies is also an active field of R&D (Ohsawa & Hirasawa, 2012). It is the subject of other chapters/subsections of this volume.

4.2.3. Action of Histamine on Mast Cells and Basophils

In 1953, Riley & West reported for the first time that the major storage site of histamine in mammalian tissues was located in the mast cells (Riley & West, 1953), usually in the cells' secretory granules. Mast cells not only release histamine upon activation through FcεRI but also via other stimuli such as cytotoxic agents, polysaccharides, lectins, anaphylatoxins, calcium and many basic compounds (such as compound 48/80) (Lagunoff, Martin, & Read, 1983). Eosinophils can also activate the mast cells to release histamine (Piliponsky *et al.*, 1999). Histamine can act both in stimulatory and inhibitory ways on immune cells; it can enhance the antigen-presentation by dendritic cells, suppress TNF-α and IL-12 and increase IL-10 production by dendritic cells and monocytes (Hsu & Boyce, 2009). Recently, histamine has been reported to induce neuronal hypertrophy and increase the mast cell density in the gastrointestinal tract

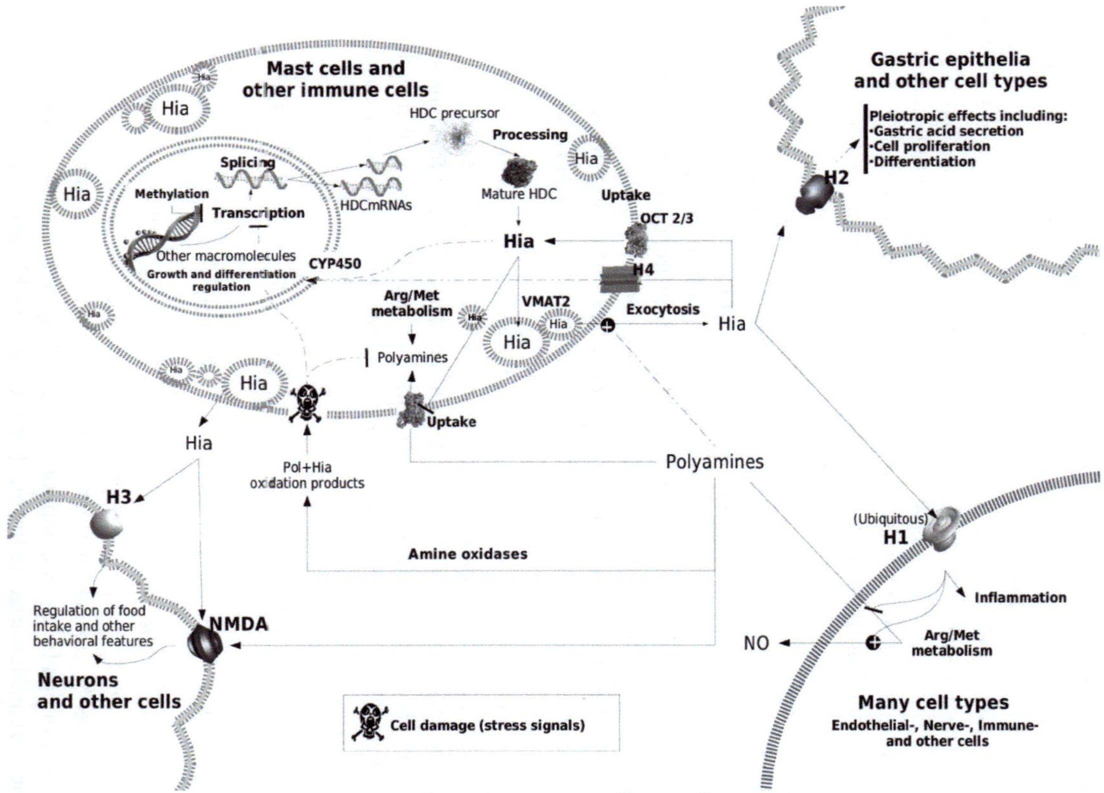

Figure 4.2 Effects of histamine on different cell types

Abbreviations: CYP450: cytochrome P450; Hia: histamine; H$_{1-4}$: histamine receptor 1-4 subtype; NMDA: N-methyl-D-aspartate receptor; NO: nitrogen oxide; OCT 2/3: organic cation transporter 2/3; Pol: polyamines; VMAT2: vesicular monoamine transporter2

(Keles *et al.*, 2012). Histamine modulates cell events and plays an important role in immune modulation and allergic inflammation through histamine receptors (Table 4.1). Histamine acts through these receptors to mediate effects that include vasodilation and vasopermeability, smooth muscle contraction of bronchial and gastrointestinal tract, secretion of gastric acid, and induction of pruritus (Jutel, Blaser, & Akdis, 2005).

Despite being the major producers and releasers of histamine, it has long been appreciated that mast cells and basophils are also affected by an autocrine action of the amine, which may play an important role in limiting their ability to further degranulate (Bourne *et al.*, 1971; Lippert *et al.*, 2004). However, owing to the difficulty in obtaining and purifying isolated primary human mast cells and basophils, a comprehensive analysis of HR expression (H_1-4R) has not yet been completed in these cells. Despite this, much of the feedback inhibition in basophils and human skin mast cells is thought to be H_2R-mediated (Lichtenstein & Gillespie 1973) although less is known regarding human lung mast cells or other mast cell types, which often display a marked degree of functional heterogeneity (Peters *et al.*, 1982, MacGlashan, 2003).

H_2R agonists (such as impromidine) but not H_1R agonists mimic the inhibitory actions of histamine on mast cells activated by compound 48/80 (an IgE-independent stimulus) and are reversed by the H_2R antagonist cimetidine (Masini *et al.*, 1982). However, the H_2R-mediated inhibitory actions of histamine caused by IgE-dependent activation, seem to be more prominent in basophils than mast cells (Kazimierczak *et al.*, 1981; Summers *et al.*, 1981; Peters *et al.*, 1982). Recently, it was reported that this receptor is involved in the early suppression of basophils to release all known major mediator classes during venom-allergen immunotherapy (Novak *et al.*, 2012). The mechanism responsible for H_2R-mediated inhibition of basophil activation is due to elevation of cyclic AMP and subsequent inhibition of extracellular calcium influx into these cells (Botana & MacGlashan, 1994; Lippert *et al.*, 2004).

Table 4.1
Effects of histamine through its specific receptors on mast cells and eosinophils.

Receptor	**Mast cell**	**Eosinophil**
H_1R	Activation ↑, IP3/[Ca^{2+}] ↑	Chemotaxis ↑
H_2R	Chemotaxis ↓	fMLPO_2^- production ↓ IL-4 mediated inflammation ↑
H_3R	Expressed in brain mast cell Auto-receptor	Chemotaxis ↑*
H_4R	Chemotaxis ↑, cAMP ↑, [Ca^{2+}] mobilization	Cell shape change chemotaxis ↑, CD11b ↑, CD54 ↑

* controversial, ↑ increase, ↓ decrease

H_1R-antagonists also lead to increased intracellular cAMP generation due to competitive antagonism with H_2R (Palacios *et al.*, 1978; Lippert, *et al.*, 2004) and some more recent generation H_1R antagonists can reduce mast cell function (Levi-Schaffer, 2009). However, H_1R expression in these mast cells and basophils is relatively low, although there is tentative evidence to suggest that it may be higher in immature mast cells (Lippert *et al.*, 2004) and guinea pig basophils, where this receptor is involved in histidine uptake (Stewart & Kay, 1980). In a mouse model of allergen-induced pulmonary inflammation, H_1R was found to have a key role in T cell chemotaxis (Bryce *et al.*, 2006). It also increases the Th1 type of cellular immune response and IFN-γ production. However, H_1R decreased the IgE production (Simons & Akdis, 2009).

Certain H_1R antagonists such as terfenadine (but not cetirizine) can inhibit IgE-dependent mediator release from basophils (Gibbs *et al.*, 1998) but only at high concentrations (μM range) and it is unlikely that these effects are H_1R specific. Nevertheless, new generation H_1R antagonists are used as anti-allergic/mast cell stabilizing drugs due to their ability to inhibit mast cell activation (Levi-Schaffer & Eliashar, 2009), even though some of them might increase cellular cAMP in competitive antagonism with H_2R. It has become increasingly apparent that species differences in histamine receptor isoforms may considerably impact on the agonist-antagonist pharmacology of various $H_{1\text{-}4}$R modulating agents, their potency as well as their specificity to a particular HR type (Schnell *et al.*, 2011; Seifert *et al.*, 2003; Seifert *et al.*, 2011).

The H_3R is expressed in brain mast cells but there have been conflicting reports regarding the actions of H_3R modulators on peripheral tissue mast cells (Bissonnette, 1996; Nemmar *et al.*, 1999; Rozniecki *et al.*, 1999). In basophils, the H_3R does not appear to affect function (Kleine-Tebbe *et al.*, 1990; Tedeschi *et al.*, 1991). These reports may reflect the problem associated with the 40% sequence homology of H_3R with H_4R (Liu *et al.*, 2001; Morse *et al.*, 2001) where agents such as α-methylhistamine (agonist) and thioperamide (antagonist) affect both receptor types.

Lippert *et al.* showed that H_3R is not expressed on human skin mast cells, in contrast to high levels of H_4R expression (Lippert *et al.*, 2004). Basophils also express H_4R (Hofstra *et al.*, 2003) but, as with human mast cells, there are no data at present to indicate a major role for this receptor in controlling mediator release from human mast cells and basophils. In mice, chemotaxis, intracellular calcium mobilization and LTB_4 have been shown to be affected by H_4R triggering, though degranulation and cAMP changes are not directly affected (Hofstra *et al.*, 2003; Rosethorne & Charlton, 2011; Takeshita *et al.*, 2003; Godot *et al.*, 2007). Even though H_4R has not been shown to substantially affect mast cell and basophil mediator release, it is thought to play a role in effector cell (such as eosinophil) recruitment to tissues affected by chronic allergic inflammation (Hofstra *et al.*, 2003). H_4R was also shown to control human mast cell precursor trafficking in the presence of CXCL12 (Godot, *et al.*, 2007) and regulate the migration of mast cells and eosinophils into guinea pig airway epithelial tissue after allergen challenge (Yu *et al.*, 2008). It would be of considerable interest if a reduction in mast cell numbers within tissues affected by allergic inflammation were demonstrated in humans due to a suppression of H_4R-mediated trafficking of mast cells and their precursors, especially given the other properties H_4R antagonists/inverse agonists on suppressing histamine-related itch (Ohsawa & Hirasawa, 2012). However, the species differences in HR isoforms and the pharmacological effects of various HR-modulating agents may be a major obstacle in the rapid development of new H_4R–blocking drugs.

4.3. Eosinophils

Eosinophils are blood-borne granulocytes that migrate into tissues in some physiological and pathological conditions where they survive for several days. Eosinophils are generated from myeloid progenitor cells in the bone marrow. The differentiation of myeloid progenitor cells into mature eosinophils requires critical transcription factors, namely GATA-1[2], PU.1[3] and C/EBP[4] (McNagny & Graf, 2002; Nerlov & Graf, 1998; Nerlov *et al.*, 1998). The development and proliferation of eosinophils is positively regulated by granulocyte-macrophage colony-stimulating factor (GM-CSF), as well as IL-3 and IL-5 cytokines (Lopez *et al.*, 1986; Lopez *et al.*, 1988; Rothenberg *et al.*, 1988; Takatsu, Takaki, & Hitoshi, 1994). Eosinophils consist of a characteristic bilobed nucleus and a cytoplasm enriched in secondary granules, which contain major basic protein (MBP), eosinophil peroxidase (EPO), eosinophil cationic protein (ECP) and eosinophil-derived neurotoxin (EDN). These cationic proteins are cytotoxic molecules that play an important role in inflammation, allergy, parasitic infection, tissue injury and tumors. Eosinophil cytoplasm also contains other 'granule' types, namely: small granules, primary granules, lipid bodies and small secretory vesicles. Eosinophils respond to diverse stimuli by degranulating and releasing these cationic proteins, lipid mediators, cytokines, chemokines, and neuromodulators (Gleich & Adolphson, 1986). Eosinophils express various cell surface markers, such as Fc receptors, CCR3, PAF receptors, CD48, 2B4, Siglec-F and histamine receptors, which are responsible for cell-cell communication via various pathways. These receptors are grouped into adhesion molecules, cytokine

2 A zinc family finger member
3 An ETS family member
4 CCAAT/enhancer-binding protein family

receptors, immunoglobulin receptors and members of the immunoglobulin superfamily, chemotactic factors, enzymes, and molecules associated with apoptosis and cellular signalling (Rothenberg & Hogan, 2006).

4.3.1. Function of Eosinophils

Eosinophils have many functions (Figure 4.3) associated with the pathogenesis of various inflammatory diseases including parasitic helminth infections (Wardlaw & Moqbel, 1992;Weller, 1994), intestinal immunity (Hogan, Waddell, & Fulkerson, 2012), allergic diseases such as asthma, allergic rhinitis, atopic dermatitis, etc. (Weller, 2009) and they can function as antigen presenting cells (APCs) since they can both process and present a variety of microbial, viral, and parasitic antigens (Shi, 2004). Eosinophils secrete various cytokines capable of promoting T-cell proliferation, activation and Th1/Th2 polarization. Murine eosinophils promote the secretion of IL-4, IL-5 and IL-13 cytokines by

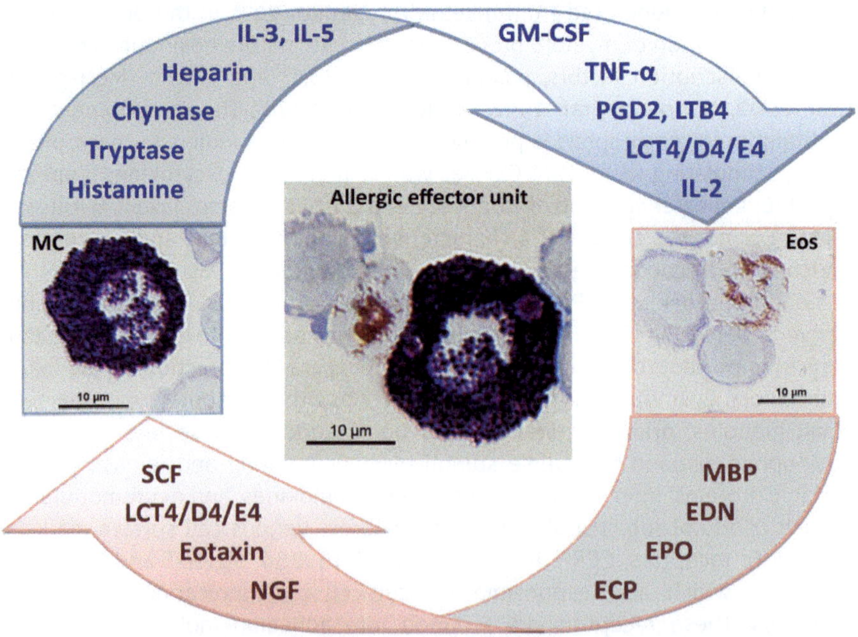

Figure 4.3 Soluble and physical interactions between mast cells and eosinophils. In this illustrative diagram, soluble interactions have been shown by boxed arrows with some of the key mediators secreted by mast cells and eosinophils responsible for mediating soluble interactions between them. At the centre, the allergic effector unit showing physical interaction between eosinophils and mast cells.

CD4+ T-cells (MacKenzie *et al.*, 2001). Human eosinophils also produce nerve growth factor (NGF), which is required for survival and functional maintenance of sympathetic neurons and also regulates immune responses (Solomon et al., 1998). However the antigen presenting capacity of eosinophils is a controversial subject (van Rijt *et al.*, 2003) as reviewed in (Rothenberg & Hogan, 2006).

4.3.2. Action of Histamine on Eosinophils

Mast cells together with eosinophils, form "the allergic effector unit" (Figure 4.3), that is considered to be the basic functional unit of allergy. Eosinophils and mast cells communicate with each other in a bidirectional manner. This interaction is both physical through cell surface receptors such as CD48, 2B4 (Elishmereni *et al.*, 2011; Minai-Fleminger *et al.*, 2010) and/or soluble through various cytokines and chemokines (Minai-Fleminger & Levi-Schaffer, 2009) (Figure 4.3). Eosinophils can regulate mast cell function and *vice versa*. Eosinophil MBP can activate human umbilical cord blood–derived mast cells to release histamine, IL-8, GM-CSF, PGD-2, and TNF-α (Piliponsky *et al.*, 2002). The mast cell protease chymase activates eosinophils and promotes the production of SCF. Eosinophil released NGF promotes mast cell survival and activation (Bullock & Johnson, 1996; Horigome, Bullock, & Johnson, 1994). NGF also acts in an autocrine manner by activating the release of EPO (Solomon *et al.*, 1998). EPO activates rat peritoneal mast cells to release histamine. The release of histamine occurs in a dose dependent manner upon incubation with MBP, EPO, and ECP (Zheutlin *et al.*, 1984). Histamine can promote superoxide (O_2^-) production in eosinophils which enhances the expression of C3b receptor and other membrane receptors (Pincus, DiNapoli, & Schooley, 1982).

Overall, histamine influences eosinophil cell shape, upregulates the expression of adhesion molecules (CD11b and CD54) and induces chemotaxis (Ling *et al.*, 2004). Histamine-induced eosinophil chemotaxis is blocked by the H_4 receptor antagonist JNJ 7777120 and the H_3/H_4 receptor antagonist thioperamide, but not by H_1 (diphenhydramine), H_2 (ranitidine) or H_3 receptor antagonists (Ling *et al.*, 2004). Moreover, selective H_1R- and H_2R agonists do not mimic the effects of histamine. In contrast, H_2R plays an inhibitory role in the activation of chemotaxis. Intriguingly, there is no inhibitory impact of H_2R on the regulation of intracellular calcium Ca^{2+} concentrations, indicating that chemotaxis is regulated independently of rises in intracellular Ca^{2+} levels (Reher *et al.*, 2012).

Synergy between histamine- and IL-4-mediated recruitment of eosinophils to the lung has been reported, as histamine is necessary to generate IL-4-driven eosinophilic inflammation, mediated via H_2R. Additionally, alveolar epithelial cells require H_2R to produce CCL24, an eosinophil recruitment factor (Swartzendruber, Byrne, & Bryce, 2012). Allergic asthma was induced in the histidine decarboxylase deficient mouse model and it was found that histamine positively controls eosinophilia (mediated by H_4 receptor) but not bronchial hypersensitivity (Ohtsu, 2010). In a mouse model of asthma, H_4R was found to influence the induction of Th2 responses by dendritic cells (Dunford *et al.*, 2006) and may play a role in mediating pruritus (Dunford *et al.*, 2007).

Based on these results, an effective treatment of allergy should be developed as a combinational therapy that targets not only H_1R but the other receptors as well, especially H_4R. This kind of therapy can interfere with both mast cells and eosinophils and reduce the overall allergic inflammation.

4.4. Monocytes and Macrophages

The mononuclear phagocytic system is generated from committed haematopoietic stem cells located in the bone marrow (Murray & Wynn, 2011). Monocytes released from bone marrow into the circulation that can further differentiate into a range of tissue macrophages and dendritic cells (Fogg *et al.*, 2006; Hume *et al.*, 2002; Shi & Pamer, 2011). Bloodstream monocytes are subdivided into subsets that differ in size, trafficking and innate immune receptor expression and in their ability to differentiate following stimulation with cytokines and/or microbial molecules (Auffray *et al.*, 2009; Geissmann *et al.*, 2003). Monocytes mediate host antimicrobial defence (Serbina *et al.*, 2008) and are also implicated in many inflammatory diseases (Woollard & Geissmann, 2010). Although several subsets of monocytes and macrophages have been identified, the individual contributions of these subsets to health and disease are not well known, and it is probable that additional, functionally distinct subsets exist (Chow *et al.*, 2011).

4.4.1. Function of Monocytes/Macrophages

When monocytes migrate from the circulation and extravasate through the endothelium, they differentiate into macrophages or dendritic cells (Murray & Wynn, 2011). Thus, the primary role of monocytes is to replenish the pool of tissue-resident macrophages and dendritic cells in steady state and in response to inflammation. Monocytes, dendritic cells and macrophages, along with neutrophils and mast cells, are 'professional' phagocytic cells which express a multitude of receptors on their surfaces that detect signals that are not normally found in healthy tissues. An initial level of macrophage activation occurs when early warning signals trigger monocyte recruitment and *in situ* activation, or when IL-4 induces *in situ* macrophage proliferation (Jenkins *et al.*, 2011). Tissue damage sensing is probably crucial at the second level of macrophage response, regardless of whether the damage is of a microbial nature. The mechanisms of tissue damage sensing have been discussed in recent reviews (Chen & Nunez,

2010; Matzinger & Kamala, 2011). Beyond the initial activation and stimulation of macrophages, cooperative actions of multiple sensors, feed forward cytokine networks and inter-organ communication increase the output of monocytes and neutrophils driving inflammatory responses. Macrophage effectors work together in cell-intrinsic and cell-extrinsic networks (Nish & Medzhitov, 2011). For example, the production of interferon-γ (IFNγ) by T helper 1 (Th1) cells requires IL-12 production from activated mononuclear phagocytes, where IFNγ then stimulates macrophages to activate the antimicrobial mechanisms (Borden *et al.*, 2007).

The activation of mononuclear phagocytes leads to the generation of the superoxide anion radical and nitric oxide (NO) by NADPH oxidase as well as inducible nitric oxide synthase (iNOS), respectively (Ambrozova *et al.*, 2010; Ambrozova *et al.*, 2011). Both the superoxide anion radical and nitric oxide generate secondary reactive oxygen species (ROS) and reactive nitrogen species (RNS). Physiologically, ROS and RNS formation is one of the essential microbicidal mechanisms in the body. Although formation of reactive species is desirable for host defence, their overproduction can cause damage to the body's own cells and tissue injury can contribute to the development of a number of serious diseases. Thus, the modulation of their production is an important target in the treatment of immune and inflammatory diseases (Lojek *et al.*, 2008; Ren & Chung, 2007).

A key component of the macrophage response is also the production of anti-inflammatory feedback mechanisms that encompass cell-intrinsic signalling feedback loops and cell-extrinsic mechanisms, such as the production of IL-10, which is an essential and non-redundant anti-inflammatory cytokine (Murray & Wynn, 2011). This part of macrophage response is the least clear and involves the final balance between chronic inflammation and re-establishment of homeostasis. The understanding of the underlying mechanisms that restore homeostasis after an inflammatory reaction underpins all research efforts related to chronic inflammatory diseases.

4.4.2. Action of Histamine on Monocytes/Macrophages

Histamine via H_1R, H_2R, H_3R, and H_4R has both proinflammatory or anti-inflammatory effects, depending on the predominance of the types of histamine receptors (Tripathi *et al.*, 2010) and histamine concentration. Histamine stimulates the release of proinflammatory cytokines and lysosomal enzymes from human macrophages.

Khan and Rai (Khan & Rai, 2007) reported that histamine differentially regulated the testicular macrophage immune responses in wall lizards. It inhibited phagocytosis and superoxide production at high concentrations (10^{-5} M) while it stimulated superoxide production and did not affect phagocytosis at low concentrations (10^{-10} M). Using selective H_1R and H_2R antagonists, pyrilamine and famotidine respectively, the H_1R subtype was responsible for mediating the inhibitory effect of histamine on testicular macrophage immune responses at high concentrations, while H_2R were involved in the stimulation at low concentrations. In contrast, Azuma and co-workers (Azuma *et al.*, 2001) found that *in vitro* treatment of macrophages with histamine resulted in inhibition of chemotaxis and histamine at 10^{-5} M markedly inhibited the production of superoxide anions from both opsonized zymosan-A and phorbol 12-myristate 13-acetate stimulated macrophages. Furthermore, histamine at a concentration range of 10^{-7} to 10^{-5} M significantly inhibited phagocytosis of *Escherichia coli* by macrophages. In addition, the H_2R-selective agonist dimaprit resulted in inhibition of macrophage chemotaxis and markedly inhibited the production of superoxide anion by phorbol 12-myristate 13-acetate-stimulated macrophages and phagocytosis of *E. coli* by macrophages. In contrast, both histamine and dimaprit caused a concentration-dependent inhibition of lipopolysaccharide-induced production of TNFα and IL-12 by macrophages. These results suggest that histamine and dimaprit may inhibit chemotaxis, phagocytosis, superoxide anion production and the production of TNFα and IL-12 by macrophages via the histamine H_2R.

Monocytes and tissue macrophages exhibit different responses to histamine treatment, indicating that differentiation of monocytes to tissue macrophages might induce changes in the expression of H_1R. Triggiani and co-workers (Triggiani *et al.*, 2007) found that human lung macrophages had a higher expression of H_1R mRNA and protein than monocytes. Moreover, they found high protein expression of H_1R in monocyte-derived macrophages in contrast to the nearly undetectable expression of this protein in monocytes. Simultaneous analysis of H_2R and H_1R mRNA expression indicated that the H_2R/H_1R ratio is approximately 100-fold higher in monocytes in comparison with monocyte-derived macrophages. Similarly, a shift from H_2R to H_1R during differentiation of histiocytic lymphoma cells U937 and human peripheral blood monocytes into macrophages was also reported (Wang *et al.*, 2000). U937 cells are widely used as a model for monocytes. Differentiation of these cells increased H_1R expression and decreased the H_2R expression. These effects may be explained by the results of Murata and coworkers (Murata *et al.*, 2005), who found H_2R mRNA in U937 cells is constitutively expressed and does not react to GM-CSF. In contrast, GM-CSF increased mRNA expression for H_1R. These U937 cells do not express H_3R (Wang *et al.*, 2000). Spontaneous and retinoic acid-induced differentiation of normal human monocytes and of leukemic THP-1 monocytes into macrophages results in a progressive loss of adenosine 3',5'-cyclic monophosphate production induced by histamine via H_2R (Mirossay *et al.*, 1994). In THP-1 cells and in HL-60 cells, retinoic acid treatment increased the abundance of mRNA expression of the H_2R gene 4-fold, suggesting transcriptional control by a retinoic acid response element. In THP-1 macrophages, histamine inhibited phorbol 12-myristate 13-acetate-induced H_2O_2 formation via the activation of H_2R. Expression of the H_2R gene, histamine accumulation, and histidine decarboxylase activity were also demonstrated in normal human monocytes/macrophages and peripheral lymphocytes.

Vignola and coworkers (Vignola *et al.*, 1994) found that histamine induced a significant increase in alveolar macrophages expressing the LFA-1, ICAM-1 and CD23b membrane markers and a significant increase in the release of fibronectin. The histamine effects were H_1R specific, since they were significantly inhibited by an H_1R antagonist pyrilamine. In contrast, Sirois and co-workers found that when alveolar macrophages from humans, Sprague Dawley rats and the rat alveolar macrophage cell line NR8383 were treated with histamine prior to their stimulation with lipopolysaccharide, histamine inhibited lipopolysaccharide-stimulated TNF release in a dose-dependent manner (Sirois *et al.*, 2000). This inhibition was mimicked by H_2R and H_3R agonists, but not by the H_1R agonist betahistine. Furthermore, these authors reported the expression of H_3R mRNA in human alveolar macrophages. Other authors (Morichika *et al.*, 2003; Takahashi *et al.*, 2002) found that histamine inhibited CD14, ICAM-1 and CD80 expression on human peripheral blood mononuclear cells induced by lipopolysaccharide and IL-18, respectively. TLR stimulation of monocytes is altered by histamine co-incubation as secretion of pro-inflammatory cytokines such as TNFα, IL-12 and IL-18 is suppressed, while IL-10 secretion is enhanced. The modulatory effects of histamine on ICAM-1 expression and cytokine production were antagonized by the H_2R antagonist famotidine but not by d-chlorpheniramine and thioperamide, and were mimicked by selective H_2R agonists but not by H_1R, H_3R and H_4R agonists, indicating the involvement of the H_2R in the action of histamine (Takahashi *et al.*, 2002, 2004; Morichika, *et al.*, 2003). Monocytes stimulated by advanced glycation end products are known to up-regulate adhesion molecule expression but this effect is inhibited by the H_2R (Zhang *et al.*, 2010). Using THP-1 cells, Tanimoto and colleagues (Tanimoto *et al.*, 2001) showed that H_1R and H_2R mRNA were present in non-differentiated monocytic THP-1 cells, whereas in TPA-differentiated THP-1 macrophages, mRNA encoding for H_1R, H_2R and H_4R (very low) was present. Nevertheless, the H_1R and H_2R mRNA were the most abundant. The monocytic THP-1 cells expressed higher amounts of H_2R mRNA than H_1R mRNA in comparison with differentiated THP-1 cells.

Dijkstra and co-workers (Dijkstra *et al.*, 2007) showed the expression of H_4R in human monocytes obtained from peripheral blood by using flow cytometry. The expression of H_4R was increased by IFN-γ. Histamine and H_4R agonists (clobenpropit and 4-methylhistamine) induced calcium mobilization in monocytes. These effects were blocked by H_4R antagonist JNJ7777120. In contrast, these agonists inhibited production of CCL2 by monocytes. Supernatants of H_4R agonist-stimulated monocytes attracted fewer monocytes in transmigration assays, resulting in a reduction of monocyte recruitment. Damaj and colleagues (Damaj *et al.*, 2007) described the protein expression of H_4R in human monocytic cell lines THP-1, U937 and human peripheral blood monocytes using flow cytometry or immunoblot assays. Their results showed much lower expression of H_4R in peripheral blood monocytes. Ohki and co-workers (Ohki *et al.*, 2007) showed co-expression of H_4R with macrophage markers CD68 and CD163 in macrophage-like cells from the human synovial tissues of patients with rheumatoid arthritis. Recently, Gschwandtner and colleagues (Gschwandtner *et al.*, 2013) reported that several H_4R agonists reduced the secretion of IL-12p70 from monocytes but the EC_{50} values were lower than those obtained with pharmacological assays.

From the available data, it is clear that histamine modulates macrophage/monocyte activity and their physiological/pathophysiological effects. It remains to be clarified which effects are receptor mediated and which histamine receptors are involved in histamine effects on macrophages.

4.5. Dendritic Cells

Dendritic cells are potent antigen-presenting cells that are present throughout the body and are central players in initiating and regulating innate and adaptive immune responses, particularly at mucosal sites. Dendritic cell survival, activation, maturation and polarization are largely influenced by local factors within their micro-environment such as microbial components, cytokines and metabolic products. Dendritic cells shape the functional polarization and differentiation of the reactive T-cells into Th1, Th2, Th9, Th17 and T_{reg} responses by producing cytokines such as IL-12, IL-18, IL-23, IL-11, IL-10 or TGFß (Akdis *et al.*, 2011). The selection of an appropriate cytokine secretion pattern by dendritic cells is dependent on a number of factors, but is particularly influenced by the binding of microbial ligands, termed pathogen-associated molecular patterns (PAMPs), to pattern recognition receptors (PRRs) such as toll-like receptors (TLRs) and C-type lectin receptors (CLRs) (O'Mahony *et al.*, 2006; Shilling *et al.*, 2007). Dendritic cells are often found in close proximity to degranulated mast cells and can secrete histamine following activation under certain conditions.

4.5.1. Action of Histamine on Dendritic Cells

While dendritic cells have been shown to express H_1R, H_2R and H_4R, it is not definitively established if the level of HR expression is altered during maturation or if different dendritic cell subsets (e.g. myeloid or plasmacytoid dendritic cells) preferentially up- or down-regulate individual HRs following specific micro-environmental cues. Dendritic cells exposed to histamine have been shown to up-regulate their antigen-presenting capacity and Th1 polarization via the H_1R

while activation of the H_2R on dendritic cells preferentially drives IL-10 secretion and, under certain circumstances, promotes Th2 polarisation (Caron *et al.*, 2001; Mazzoni *et al.*, 2001; van der Pouw Kraan *et al.*, 1998). H_2R activation of human plasmacytoid dendritic cells leads to a significant down-regulation of IFNγ and TNFα secretion following CpG stimulation (Mazzoni *et al.*, 2003). In addition, accumulation of plasmacytoid dendritic cells and $CD11b^+$ dendritic cells, but not $CD8^+$ dendritic cells, in draining lymph nodes is H_2R dependent (Dawicki *et al.*, 2010). Furthermore, histamine reduces the production of NADPH-oxidase-derived oxygen radicals by several types of myeloid cells, an effect mediated by the H_2R (Hellstrand *et al.*, 1994). However, Gschwandtner *et al.*, (2011) found that human plasmacytoid dendritic cells from patients with psoriasis had high levels of the H_4R compared to controls or those from atopic dermatitis patients. Histamine treatment reduced cytokine production from these cells and was more marked in those from patients with psoriasis.

In contrast to other dendritic cell subsets, Langerhans cells within the epidermis seem to be regulated uniquely by histamine as they do not express H_1R or H_2R (Ohtani *et al.*, 2003). However, Langerhans cells do express H_4R and human inflammatory dendritic epidermal cells can express H_4R following exposure to IFNγ, which results in the downregulation of CCL2 and IL-12 secretion in an H_4R-dependent mechanism (Gschwandtner *et al.*, 2010; Dijkstra, *et al.*, 2008). In contrast, murine studies suggest that inhibition of H_4R on dendritic cells leads to decreased cytokine and chemokine production, which limits their ability to induce Th2 responses. Simon *et al.*, 2011 found an increased antigen presentation capacity in dendritic cells from $H_4R^{-/-}$ mice. These mice had reduced mRNA expression of cytokines such as IFNγ and IL-10. Recently, dendritic cell H_1R expression was shown to be required for IFNγ production by CD8 cells resulting in atopic dermatitis in a murine model (Vanbervliet *et al.*, 2011). Interestingly, IL-17^+CD8 cells were induced by dendritic cells lacking H_1R. Certain bacterial strains present within the human diet express the HDC gene and have been shown to release histamine (Coton *et al.*, 2010). However, the *in vivo* homeostatic consequences of histamine release by these bacteria has not yet been determined, although it is likely that histamine-producing bacteria could impact the dendritic cell response to the bacteria itself and other bacterial strains through a bystander effect.

Due to their potent ability to influence innate activation and adaptive polarization, dendritic cells are key cellular players in orchestrating protective and tolerogenic immune responses. Inappropriate dendritic cell PRR signalling has been associated with multiple inflammatory diseases, particularly those of tissues exposed to the external environment such as the gastrointestinal tract, the lungs and the skin. Histamine and its four receptors represent a complex system of immunoregulation with distinct dendritic cell effects dependent on receptor subtype expression (O'Mahony *et al.*, 2011). Strategies that promote dendritic cell H_2R expression and activity could improve mucosal immuoregulatory activity and protect against allergic sensitization and inflammatory disorders.

4.6. Polymorphonuclear Neutrophils

Polymorphonuclear neutrophils (or neutrophils) are the most abundant leukocytes in blood and constitute the first line of defence against bacteria and fungi. During episodes of infection, neutrophils leave the blood circulation and migrate to inflamed tissues (extravasation) where they clear the pathogens by a process called phagocytosis. The recruitment of neutrophils to sites of inflammation is a complex process orchestrated by soluble mediators (formyl peptides, chemokines, pro-inflammatory cytokines, complement split products) and adhesion receptors (selectins and β2 integrins) (Ley *et al.*, 2007). The soluble mediators, which are produced during episodes of infection, induce expression of β2 integrins on the membrane surface of neutrophils (Sengeløv, 1993), the main adhesion receptors expressed in neutrophils. The soluble mediators also switch the β2 integrins from a low to a high affinity ligand binding conformation (inside-out signalling) (Diamond & Springer, 1993). This change of conformation in β2 integrins is required for strong attachment to the endothelium, migration along the endothelial layer and the subsequent transendothelial migration. In infected tissues, neutrophils migrate towards the pathogens and capture them. The capture of opsonised microorganisms primarily involves two receptors: the β2 integrins and the Fcγ receptors. The microoganisms are trapped inside the phagosome, which fuses with intracellular granules. The release of the granule content (proteases and anti-microbial peptides) into the phagosome, coupled with the production of reactive oxygen species, is responsible for the destruction of the microorganisms (Sheppard *et al.*, 2005).

4.6.1. Action of Histamine on Neutrophils

Histamine H_1 binding sites are present in human neutrophils (Wescott & Kaliner, 1983). [^{3}H]pyrilamine, an H_1R antagonist, binds human neutrophils in a specific, saturable and reversible fashion. Furthermore, the binding of [^{3}H]pyrilamine is competed (by order of potency) by H_1 antagonists, histamine, and H_2 antagonists. It was estimated that human neutrophils express a large number of H_1R (265 x 10^3/cell) and have a homogeneous population of H_1R, with a moderate affinity for histamine (K_D = 52 nM). Another study reported presence of H_2- but not H_1-binding sites in neutrophils. A bioactive fluorescent derivative of histamine was shown to bind to the membrane surface of human neutrophils and histamine, cimetidine (an H_2R antagonist) but not diphenhydramine (H_1R antagonist) competed with the fluorescent derivative of histamine for binding sites (Petty & Francis, 1986).

Little information is available regarding expression of histamine receptors in neutrophils. There is a lack of information regarding expression of H_1R, H_2R or H_3R in neutrophils. There is controversy as to whether the H_4R is present in human neutrophils. The mRNA encoding for H_4R was found in human neutrophils (Oda *et al.*, 2000; Zhu *et al.*, 2001) and in HL-60 cells differentiated into granulocytes (Van Rijn *et al.*, 2006). However, other investigators did not detect the H_4R messenger RNA in neutrophils (Ling *et al.*, 2004).

There is consistent support for a role of histamine in the inhibition of antimicrobial functions of human neutrophils. Histamine inhibits neutrophil degranulation induced by the potent chemoattractant fMLP (Seligmann *et al.*, 1983; Burde *et al.*, 1989) or by zymosan particles (Busse & Sosman, 1976). Histamine also blunts fMLP-induced superoxide production in neutrophils (Seligmann *et al.*, 1983; Burde *et al.*, 1989). These effects of histamine are mediated by H_2R but not by H_1R. Indeed, antagonists of H_2R such as metiamide (Busse & Sosman, 1976), cimetidine (Seligmann *et al.*, 1983), ranitidine (Zimmerman & Millard, 1989) and famotidine (Burde *et al.*, 1989) prevented these inhibitory effects of histamine. In contrast, chlorpheniramine, an antagonist of the H_1R did not prevent histamine blocking degranulation induced by zymosan particles (Busse & Sosman, 1976). Additional evidence for the involvement of H_2R in the negative regulation of neutrophil functions is provided by studies showing that inhibition of neutrophil degranulation by histamine is also reproduced by guanidine-type H_2-agonists structurally derived from impromidine (Burde *et al.*, 1989). The regulatory effect of the H_2R on neutrophil functions may be mediated through activation of adenylyl cyclase and increased production of the second messenger cAMP (Burde *et al.*, 1989; Busse & Sosman, 1976). It remains to be investigated whether the H_4R plays a role in the regulation of neutrophil antimicrobial functions.

Neutrophils stimulated with fMLP synthesize and release leukotrienes which are potent chemoattractants for neutrophils. Furthermore, histamine blocks fMLP-induced production of leukotrienes (Flamand *et al.*, 2004). It is therefore plausible that the ability of histamine to block neutrophil functions is due to reduced synthesis of potent activators of neutrophil functions. Based on the use of pharmacological antagonists of H_2R (cimetidine, ranitidine, and tiotidine) and H_4R (thioperamide), it has been shown that H_2R but not H_4R is involved in the inhibition of fMLP-induced biosynthesis of leukotrienes by histamine (Flamand *et al.*, 2004). Histamine also has profound effects on the oxidative burst and release of reactive oxygen species from neutrophils; however, these observations could be considered controversial (for full discussion see: Číž & Lojek, 2013).

In addition to its regulatory role of antimicrobial functions, histamine has been demonstrated to regulate neutrophil adhesion and chemotaxis. Histamine caused limited inhibition of fMLP-dependent chemotaxis while stimulating chemokinesis (Seligmann *et al.*, 1983). The ability of histamine to prevent chemotaxis may be explained by the fact that the vasoamine diminishes expression of β2 integrins on the membrane surface (Francis *et al.*, 1991). In contrast, by using a mouse mast-cell-dependent model of zymosan-induced peritonitis, it was proposed that histamine, by acting on H_4R, controls neutrophil chemotaxis. Indeed, the H_4R antagonist A-940894 blocks neutrophil influx in the peritoneum (Strakhova *et al.*, 2009). However, the H_4R may indirectly control neutrophil migration by regulating expression levels of chemoattactants and chemokines.

Histamine could play a key role in the resolution of inflammation. Indeed, in response to engagement of the TLR 4 (which binds gram negative bacteria), neutrophils produce histamine (Smuda *et al.*, 2011). When pathogens have been cleared, histamine could limit neutrophil degranulation and reactive oxygen species production in order to avoid tissue damage.

4.7. Conclusions

The role of histamine in allergic reactions is undisputed. However, our increasing knowledge about this mediator and its functions places it central stage in the orchestration of the innate immune response and leading to the adaptive response. Studies have been complicated because of the differences in the receptors between species. Cellular responses also vary depending on the original tissue from which the cells were isolated, which is of special relevance for mast cells. The movement of cells from the blood stream out into the tissues, airways or gut can also modulate the expression of receptors and receptor subtypes. Finally an area that has received less attention, the underlying disease state may also cause modifications to receptor expression and/or the response to stimuli.

From the data presented above, further work is necessary to elucidate the different HRs on inflammatory cells such as basophils, mast cells and neutrophils. The complex interplay between the inflammatory cells also requires further study. Using novel compounds with actions at more than one histamine receptor will provide both interesting scientific knowledge and possibly better treatments.

Acknowledgements

We thank the following for support: COST Action BM0806, Grant SAF2011-26518, BIO-267, CIBERER is an initiative of ISCIII (Spain), grant no. LD11010 of the MEYS of the Czech Republic. Thanks are due to Dr. R. Montañez for preparing figure 4.2.

References

Abrighach H., Fajardo I., Sánchez-Jiménez F. *et al.* (2010) Exploring polyamine regulation by nascent histamine in a human-transfected cell model. *Amino Acids, 38(2)*, 561-73.

Aichberger K. J., Mayerhofer M., Vales A. *et al.* (2006) The CML-related oncoprotein BCR/ABL induces expression of histidine decarboxylase (HDC) and the synthesis of histamine in leukemic cells. *Blood, 108(10)*, 3538-47.

Akdis, C. A., & Simons, F. E. (2006). Histamine receptors are hot in immunopharmacology. *Eur J Pharmacol, 533*(1-3), 69-76.

Akdis, M., Burgler, S., Crameri, R. *et al.* (2011) Interleukins, from 1 to 37, and interferon-gamma: Receptors, functions, and roles in diseases. *J Allergy Clin Immunol, 127*,701-721.

Ambrozova, G., Pekarova, M., & Lojek, A. (2010). Effect of polyunsaturated fatty acids on the reactive oxygen and nitrogen species production by raw 264.7 macrophages. *Eur J Nutr, 49*, 133-139.

Ambrozova, G., Pekarova, M., & Lojek, A. (2011). The effect of lipid peroxidation products on reactive oxygen species formation and nitric oxide production in lipopolysaccharide-stimulated RAW 264.7 macrophages. *Toxicol In Vitro, 25*, 145-152.

Auffray, C., Sieweke, M. H., & Geissmann, F. (2009). Blood monocytes: development, heterogeneity, and relationship with dendritic cells. *Annu Rev Immunol, 27*, 669-692.

Azuma, Y., Shinohara, M., Wang, P. L. *et al.* (2001). Histamine inhibits chemotaxis, phagocytosis, superoxide anion production, and the production of TNFalpha and IL-12 by macrophages via H_2-receptors. *Int Immunopharmacol, 1*, 1867-1875.

Babbar N., Murray-Stewart T., & Casero R. A. Jr. (2007) Inflammation and polyamine catabolism: the good, the bad and the ugly. *Biochem Soc Trans, 35 (Pt 2)*, 300-4.

Belting M., Mani K., Jönsson M. *et al.* (2003) Glypican-1 is a vehicle for polyamine uptake in mammalian cells: a pivotal role for nitrosothiol-derived nitric oxide. *J Biol Chem, 278(47)*, 47181-9.

Bissonnette, E. Y. (1996). Histamine inhibits tumor necrosis factor alpha release by mast cells through H2 and H3 receptors. *Am J Respir Cell Mol Biol*, 14,620– 626.

Borden, E. C., Sen, G. C., Uze, G. *et al.* (2007). Interferons at age 50: past, current and future impact on biomedicine. *Nat Rev Drug Discov, 6*, 975-990.

Botana, L.M. & MacGlashan, D.W. (1994). Effect of cAMP-elevating drugs on stimulus-induced cytosolic calcium changes in human basophils. *J Leuko Biol*, 55,798-804.

Bourne, H.R., Melmon, K.L. & Lichtenstein, L.M. (1971). Histamine augments leukocyte adenosine 3´,5´-monophosphate and blocks antigenic histamine release. *Science*, 173,743-745.

Bryce, P. J., Mathias, C. B., Harrison, K. L., *et al.* (2006). The H_1 histamine receptor regulates allergic lung responses. *J Clin Invest, 116*(6), 1624-1632.

Bullock, E. D., & Johnson, E. M., Jr. (1996). Nerve growth factor induces the expression of certain cytokine genes and bcl-2 in mast cells. Potential role in survival promotion. *J Biol Chem, 271*(44), 27500-27508.

Bunikowski, R., Mielke, M., Skarabis, H., *et al.* (1999). Prevalence and role of serum IgE antibodies to the Staphylococcus aureus-derived superantigens SEA and SEB in children with atopic dermatitis. *J Allergy Clin Immunol, 103*(1 Pt 1), 119-124.

Burde, R., Seifert, R., Buschauer, A. *et al.* (1989). Histamine inhibits activation of human neutrophils and HL-60 leukemic cells via H2-receptors. *Naunyn Schmiedebergs Arch Pharmacol, 340*, 671-678.

Busse, W.W., Cooper, W., & Anderson, C. (1980). Dimaprit inhibition of zymosan-stimulated beta-glucuronidase release from human granulocytes. *Agents Actions, 10*, 15-18.

Busse, W. W., & Sosman, J. (1976). Histamine inhibition of neutrophil lysosomal enzyme release: an H_2 histamine receptor response. *Science, 194*, 737-738.

Caron, G., Delneste, Y., Roelandts, E. *et al.* (2001) Histamine induces CD86 expression and chemokine production by human immature dendritic cells. *J Immunol 166*, 6000-6.

Chen, G. Y., & Nunez, G. (2010). Sterile inflammation: sensing and reacting to damage. *Nat Rev Immunol, 10*, 826-837.

Chow, A., Brown, B. D., & Merad, M. (2011). Studying the mononuclear phagocyte system in the molecular age. *Nat Rev Immunol, 11*, 788-798.

Cíž, M. & Lojek, A. (2013) Modulation of neutrophil oxidative burst via histamine receptors. *Br J Pharmacol, 170*, 17-22.

Code, C. F. (2008). Histamine and Gastric Secretion *Ciba Foundation Symposium - Histamine* (pp. 189-219): John Wiley & Sons, Ltd.

Colwell C. S., & Levine M. S. (1997) Histamine modulates NMDA-dependent swelling in the neostriatum. *Brain Res, 766(1-2)*, 205-12.

Coton, M., Romano, A., Spano, G. *et al.* (2010) Occurrence of biogenic amine-forming lactic acid bacteria in wine and cider. *Food Microbiol* 27:1078-85.

Damaj, B. B., Becerra, C. B., Esber, H. J. *et al.* (2007). Functional expression of H4 histamine receptor in human natural killer cells, monocytes, and dendritic cells. *J Immunol, 179*, 7907-7915.

Dawicki, W., Jawdat, D.W., Xu, N. *et al.* (2010) Mast cells, histamine, and IL-6 regulate the selective influx of dendritic cell subsets into an inflamed lymph node. *J Immunol*, 184, 2116-2123.

Diamond, M. S., & Springer, T. A. (1993). A subpopulation of Mac-1 (CD11b/CD18) molecules mediates neutrophil adhesion to ICAM-1 and fibrinogen. *J Cell Biol, 120*, 545-556.

Dijkstra, D., Leurs, R., Chazot, P. *et al.* (2007). Histamine downregulates monocyte CCL2 production through the histamine H4 receptor. *J Allergy Clin Immunol, 120*, 300-307.

Dijkstra D, Stark H, Chazot PL, *et al.* (2008) Human inflammatory dendritic epidermal cells express a functional histamine H4 receptor. *J Invest Dermatol,* 128, 1696-1703.

Dunford, P. J., O'Donnell, N., Riley, J. P., *et al.* (2006). The histamine H4 receptor mediates allergic airway inflammation by regulating the activation of CD4+ T cells. *J Immunol, 176*(11), 7062-7070.

Dunford, P. J., Williams, K. N., Desai, P. J., *et al.* (2007). Histamine H4 receptor antagonists are superior to traditional antihistamines in the attenuation of experimental pruritus. *J Allergy Clin Immunol, 119*(1), 176-183.

Elishmereni, M., Alenius, H. T., Bradding, P., *et al.* (2011). Physical interactions between mast cells and eosinophils: a novel mechanism enhancing eosinophil survival in vitro. *Allergy, 66*(3), 376-385.

Eller, K., & Rosenkranz, A. R. (2012). Mast Cells: Subordinates or Masterminds in Autoimmunity? *J Am Soc Nephrol. 23(12)*:1913-4.

Fajardo I., Urdiales J. L., Medina M. A. *et al.* (2001a) Effects of phorbol ester and dexamethasone treatment on histidine decarboxylase and ornithine decarboxylase in basophilic cells. *Biochem Pharmacol, 61(9),* 1101-6.

Fajardo I., Urdiales J. L., Paz J. C. *et al.* (2001b) Histamine prevents polyamine accumulation in mouse C57.1 mast cell cultures. *Eur J Biochem, 268(3),* 768-73.

Falcone, F.H., Knol, E.F. & Gibbs, B.F. (2011). The role of basophils in the pathogenesis of allergic disease. *Clin Exp Allergy,* 41,939-947.

Falcone, F.H., Pritchard, D.I. & Gibbs, B.F. (2001). Do basophils play a role in immunity against parasites? *Trends Parasitol,* 17,126-129.

Flamand, N., Plante, H., Picard, S., *et al.* (2004). Histamine-induced inhibition of leukotriene biosynthesis in human neutrophils: involvement of the H_2 receptor and cAMP. *Br J Pharmacol, 141,* 552-561.

Fogg, D. K., Sibon, C., Miled, C. *et al.* (2006). A clonogenic bone marrow progenitor specific for macrophages and dendritic cells. *Science, 311,* 83-87.

Francis, J.W., Todd, R.F., Boxer, L.A., *et al.* (1991). Histamine inhibits cell spreading and C3bi receptor clustering and diminishes hydrogen peroxide production by adherent human neutrophils. *J Cell Physiol, 147,* 128-137

Friedman, S. J., Schroeter, A. L., & Homburger, H. A. (1985). IgE antibodies to Staphylococcus aureus. Prevalence in patients with atopic dermatitis. *Arch Dermatol, 121*(7), 869-872.

Furuta K., Ichikawa A., Nakayama K. *et al.* (2006) Membrane orientation of the precursor 74-kDa form of L-histidine decarboxylase. *Inflamm Res, 55(5)*, 185-91.

Galli, S. J., Tsai, M., & Piliponsky, A. M. (2008). The development of allergic inflammation. *Nature, 454*(7203), 445-454.

García-Faroldi G., Correa-Fiz F., Abrighach H. *et al.* (2009a)Polyamines affect histamine synthesis during early stages of IL-3-induced bone marrow cell differentiation. *J Cell Biochem, 108(1)*, 261-71.

García-Faroldi G., Sánchez-Jiménez F., & Fajardo I. (2009b) The polyamine and histamine metabolic interplay in cancer and chronic inflammation. *Curr Opin Clin Nutr Metab Care, 12(1)*, 59-65.

García-Faroldi G., Rodríguez C. E., Urdiales J. L. *et al.* (2010) Polyamines are present in mast cell secretory granules and are important for granule homeostasis. *PLoS One, 5(11)*, e15071.

Garcia-Montero A. C., Jara-Acevedo M., Teodosio C. *et al.* (2006) KIT mutation in mast cells and other bone marrow hematopoietic cell lineages in systemic mast cell disorders: a prospective study of the Spanish Network on Mastocytosis (REMA) in a series of 113 patients. *Blood, 108(7)*, 2366-72.

Gaudy-Marqueste, C. (2010). Antihistamines. In L. Misery & S. Ständer (Eds.), *Pruritus* (pp. 276-288). New York ; London: Springer.

Geissmann, F., Jung, S., & Littman, D. R. (2003). Blood monocytes consist of two principal subsets with distinct migratory properties. *Immunity, 19*, 71-82.

Gibbs, B.F., Vollrath, I.B., Albrecht, C. *et al.* (1998). Inhibition of interleukin-4 and interleukin-13 release from immunologically activated human basophils due to the actions of anti-allergic drugs. *Naunyn Schmiedebergs Archiv Pharmacol,* 357,573-578.

Gibbs, B.F., Wierecky, J., Welker, P. *et al.* (2001). Human skin mast cells rapidly release preformed and newly generated TNF-alpha and IL-8 following stimulation with anti-IgE and other secretagogues. *Exp Dermatol,* 10,312-320.

Gleich, G. J., & Adolphson, C. R. (1986). The eosinophilic leukocyte: structure and function. *Adv Immunol, 39*, 177-253.

Godot, V., Arock, M., Garcia, G. *et al.* (2007). H4 histamine receptor mediates optimal migration of mast cell precursors to CXCL12. *J Allergy Clin Immunol*, 120, 827–834.

Gschwandtner, M., Koether, B., Werfel, T. *et al.* (2013) Profiling of histamine H4 receptor agonists in native human monocytes. *Br J Pharmacol*, 170, 136-143.

Gschwandtner, M., Mommert, S., Köther, B. *et al.* (2011) The histamine H4 receptor is highly expressed on plasmacytoid dendritic cells in psoriasis and histamine regulates their cytokine production and migration. *J Invest Dermatol*, 131(8):1668-76.

Gschwandtner, M., Rossbach, K., Dijkstra, D. *et al.* (2010) Murine and human Langerhans cells express a functional histamine H4 receptor: modulation of cell migration and function. *Allergy 65*, 840-9.

Hellstrand, K., Asea, A., Dahlgren, C. *et al.* (1994) Histaminergic regulation of NK cells. Role of monocyte-derived reactive oxygen metabolites. *J Immunol 153*, 4940-7.

Hill, S. J., Ganellin, C. R., Timmerman, H., et al. (1997). International Union of Pharmacology. XIII. Classification of histamine receptors. *Pharmacol Rev, 49*(3), 253-278.

Hofstra, C.L., Desai, P.J., Thurmond, R.L. *et al.* (2003). Histamine H4 receptor mediates chemotaxis and calcium mobilization of mast cells. *J Pharmacol Exp Ther*, 305, 1212–1221.

Hogan, S. P., Waddell, A., & Fulkerson, P. C. (2013). Eosinophils in infection and intestinal immunity. *Curr Opin Gastroenterol.* 29(1), 7-14.

Horigome, K., Bullock, E. D., & Johnson, E. M., Jr. (1994). Effects of nerve growth factor on rat peritoneal mast cells. Survival promotion and immediate-early gene induction. *J Biol Chem, 269*(4), 2695-2702.

Hsu, F. I., & Boyce, J. A. (2009). Biology of Mast Cells and Their Mediators. In N. F. Adkinson Jr, W. W. Busse, B. S. Bochner, S. T. Holgate, F. E. R. Simons & R. F. Lemanske (Eds.), *Middleton's allergy: principles & practice* (7th ed., Vol. 2, pp. 311-328). Philadelphia, Pa.; [Edinburgh]: Mosby Elsevier.

Hume, D. A., Ross, I. L., Himes, S. R. *et al.* (2002). The mononuclear phagocyte system revisited. *J Leukoc Biol, 72*, 621-627.

Jacobsen, E. A., Helmers, R. A., Lee, J. J., et al. (2012). The expanding role(s) of eosinophils in health and disease. *Blood, 120*(19), 3882-3890.

Jenkins, S. J., Ruckerl, D., Cook, P. C. *et al.* (2011). Local macrophage proliferation, rather than recruitment from the blood, is a signature of Th2 inflammation. *Science, 332*, 1284-1288.

Jutel, M., Blaser, K., & Akdis, C. A. (2005). Histamine in allergic inflammation and immune modulation. *Int Arch Allergy Immunol, 137*(1), 82-92.

Kahana C. (2009) Regulation of cellular polyamine levels and cellular proliferation by antizyme and antizyme inhibitor. *Essays Biochem, 46*, 47-61.

Kalesnikoff, J., & Galli, S. J. (2008). New developments in mast cell biology. *Nat Immunol, 9*(11), 1215-1223.

Kawahara T. (2012) Establishment and characterization of mouse bone marrow-derived mast cell hybridomas. *Exp Cell Res, 318(18)*, 2385-96.

Kazimierczak, W., Szczepaniak, K. & Bańkowska, K. (1981). A modulation of the anaphylactic basophil histamine release by selective H2 histamine agonists. *Agents Actions*, 11,96-98.

Keles, N., Yavuz Arican, R., Coskun, M., *et al.* (2012). Histamine induces the neuronal hypertrophy and increases the mast cell density in gastrointestinal tract. *Exp Toxicol Pathol, 64*(7-8), 713-716.

Khan, U. W., & Rai, U. (2007). Differential effects of histamine on Leydig cell and testicularmacrophage activities in wall lizards: precise role of H1/H2 receptor subtypes. *J Endocrinol, 194*, 441-448.

Kleine-Tebbe, J., Schramm, J., Bolz, M. *et al.* (1990). Influence of histamine H3-antagonists on human leukocytes. *Agents Actions*, 30,137-139.

Knight, P.A., Brown, J.K. & Pemberton, A.D. (2008). Innate immune response mechanisms in the intestinal epithelium: potential roles for mast cells and goblet cells in the expulsion of adult Trichinella spiralis. *Parasitology*, 135,655-670.

Krauth M. T., Agis H., Aichberger K. J. *et al.* (2006) Immunohistochemical detection of histidine decarboxylase in neoplastic mast cells in patients with systemic mastocytosis. *Hum Pathol, 37(4)*, 439-47.

LaBella, F. S., & Brandes, L. J. (2000). Interaction of histamine and other bioamines with cytochromes P450: implications for cell growth modulation and chemopotentiation by drugs. *Sem Canc Biol, 10*(1), 47–53.

Lagunoff, D., Martin, T. W., & Read, G. (1983). Agents that release histamine from mast cells. *Annu Rev Pharmacol Toxicol, 23*, 331-351.

Lantz, C.S., Boesiger, J., Song, C.H. *et al.* (1998). Role for interleukin-3 in mast-cell and basophil development and in immunity to parasites. *Nature*, 392,90-93.

Levi-Schaffer, F & Eliashar, R. (2009). Mast cell stabilizing properties of antihistamines. *J Invest Dermatol*, 129,2549-2551.

Ley, K., Laudanna, C., Cybulsky, M. I., *et al.* (2007). Getting to the site of inflammation: the leukocyte adhesion cascade updated. *Nat Rev Immunol, 7*, 678-689.

Lichtenstein, L. M. & Gillespie, E. (1973). Inhibition of histamine release by histamine controlled by H2 receptor. *Nature*, 244,287-288.

Ling, P., Ngo, K., Nguyen, S., *et al.* (2004). Histamine H4 receptor mediates eosinophil chemotaxis with cell shape change and adhesion molecule upregulation. *Br J Pharmacol, 142*(1), 161-171.

Lippert, U., Artuc, M., Grützkau, A. *et al.* (2004). Human skin mast cells express H2 and H4, but not H3 receptors. *J Invest Dermatol*, 123, 116-123.

Liu, C., Ma, X., Jiang, X. *et al.* (2001). Cloning and pharmacological characterization of a fourth histamine receptor (H4) expressed in bone marrow. *Molecular Pharmacology*, 59,420–426.

Lojek, A., Pecivova, J., Macickova, T. *et al.* (2008). Effect of carvedilol on the production of reactive oxygen species by HL-60 cells. *Neuro Endocrinol Lett, 29*, 779-783.

Lopez, A. F., Begley, C. G., Williamson, D. J., et al. (1986). Murine eosinophil differentiation factor. An eosinophil-specific colony-stimulating factor with activity for human cells. *J Exp Med, 163*(5), 1085-1099.

Lopez, A. F., Sanderson, C. J., Gamble, J. R., et al. (1988). Recombinant human interleukin 5 is a selective activator of human eosinophil function. *J Exp Med, 167*(1), 219-224.

Lundequist, A., & Pejler, G. (2010). Biological implications of preformed mast cell mediators. *Cell Mol Life Sci, 68*(6), 965–975.

MacGlashan, D. (2003). Histamine: A mediator of inflammation. *J Allergy Clin Immunol*, 112 (4 Suppl),S53-S59.

MacKenzie, J. R., Mattes, J., Dent, L. A., *et al.* (2001). Eosinophils promote allergic disease of the lung by regulating CD4(+) Th2 lymphocyte function. *J Immunol, 167*(6), 3146-3155.

Martinez-Mir, M. I., Pollard, H., Moreau, J., et al. (1990). Three histamine receptors (H1, H2 and H3) visualized in the brain of human and non-human primates. *Brain Res, 526*(2), 322-327.

Masini, E., Blandina, P., Brunelleschi, S. *et al.* (1982). Evidence for H2-receptor-mediated inhibition of histamine release from isolated rat mast cells. *Agents Actions*, 12,85-88.

Matzinger, P., & Kamala, T. (2011). Tissue-based class control: the other side of tolerance. *Nat Rev Immunol, 11*, 221-230.

Mazzoni, A., Leifer, C.A., Mullen, G.E. *et al.* (2003) Cutting edge: histamine inhibits IFN-alpha release from plasmacytoid dendritic cells. *J Immunol, 170*, 2269-73.

Mazzoni, A., Young, H.A., Spitzer, J.H. *et al.* (2001) Histamine regulates cytokine production in maturing dendritic cells, resulting in altered T cell polarization. *J Clin Invest, 108*, 1865-73.

McNagny, K., & Graf, T. (2002). Making eosinophils through subtle shifts in transcription factor expression. *J Exp Med, 195*(11), F43-47.

Medina M. A., Correa-Fiz F., Rodríguez-Caso C. *et al.* (2005) A comprehensive view of polyamine and histamine metabolism to the light of new technologies. *J Cell Mol Med, 9(4)*, 854-64.

Medina V., Croci M., Crescenti E. *et al.* (*2008*) The role of histamine in human mammary carcinogenesis: H3 and H4 receptors as potential therapeutic targets for breast cancer treatment. *Cancer Biol Ther, 7(1)*, 28-35.

Melgarejo E., Medina M. A., Sánchez-Jiménez F. *et al.* (2007) (-)-Epigallocatechin-3-gallate interferes with mast cell adhesiveness, migration and its potential to recruit monocytes. *Cell Mol Life Sci, 64(19-20)*, 2690-701.

Melgarejo E., Medina M. A., Sánchez-Jiménez F., *et al.* (2009) Epigallocatechin gallate reduces human monocyte mobility and adhesion in vitro. Br J Pharmacol, 158(7), 1705-12.

Melgarejo E., Medina M. A., Sánchez-Jiménez F. *et al.* (2010a) Targeting of histamine producing cells by EGCG: a green dart against inflammation? *J Physiol Biochem, 66(3)*, 265-70.

Melgarejo E., Urdiales J. L., Sánchez-Jiménez F. *et al.* (2010b) Targeting polyamines and biogenic amines by green tea epigallocatechin-3-gallate. *Amino Acids, 38(2)*, 519-23.

Metcalfe, D.D. (2008). Mast cells and mastocytosis. *Blood*, 112,946-956.

Minai-Fleminger, Y., Elishmereni, M., Vita, F., *et al.* (2010). Ultrastructural evidence for human mast cell-eosinophil interactions in vitro. *Cell Tissue Res, 341*(3), 405-415.

Minai-Fleminger, Y., & Levi-Schaffer, F. (2009). Mast cells and eosinophils: the two key effector cells in allergic inflammation. *Inflamm Res, 58*(10), 631-638.

Mirossay, L., Chastre, E., Callebert, J. *et al.* (1994). Histamine H2 receptors and histidine decarboxylase in normal and leukemic human monocytes and macrophages. *Am J Physiol, 267*, R602-611.

Morichika, T., Takahashi, H. K., Iwagaki, H. *et al.* (2003). Effect of prostaglandin E2 on intercellular adhesion molecule-1 and B7 expre ssion in mixed lymphocyte reaction. *Transplantation, 75*, 2100-2105.

Morse, K.L., Behan, J., Laz, T.M. *et al.* (2001). Cloning and characterization of a novel human histamine receptor. *J Pharmacol Exp Ther*, 296,1058–1066.

Motala, C., Potter, P. C., Weinberg, E. G., *et al.* (1986). Anti-Staphylococcus aureus-specific IgE in atopic dermatitis. *J Allergy Clin Immunol, 78*(4 Pt 1), 583-589.

Moya-Garcia A. A., Medina M. A., & Sánchez-Jiménez F. (2005) Mammalian histidine decarboxylase: from structure to function. *Bioessays, 27(1)*, 57-63.

Murata, Y., Tanimoto, A., Wang, K. Y. *et al.* (2005). Granulocyte macrophage-colony stimulating factor increases the expression of histamine and histamine receptors in monocytes/macrophages in relation to arteriosclerosis. *Arterioscler Thromb Vasc Biol, 25*, 430-435.

Murray, P. J., & Wynn, T. A. (2011). Protective and pathogenic functions of macrophage subsets. *Nat Rev Immunol, 11*, 723-737.

Mutti F. G., Gullotti M., Casella L. *et al.* (2011) A new chiral, poly-imidazole N8-ligand and the related di- and tri-copper(II) complexes: synthesis, theoretical modelling, spectroscopic properties, and biomimetic stereoselective oxidations. *Dalton Trans, 40(20)*, 5436-57.

Nemmar, A., Delaunois, A., Beckers, J.F. *et al.* (1999). Modulatory effect of imetit, a histamine H3 receptor agonist, on C-fibers, cholinergic fibers and mast cells in rabbit lungs in vitro. *Eur J Pharmacol*, 371,23–30.

Nerlov, C., & Graf, T. (1998). PU.1 induces myeloid lineage commitment in multipotent hematopoietic progenitors. *Genes Dev, 12*(15), 2403-2412.

Nerlov, C., McNagny, K. M., Doderlein, G., *et al.* (1998). Distinct C/EBP functions are required for eosinophil lineage commitment and maturation. *Genes Dev, 12*(15), 2413-2423.

Nish, S., & Medzhitov, R. (2011). Host defense pathways: role of redundancy and compensation in infectious disease phenotypes. *Immunity, 34*, 629-636.

Novak, N., Mete, N., Bussmann, C. *et al.* (2012). Early suppression of basophil activation during allergen-specific immunotherapy by histamine receptor 2. *J Allergy Clin Immunol*, 130,1153-1158.

Oda, T., Morikawa, N., Saito, Y., *et al.* (2000). Molecular cloning and characterization of a novel type of histamine receptor preferentially expressed in leukocytes. *J. Biol. Chem., 275*, 36781-36786.

Ohki, E., Suzuki, M., Aoe, T. *et al.* (2007). Expression of histamine H4 receptor in synovial cells from rheumatoid arthritic patients. *Biol Pharm Bull, 30*, 2217-2220.

Ohsawa, Y. & Hirasawa, N. (2012). The antagonism of histamine H1 and H4 receptors ameliorates chronic allergic dermatitis via anti-pruritic and anti-inflammatory effects in NC/Nga mice. *Allergy*, 67,1014-1022.

Ohtani, T., Aiba, S., Mizuashi, M. *et al.* (2003) H1 and H2 histamine receptors are absent on Langerhans cells and present on dermal dendritic cells. *J Invest Dermatol* 121, 1073-9.

Ohtsu, H. (2010). Histamine Synthesis and Lessons Learned from Histidine Decarboxylase Deficient Mice. In R. Thurmond (Ed.), *Histamine in Inflammation* (Vol. 709, pp. 21-31): Springer US.

Olmo M. T., Sánchez-Jiménez F., Medina M. A. *et al.* (2002) Spectroscopic analysis of recombinant rat histidine decarboxylase. *J Biochem, 132(3)*, 433-9.

O'Mahony, L., Akdis, M. & Akdis, C.A. (2011) Regulation of the immune response and inflammation by histamine and histamine receptors. *J Allergy Clin Immunol 128*, 1153-62.

O'Mahony, L., O'Callaghan, L., McCarthy, J. *et al.* (2006) Differential cytokine response from dendritic cells to commensal and pathogenic bacteria in different lymphoid compartments in humans. *Am J Physiol Gastrointest Liver Physiol* 290, G839-45.

Palacios, J.M., Garbarg, M., Barbin, G. *et al.* (1978). Pharmacological characterization of histamine receptors mediating the stimulation of cyclic AMP accumulation in slices from guinea-pig hippocampus. *Mol Pharmacol*, 14,971–982.

Parsons, M. E., & Ganellin, C. R. (2006). Histamine and its receptors. *Br J Pharmacol, 147 Suppl 1*, S127-135.

Pegg AE. (2009) Mammalian polyamine metabolism and function. *IUBMB Life, 61(9):* 880-94.

Peters, S.P., Schulman, E.S., Schleimer, R.P. *et al.* (1982). Dispersed human lung mast cells. Pharmacologic aspects and comparison with human lung tissue fragments. *Am Rev Resp Dis*, 126,1034-1039.

Petty, R., & Francis J. W. (1986). Polymorphonuclear leukocyte histamine receptors; occurrence in cell surface clusters and their redistribution during locomotion. *Proc Natl Acad Sci*, 83, 4332-4335.

Pilbák S., Farkas Ö., & Poppe L. (2012) Mechanism of the tyrosine ammonia lyase reaction-tandem nucleophilic and electrophilic enhancement by a proton transfer. *Chemistry, 18(25)*, 7793-802.

Piliponsky, A. M., Chen, C. C., Grimbaldeston, M. A., *et al.* (2010). Mast cell-derived TNF can exacerbate mortality during severe bacterial infections in C57BL/6-KitW-sh/W-sh mice. *Am J Pathol, 176*(2), 926-938.

Piliponsky, A. M., Gleich, G. J., Bar, I., *et al.* (2002). Effects of eosinophils on mast cells: a new pathway for the perpetuation of allergic inflammation. *Mol Immunol, 38*(16-18), 1369.

Piliponsky, A. M., Pickholtz, D., Gleich, G. J., *et al.* (1999). Eosinophils activate mast cells to release histamine. *Int Arch Allergy Immunol, 118*(2-4), 202-203.

Pincus, S. H., DiNapoli, A. M., & Schooley, W. R. (1982). Superoxide production by eosinophils: activation by histamine. *J Invest Dermatol, 79*(1), 53-57.

Pino-Ángeles A., Reyes-Palomares A., Melgarejo E. *et al.* (2012) Histamine: an undercover agent in multiple rare diseases? *J Cell Mol Med, 16(9)*, 1947-60.

Poulin R., Casero R. A. Jr, & Soulet D. (2012) Recent advances in the molecular biology of metazoan polyamine transport. *Amino Acids, 42(2-3)*, 711-23.

Raible, D. G., Lenahan, T., Fayvilevich, Y., *et al.* (1994). Pharmacologic characterization of a novel histamine receptor on human eosinophils. *Am J Respir Crit Care Med, 149*(6), 1506-1511.

Reber, L. L., Marichal, T., & Galli, S. J. (2012). New models for analyzing mast cell functions in vivo. *Trends Immunol, 33*(12), 613-625.

Reher, T. M., Brunskole, I., Neumann, D., *et al.* (2012). Evidence for ligand-specific conformations of the histamine H(2)-receptor in human eosinophils and neutrophils. *Biochem Pharmacol, 84*(9), 1174-1185.

Reher, T. M., Neumann, D., Buschauer, A., *et al.* (2012). Incomplete activation of human eosinophils via the histamine H4-receptor: evidence for ligand-specific receptor conformations. *Biochem Pharmacol, 84*(2), 192-203.

Ren, J., & Chung, S. H. (2007). Anti-inflammatory effect of alpha-linolenic acid and its mode of action through the inhibition of nitric oxide production and inducible nitric oxide synthase gene expression via NF-kappaB and mitogen-activated protein kinase pathways. *J Agric Food Chem, 55*, 5073-5080.

Riley, J. F., & West, G. B. (1953). The presence of histamine in tissue mast cells. *J Physiol, 120*(4), 528-537.

Ringvall M., Rönnberg E., Wernersson S. *et al.* (2008) Serotonin and histamine storage in mast cell secretory granules is dependent on serglycin proteoglycan. *J Allergy Clin Immunol, 121(4)*, 1020-6.

Rodríguez-Caso C., Rodríguez-Agudo D., Moya-García A. A. *et al.* (2003) Local changes in the catalytic site of mammalian histidine decarboxylase can affect its global conformation and stability. *Eur J Biochem, 270(21)*, 4376-87.

Rönnberg E., Calounova G., & Pejler G. (2012) Mast cells express tyrosine hydroxylase and store dopamine in a serglycin-dependent manner. *Biol Chem, 393(1-2)*, 107-12.

Rosenberg, H. F., Dyer, K. D., & Foster, P. S. (2013). Eosinophils: changing perspectives in health and disease. *Nat Rev Immunol.* 13(1), 9-22

Rosethorne, E. M., & Charlton, S. J. (2011). Agonist-biased signaling at the histamine H4 receptor: JNJ7777120 recruits beta-arrestin without activating G proteins. *Mol Pharmacol, 79*(4), 749-757.

Rothenberg, M. E., & Hogan, S. P. (2006). The eosinophil. *Annu Rev Immunol, 24*, 147-174.

Rothenberg, M. E., Pomerantz, J. L., Owen, W. F., Jr., *et al.* (1988). Characterization of a human eosinophil proteoglycan, and augmentation of its biosynthesis and size by interleukin 3, interleukin 5, and granulocyte/macrophage colony stimulating factor. *J Biol Chem, 263*(27), 13901-13908.

Rozniecki, J.J., Letourneau, R., Sugiultzoglu, M. *et al.* (1999). Differential effect of histamine 3 receptor-active agents on brain, but not peritoneal, mast cell activation. *J Pharmacol Exp Ther*, 290, 1427-1435.

Ruiz-Chica, J., Medina, M. A., Sánchez-Jiménez, F. *et al.* (2001a). Raman study of the interaction between polyamines and a GC oligonucleotide. *Biochem Biophys Res Comm, 285*(2), 437–446.

Ruiz-Chica, J., Medina, M. A., Sánchez-Jiménez, F. *et al.* (2001b). Fourier transform Raman study of the structural specificities on the interaction between DNA and biogenic polyamines. *Biophys J, 80*(1), 443–454.

Ruiz-Chica, J., Medina, M. A., Sánchez-Jiménez, F. *et al.* (2003). Raman spectroscopy study of the interaction between biogenic polyamines and an alternating AT oligodeoxyribonucleotide. *Biochim Biophys Acta, 1628*(1), 11–21.

Ruiz-Chica, A. J., Soriano, A., Tuñón, I. *et al.* (2006). FT-Raman and QM/MM study of the interaction between histamine and DNA. *Chem Phys, 324*(2-3), 579–590.

Ruiz-Pérez M. V., Pino-Ángeles A., Medina M. A. *et al.* (2012) Structural perspective on the direct inhibition mechanism of EGCG on mammalian histidine decarboxylase and DOPA decarboxylase. *J Chem Inf Model, 52(1)*, 113-9.

Sánchez-Jiménez F., Montañez R., Correa-Fiz F. *et al.* (2007) The usefulness of post-genomics tools for characterization of the amine cross-talk in mammalian cells. *Biochem Soc Trans, 35(Pt 2)*, 381-5.

Schneider E., de-Moraes M. L., & Dy M. (2010) Histamine, Immune Cells and Autoimmunity. In R. L. Thurmond (ed), *Histamine in Inflammation.* (pp. 81-94). Berlin: Landes Bioscience and Springer Science.

Schnell, D., Brunskole, I., Ladova, K. *et al.* (2011). Expression and functional properties of canine, rat, and murine histamine H_4 receptors in Sf9 insect cells. *Naunyn Schmiedebergs Archives Pharmacol*, 383,457-470.

Schroeder, J. T. (2011). Basophils: Emerging roles in the pathogenesis of allergic disease. *Immunol Rev*, 242, 144-160.

Seifert, R., Schneider, E.H., Dove, S., *et al.* (2011). Paradoxical stimulatory effects of the "standard" histamine H4-receptor antagonist JNJ7777120: the H4 receptor joins the club of 7 transmembrane domain receptors exhibiting functional selectivity. *Mol Pharmacol*, 79,631-638.

Seifert, R., Wenzel-Seifert, K., Burckstummer, T., *et al.* (2003). Multiple differences in agonist and antagonist pharmacology between human and guinea pig histamine H1-receptor. *J Pharmacol Exp Ther*, 305, 1104-1115.

Seligmann, B. E., Fletcher, M. P., & Gallin J. I. (1983). Histamine modulation of human neutrophil oxidative metabolism, locomotion, degranulation, and membrane potential changes. *J Biol Chem* 130, 1902-1909.

Sengeløv, H., Kjeldsen, L., Diamond, M. S., *et al.* (1993). Subcellular localization and dynamics of Mac-1 (alpha m beta 2) in human neutrophils. *J Clin Invest*, 92, 1467-1476.

Serbina, N. V., Jia, T., Hohl, T. M. *et al.* (2008). Monocyte-mediated defense against microbial pathogens. *Annu Rev Immunol, 26*, 421-452.

Shen, T., Kim, S., Do, J.S. *et al.* (2008). T cell-derived IL-3 plays key role in parasite infection-induced basophil production but is dispensable for in vivo basophil survival. *Int Immunol*, 20,1201-1209.

Sheppard, F. R., Kelher, M. R., Moore, E.E., *et al.* (2005). Structural organization of the neutrophil NADPH oxidase: phosphorylation and translocation during priming and activation. *J. Leuk. Biol.*, 78, 1025-1042.

Shi, C., & Pamer, E. G. (2011). Monocyte recruitment during infection and inflammation. *Nat Rev Immunol, 11*, 762-774.

Shi, H. Z. (2004). Eosinophils function as antigen-presenting cells. *J Leukoc Biol, 76*(3), 520-527.

Shilling DA, Smith MJ, Tyther R, *et al.* (2007) Salmonella typhimurium stimulation combined with tumour-derived heat shock proteins induces potent dendritic cell anti-tumour responses in a murine model. *Clin Exp Immunol*, 149, 109-116.

Simon, T., Laszlo, V., Lang, O. *et al.* (2011) Histamine regulates relevant murine dendritic cell functions via H4 receptor. *Front Biosci (Elite Ed).* 2011 Jun 1;3:1414-24.

Simons, F. E. R., & Akdis, C. A. (2009). Histamine and H1-Antihistamines. In N. F. Adkinson Jr, W. W. Busse, B. S. Bochner, S. T. Holgate, F. E. R. Simons & R. F. Lemanske (Eds.), *Middleton's allergy: principles & practice* (7th ed., Vol. 2, pp. 1517-1547). Philadelphia, Pa.; [Edinburgh]: Mosby Elsevier.

Sirois, J., Menard, G., Moses, A. S. *et al.* (2000). Importance of histamine in the cytokine network in the lung through H2 and H3 receptors: stimulation of IL-10 production. *J Immunol, 164*, 2964-2970.

Smuda, C., Wechsler, J. B., & Bryce, P. J. (2011). TLR-induced activation of neutrophils promotes histamine production via a PI3 Kinase-dependent mechanism. *Immunol Lett*, 141, 102-108.

Solomon, A., Aloe, L., Pe'er, J., *et al.* (1998). Nerve growth factor is preformed in and activates human peripheral blood eosinophils. *J Allergy Clin Immunol, 102*(3), 454-460.

Stewart, J. & Kay, A.B. (1980). Histamine increases histidine uptake by basophils. *Clin Exp Immunol*, 40,423-426.

Strakhova, M.I., Cuff, C.A., Manelli, A.M., *et al.* (2009). In vitro and in vivo characterization of A-940894: a potent histamine H4 receptor antagonist with anti-inflammatory properties. *Brit J Pharmacol*, 157, 44-54.

Summers, R., Sigler, R., Shelhamer, J.H. *et al.* (1981). Effects of infused histamine on asthmatic and normal subjects: comparison of skin test responses. *J Allergy Clin Immunol*, 67,456-464.

Suzuki S., Tanaka S., Nemoto K. *et al.* (1998) Membrane targeting and binding of the 74-kDa form of mouse L-histidine decarboxylase via its carboxyl-terminal sequence. *FEBS Lett, 437(1-2)*, 44-8.

Swartzendruber, J. A., Byrne, A. J., & Bryce, P. J. (2012). Cutting edge: histamine is required for IL-4-driven eosinophilic allergic responses. *J Immunol, 188*(2), 536-540.

Takatsu, K., Takaki, S., & Hitoshi, Y. (1994). Interleukin-5 and its receptor system: implications in the immune system and inflammation. *Adv Immunol, 57*, 145-190.

Takahashi, H.K., Iwagaki, H., Mori, S., *et al.* (2004) Histamine inhibits lipopolysaccharide-induced interleukin (IL)-18 production in human monocytes. *Clin Immunol*, 112, 30-34.

Takahashi, H. K., Yoshida, A., Iwagaki, H. *et al.* (2002). Histamine regulation of interleukin-18-initiating cytokine cascade is associated with down-regulation of intercellular adhesion molecule-1 expression in human peripheral blood mononuclear cells. *J Pharmacol Exp Ther, 300*, 227-235.

Takeshita, K., Sakai, K., Bacon, K.B. *et al.* (2003). Critical role of histamine H4 receptor in leukotriene B4 production and mast cell dependent neutrophil recruitment induced by zymosan in vivo. *J Pharmacol Exp Ther*, 307, 1072–1078.

Tanimoto, A., Murata, Y., Nomaguchi, M. *et al.* (2001). Histamine increases the expression of LOX-1 via H2 receptor in human monocytic THP-1 cells. *FEBS Lett, 508*, 345-349.

Tedeschi, A., Lorini, M., Arquati, M. *et al.* (1991). Regulation of histamine release from human basophil leucocytes: role of H1, H2 and H3 receptors. *Allergy*, 46,626-631.

Thakurdas, S. M., Melicoff, E., Sansores-Garcia, L., et al. (2007). The mast cell-restricted tryptase mMCP-6 has a critical immunoprotective role in bacterial infections. *J Biol Chem, 282*(29), 20809-20815.

Triggiani, M., Petraroli, A., Loffredo, S. *et al.* (2007). Differentiation of monocytes into macrophages induces the upregulation of histamine H1 receptor. *J Allergy Clin Immunol, 119*, 472-481.

Tripathi, T., Shahid, M., Sobia, F. *et al.* (2010). Immune Regulation by Various Facets of Histamine in Immunomodulation and Allergic Disorders. In M. Shahid, N. Khardori, R. A. Khan & T. Tripathi (Eds.), *Biomedical Aspects of Histamine: Current Perspectives* (pp. 133-147). New York, NY: Springer.

Urdiales J. L., Medina MA, & Sánchez-Jiménez, F. (2001). Polyamine metabolism revisited. *Eur J Gastroent Hepat, 13*(9), 1015–1019.

Vanbervliet, B., Akdis, M., Vocanson, M. *et al.* (2011) Histamine receptor H1 signaling on dendritic cells plays a key role in the IFN-gamma/IL-17 balance in T cell-mediated skin inflammation. *J Allergy Clin Immunol* 127, 943-53.

van der Pouw Kraan, T.C., Snijders, A., Boeijem L.C. *et al.* (1998) Histamine inhibits the production of interleukin-12 through interaction with H2 receptors. *J Clin Invest* 102, 1866-73.

Van Rijn, R.M., Chazot, P.L., Shenton, FC., *et al.* (2006). Oligomerization of the recombinant and endogenously expressed human histamine H4 receptors. *Mol. Pharmacol.*, 70, 604-615.

van Rijt, L. S., Vos, N., Hijdra, D., et al. (2003). Airway eosinophils accumulate in the mediastinal lymph nodes but lack antigen-presenting potential for naive T cells. *J Immunol, 171*(7), 3372-3378.

Vignola, A. M., Chanez, P., Paul-Lacoste, P., *et al.* (1994). Phenotypic and functional modulation of normal human alveolar macrophages by histamine. *Am J Respir Cell Mol Biol, 11*, 456-463.

Wang, K. Y., Arima, N., Higuchi, S. *et al.* (2000). Switch of histamine receptor expression from H2 to H1 during differentiation of monocytes into macrophages. *FEBS Lett, 473*, 345-348.

Wardlaw, A. J., & Moqbel, R. (1992). The eosinophil in allergic and helminth-related inflammatory responses. In R. Moqbel (Ed.), *Allergy and immunity to helminths: common mechanisms or divergent pathways.* (pp. 154-186). London: Taylor & Francis.

Weller, P. F. (1994). Eosinophils: structure and functions. *Curr Opin Immunol, 6*(1), 85-90.

Weller, P. F. (2009). Eosinophilia and Eosinophil-Related Disorders. In N. F. Adkinson Jr, W. W. Busse, B. S. Bochner, S. T. Holgate, F. E. R. Simons & R. F. Lemanske (Eds.), *Middleton's allergy: principles & practice* (7th ed., Vol. 2, pp. 859-877). Philadelphia, Pa.; [Edinburgh]: Mosby Elsevier.

Wescott, S., & Kaliner, M. (1983). Histamine H-1 binding site on human polymorphonuclear leukocytes. *Inflammation*,7, 291-300.

Woollard, K. J., & Geissmann, F. (2010). Monocytes in atherosclerosis: subsets and functions. *Nat Rev Cardiol, 7*, 77-86.

Woska J. R. Jr, & Gillespie M. E. (2012) SNARE complex-mediated degranulation in mast cells. *J Cell Mol Med, 16(4)*, 649-56.

Yanagihara, Y., Kajiwara, K., Basaki, Y., *et al.* (1998). Cultured basophils but not cultured mast cells induce human IgE synthesis in B cells after immunologic stimulation. *Clin Exp Immunol*, 111,136-143.

Yu, S., Stahl, E., Li, Q. *et al.* (2008). Antigen inhalation induces mast cells and eosinophils infiltration in the guinea pig esophageal epithelium involving histamine-mediated pathway. *Life Sci*, 82, 324–330.

Zhang, J., Takahashi, H.K., Liu, K. *et al.* (2010) Histamine inhibits adhesion molecule expression in human monocytes, induced by advanced glycation end products, during the mixed lymphocyte reaction. *Br J Pharmacol* 160, 1378-86.

Zheutlin, L. M., Ackerman, S. J., Gleich, G. J., et al. (1984). Stimulation of basophil and rat mast cell histamine release by eosinophil granule-derived cationic proteins. *J Immunol, 133*(4), 2180-2185.

Zhu, Y., Michalovich, D., Wu, H-L., *et al.* (2001). Cloning, expression, and pharmacological characterization of a novel human histamine receptor. *Mol Pharmacol*, 59, 434-441.

Zimmerman, J., & Mallard, J. (1989). H2-antagonist inhibition of human neutrophil superoxide anion synthesis. *Clin Pharmacol Ther*, 45, 487-494.

Chapter 5

Histamine in Asthmatic and Fibrotic Lung Disorders

Emanuela Masini[1*], Laura Lucarini[1], Anita Sydbom[2], Barbro Dahlén[3], Sven-Erik Dahlén[2]

[1]Dept of NEUROFARBA, Section of Pharmacology, University of Florence, Florence, Italy

[2]Institute of Environmental Medicine, [3]Department of Medicine at Karolinska University Hospital Huddinge, both belonging to The Centre for Allergy Research at Karolinska Institutet, Stockholm Sweden

* Corresponding author:
Emanuela Masini, Department of NEUROFARBA, Section of Pharmacology, V. le G. Pieraccini n.6, 50139 Florence, Italy, email: emanuela.masini@unifi.it

Abstract

Histamine has long been known to mediate inflammatory and allergic responses acting predominately through H_1 receptors and H_1 receptor antagonists have been used to treat allergic rhinitis for many years. Since its discovery, histamine H_4 receptor has been identified as a potential drug target for the treatment of allergic and inflammatory lung disease and has been identified as a potential modulator of allergy and inflammation in animal models. The use of the H_4 receptor (H_4R) antagonist JNJ7777120 in mouse and guinea pig models of allergic bronchoconstriction highlighted the involvement of this receptor in asthma. Histamine released from tissue mast cells by the cross-linking of antigen with IgE is deaminated by amine oxidases, enzymes widely distributed among living organisms and a histaminase, purified from the pea seedlings has been shown to exert a protective effect in an experimentally animal model of asthma-like reaction in guinea pig. In an animal model of fibrosis, H_4R antagonists demonstrated anti-inflammatory and anti-fibrotic properties, suggesting a possible therapeutic potential in the treatment of Th2-mediated diseases, including pulmonary fibrosis.

The combined effects of antihistamines and antileukotrienes inhibit both phases of the allergen induced bronchoconstriction in subjects with asthma. Finally, available data support H_1 antihistamines as part of the treatment of asthma especially in combination with other drugs, however bronchoprovocation trials or treatment studies with histamine H_4 antagonists are not available. In view of the interesting profile of activities of H_4 antagonists, it is possible to hypothesize that this class of drugs may add further clinical benefit by causing a broader inhibition of histamine effects.

5.1. Introduction

Since its discovery at the turn of the century, the Histamine H_4R (Nakamura *et al*, 2000; Oda *et al*, 2000; Cogé *et al*, 2001; Liu *et al*, 2001 ; Morse *et al*, 2001; Nguyen *et al*, 2001; Zhu *et al*, 2001) has been a potential drug target for the treatment of inflammatory diseases.

Asthma is a heterogeneous airway inflammatory disease associated with the involvement of T helper cell, recruitment of mast cells and eosinophils into airways, production of cytokines (IL-4, IL-5 and IL-13) and other mediators leading to chronic inflammation as well as remodelling (Bousquet *et al*, 2000; Busse *et al*, 2001, Holgate, 2012). The Histamine H_4R belongs to the GPCR family ($G\alpha_{i/o}$) and is expressed on many cells of the immune system, especially those associated with the pathology of asthma, such as dendritics cells, eosinophils, mast cells, monocytes, neutrophils, T-lymphocytes, NK-cells and also on fibroblasts (Zampeli & Tiligada, 2009). Many cells involved in the asthmatic inflammatory response express both H_1 and H_4R (see review by Thurmond *et al*, 2008). Histamine is also found in increased levels in bronchoalveolar lavage from asthma patients and this is associated with worsening of lung function. In mice, genetic deletion of histamine-forming enzyme L-histidine decarboxylase or of H_1 receptor provides beneficial effects in experimental asthma (Ohtsu *et al.*, 2001; Miyamoto *et al.*, 2006). In animal models, the H_4R has been identified as a potential modulator of the recruitment of T cell into the lungs after allergen exposure (Dunford *et al.*, 2006) and by cytokines amplifying the allergic symptoms and thus leading to chronic inflammation (Cowden *et al.*, 2010). This also leads to the activation of eosinophils and increased expression of adhesion molecules.

Evidence for the involvement of Histamine H_4R in asthma was shown with the use of the H_4R antagonist JNJ7777120, which inhibits T-cell infiltration to the lung and decreases Th2 cytokines in a mouse model (Cowden *et al*, 2010; Thurmond *et al*, 2004).

Although many studies in different animal models point to the role of Histamine H_4R in the management of allergic disorders and asthma, there are no real human studies yet concerning asthma.

Several compounds have been produced by different pharmaceutical companies trying to address this problem. There are more than 13 pharmaceutical companies with patent applications for substances manipulating the H_4R (Kiss *et al*, 2012). Among these companies, Johnson & Johnson is the most active company with 54 patent applications, and holds the patent for the most studied Histamine H_4 antagonist, JNJ7777120. Pfizer has also developed two other H_4R antagonists, PF-3893787 and INCB38579 by IncyteCor, where both compounds have shown promising results in preclinical animal models. The first Histamine H_4R antagonist to enter clinical trial Phase I was the Palau Pharma´s H_4R antagonist UR-65318 and last year this compound entered the Phase IIa studies for patients suffering from chronic allergic asthma (Salcedo *et al*, 2013). Recently, another H_4 antagonist from Jansen Research & Development, JNJ 39758979, has been published which was well tolerated in Phase I human clinical trials and has now progressed into Phase II (Savall, 2013). The results from these clinical studies have yet to be published.

5.2. Histamine and Histaminergic H_4R Ligands in Animal Models of Allergic Asthma and Pulmonary Fibrosis

Asthma is a heterogeneous chronic condition characterized by widespread, variable and reversible airflow obstruction which is released either spontaneously or with pharmacological treatment. The underlying patho-physiological feature of asthma is increased airway responsiveness which develops on a basis of diffuse bronchial inflammation and functional changes in the airway smooth muscle. The prevalence of asthma is increasing worldwide (Peebles & Hartert, 2002; Myers, 2000), but medications can be highly effective in reducing the burden of asthma for many people and clinical studies show that proper treatment of asthma with appropriate medications can reduce deaths, hospitalizations, and symptoms. Over the last few decades, leukotriene modifiers have emerged as one of the few pharmacological options for asthma that specifically target a pathway of pathogenesis (Drazen *et al.*, 1999; Kemp, 2003). Moreover, anti-IgE therapy has been introduced in the last ten years for the treatment of moderate to severe allergic asthma in adults and adolescent (Johansson & Buhl, 2006). Histamine has long been known to mediate inflammatory and allergic responses acting predominately through H_1 receptor and H_1 receptor antagonists have been used to treat allergic rhinitis for many years (Hill *et al.*, 1997).

Direct evidence for the role of free radicals in asthma comes from human studies that demonstrate that the levels of antioxidants in airway of asthmatics are higher than normal (Comhair *et al.*, 1999) and bronchial obstruction in asthma is associated with an increased production of oxygen free radicals in airway lumen from inflammatory cells (Calhoun *et al.*, 1992). It has been reported that inflammatory cells produce large amount of superoxide anion and the scavenging of superoxide by superoxide dismutase (SOD) mimetics results in clear protection against ovalbumin-induced bronchospasm, lung inflammation and prostaglandin production in sensitized guinea pigs (Masini *et al.*, 2005). In another study, epigallocatechin-3-gallate (EGCG), an antioxidant molecule that enhances constitutive nitric oxide synthase (NOS) activity, reduced leukocyte infiltration as measured by myeloperoxidase (MPO) activity and eosinophilic accumulation, evaluated as eMBP-positive cells, and reduced acute bronchospasm in ovalbumin-challenged sensitized guinea pig (Bani *et al.*, 2006).

Infiltration of inflammatory cells, especially mast cells and eosinophils, is thus a prominent feature of asthmatic lungs, and acute airway response to allergens is known to depend on eosinophils and mast cells (Wardlaw *et al.*, 1988). It is improbable that only these cell types are responsible for the initiation and perpetuation of bronchoconstriction and airway inflammation, however, mast cell has long been considered to be of paramount importance in the patho-physiology of asthma (Rossi & Olivieri, 1997). In fact, the release of IgE-dependent mediators from mast cells, such as histamine, plays a central role in the pathogenesis of the disease (Galli, 1993; Bradding & Holgate, 1996; Barnes & Page, 2001). The release of histamine from mast cells infiltrating airway smooth muscle elicited by the cross-linking of antigen with IgE bound to specific receptors of cell membrane is considered of paramount importance not only in the early phase of asthmatic response, but also in the late one (Brightling *et al.*, 2002). Mast cell mediators cause bronchoconstriction and smooth muscle cell proliferation and can recruit other inflammatory cells, thereby initiating a vicious cycle that amplifies the pathological features of the disease. Studies in transgenic mast cell-deficient mice have shown that these cells act as a local amplifier of bronchoconstriction (Williams & Galli, 2000). The number of mast cells in the smooth muscle of patients with asthma is inversely correlated with the degree of airway hyperresponsiveness (Brigthling *et al.*, 2002) and endobronchial biopsies show increased mast cell degranulation in asthmatic patients compared with non asthmatic subjects (Pesci *et al.*, 1993). The levels of prostaglandin D_2, the major cyclooxygenase product generated by activated mast cells during allergic response (Bochenek *et al.*, 2004), are elevated in the urine of asthmatic patients after antigen exposure in both the early and late phases (O'Sullivan, 1999).

Histamine released from tissue mast cells by the cross-linking of antigen with IgE is oxidatively deaminated by amine oxidases (AO_s), enzymes widely distributed among living organisms (Mondovì, 1985), and is involved not only in the metabolism of histamine but of other primary amines released during anaphylactic reaction. Amine oxidases can be divided in two classes, depending whether the prosthetic group is a flavin adenine dinucleotide (FAD) or 2,4,5 trihydroxyphenylalaninequinone (TPQ), a cofactor derived from the post-translactional oxidation of a tyrosine residue (Klinmann & Mu, 1994). This second group of enzymes (E.C.1.4.3.6) contains copper ($CuAO_s$) and TPQ and within the Cu-TPQ class, plant enzymes are in general more efficient than animal ones, probably because they also function through a radical mechanism not present in the enzymes from animal sources. (Bellelli *et al.*, 2000). Recently, a copper amine oxidase (histaminase) purified from the pea seedling has been shown to exert a protective effect on cardiac anaphylactic response in guinea pig (Masini *et al.*, 2002) and a protective effect of histaminase, purified from the pea seedling, has been demonstrated in an experimentally animal model of asthma-like reaction in guinea pig (Masini *et al.*, 2004). This enzyme has a therapeutic application in various histamine-related affections, such as food histaminosis and inflammatory bowel disease and may be involved in allergic and anaphylactic responses (Mondovì *et al.*, 2013). In fact, the pre-treatment of animals with histaminases, i.p. or as aerosol solution, resulted in a marked reduction of breathing abnormalities and prevention of respiratory failure (Mondovì *et al.*, 2013).

5.3. Histamine H_4R Ligands in Animal Models of Asthma

The histaminergic system has over the last five decades proved to be a rich source of drugs. Histamine H_1 and H_2 receptor antagonists are widely used for the treatment of allergies and peptic ulcers, respectively, while H_3 receptor antagonists could have a therapeutic use, in a near future, in dementia, obesity and psychotic and sleep disorders (Hill *et al.*, 1997). Most importantly, growing attention is directed towards the validation of H_4R ligands in some inflammatory and immunological diseases, such as asthma and pulmonary fibrosis. Over the years, studies on the effect of histamine H_4R ligands in animal models of immune and allergic diseases have produced a large amount of information, albeit with some conflicting reports. These different responses might arise from the differential expression of histamine receptors as well as interspecies differences in the action of these molecules (Zampeli & Tiligada, 2009). Recent results obtained in animal models of airway inflammation and fibrosis are reported here.

Many animal models are available for studying human diseases from mice to primates, but we must consider that these models are not perfect replicas of human disease and may not be predictive of human outcome. Nonetheless, the growing knowledge in genomics, bioinformatics, immunology and molecular pharmacology has helped us to better understand the various animal models and apply the information to human conditions.

The widespread use of the mouse asthma model is likely due to the availability of specific molecular, immunological and genetic tools for exploring the inflammatory mechanisms contributing to the underlying pathophysiology of asthma. The H_4R is present in the lung, bronchial epithelium, smooth muscle cells and microvascular endothelial cells (Gantner *et al.*, 2002) and can thus theoretically contribute to the airway disease pathobiology. The H_4R mediates the synergistic sequential action of histamine and CXCL12, a chemokine constitutively expressed in skin and airway epithelium, and the migration of mast cells in the mucosal epithelium in response to the allergens (Thurmond *et al.*, 2004; Godot *et al.*, 2007). Accordingly, previous experiments with H_4R-deficient($H_4R^{-/-}$) mice and oral gavage administration of selective H_4R antagonists in a murine model showed a reduction of lung inflammation upon allergen stimulation (Dunford *et al.*, 2006). This demonstrated the role of the H_4R in modulating Th2 allergic responses, where the H_4R influences $CD4^+$ T cell activation attributed to a decreased production of IL-4, IL-5 and IL-13 and chemokines from dendritic cells (Dunford *et al.*, 2006). These cytokines have individually been targeted for potential therapeutic benefit in human disease. IL-4 is the main driver of isotype switching in B cells to produce high affinity IgE and it is an important factor in Th2-cell development. A similar effect is exploited by IL-5, which is an activation factor for eosinophils, whilst IL-13 is a mediator of remodelling, airway hyperreactivity and goblet cell hyperplasia.

In the mouse model of acute lung inflammation and hyper-responsiveness, dendritic cells seem to be the target of the effects of H_4R antagonists, where dendritic cells isolated from H_4R knockout mice or treated with H_4R antagonists were unable to properly stimulate $CD4^+$ T cells under Th2-polarizing conditions (Zhang *et al.*, 2007). In the same murine model of asthma, H_4R antagonists, given only during the sensitization stage, had an inhibitory effect upon subsequent airway inflammation and T-cell cytokine production, indicating that T-cell activation was impaired at the priming stage. In another study, inhibition of airway resistance and inflammation mediated through the recruitment of T regulatory (Treg) cells was observed with 4-methylhistamine, a selective H_4R agonist, given intratracheally to asthmatic mice (Morgan *et al.*, 2007). The authors attributed their different results to the local administration of the compound and to the resulting concentration gradient within the lung that would allow the migration of the Treg cells and the immune suppressive response (Morgan *et al.*, 2007).

The allergic guinea pig model, while validated many decades ago, has continued to be used to investigate the pathophysiology of asthma for many reasons. The guinea pig is a particular suitable model for studying histaminergic system and allows the reproduction of the different syndromes presented by human asthma. In fact, a model of occupational asthma has been validated using toluene-2,4-diisocyanate (Nabe *et al.*, 2005) and a model of cough variant asthma has been developed (Bani *et al.*, 1997). Guinea pigs are sensitized to ovalbumin and bronchial responsiveness, as well as the cough reflex response, which is measured after antigen inhalation. This animal model has been widely used to evaluate the anti-asthmatic effect of different classes of compounds. For instance, a phosphodiesterase 4 inhibitor and its active metabolite reduced allergen-induced eosinophilia and airway hyper-responsiveness (AHR) in a guinea pig model (Billah *et al.*, 2002). Cyclosporine (Xie *et al.*, 2002) and FK-506 (Morishita *et al.*, 2005), administered by inhalation, inhibited AHR and cellular influx following allergen challenge, and similar effects were observed with a derivative of methotrexate. The role of neurokinins and their receptors has been extensively studied in the guinea pig model, with the evidence for a complex interplay between the three receptor subtypes (Schuiling *et al.*, 1999).

Masini and coworkers have utilized the guinea pig model for several studies. First, we studied the effect of the hormone relaxin, which was found to inhibit mast cell histamine release (Masini *et al.*, 1994) and to counteract the respiratory and histopathological abnormalities induced by inhaled antigen (Bani *et al.*, 1997). Relaxin was also found to promote dilatation of alveolar blood capillaries and to reduce the thickness of the air blood barrier. In the following years, using this model of allergic asthmatic response, we investigated several antioxidants or SOD mimetic drugs (Masini *et al.*, 2005; Bani *et al.*, 2006) as well as molecules that interfere with the production of pro-inflammatory mediators, such as ceramide (Masini *et al.*, 2008).

The recent discovery of the histamine H_4R, functionally expressed on many cell types associated with asthma pathology (eosinophils, basophils, mast cells, dendritic cells and $CD8^+$ T cells (Thurmond *et al.*, 2008)), prompted an investigation in the guinea pig asthma model for the ability of an H_4R antagonist to modify antigen-induced oxidative stress, airway inflammation, bronchoconstriction, as well as cytokine and prostaglandin production. Ovalbumin challenge-induced cough, dyspnoea, bronchoconstriction and alveolar space dilatation were observed along with increased leukocyte infiltration and production of PGD_2, LTB_4 and TNFα in lung tissue and bronchoalveolar lavage (BAL) fluid. Compound JNJ7777120 (JNJ), a selective H_4R antagonist, counteracts the functional, histopatological and biochemical changes induced by antigen challenge (Somma *et al.*, 2013). The authors tried to find a possible explanation for the protective mechanism of JNJ, evaluating the possible interaction of this molecule with lipocortin-1 (LC-1), a glucocorticoid-modulated protein, and initially characterized its ability to inhibit prostanoid release (Cirino *et al.*, 1987). LC-1 inhibits the activity of cytoplasmic phospholipase A_2, which plays a key role in the production of inflammatory lipid mediators, prostaglandins and leukotrienes and inhibits the extravasation of leukocytes (D'Acquisto *et al.*, 2008). Given that histamine H_4R are expressed in dendritic, $CD4^+$ and $CD8^+$ T cells (Zampeli &Tiligada, 2009), it seemed logical to assume that JNJ could influence the complex cytokine interplay involved in the pathogenesis of asthma (Figure 5. 1).

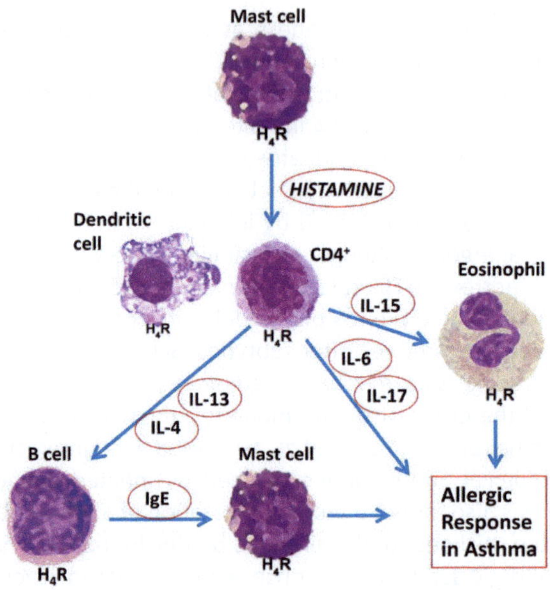

Figure 5.1 The complex cytokine interaction in allergic response in asthma.

In fact, antigen-challenge caused leukocyte infiltration in lung tissue, especially eosinophils, which produce large amounts of free radicals and are responsible for airway remodelling (Jarjour & Calhoun, 1994). In addition, superoxide produced during the asthmatic response can promote the expression of genes encoding the pro-inflammatory cytokines IL-1, IL-6 and TNFα, which can cause endothelial cell damage (Ndengele *et al.*, 2005). Moreover, superoxide promotes mast cell degranulation and histamine release (Masini *et al.*, 2005), thus amplifying the inflammatory response. Notably, JNJ prevents lung mast cell degranulation and chemotaxis of eosinophils, which are involved in adverse airway remodelling through the release of transforming growth factor-β (TGF-β), which shifts stromal cells towards the myofibroblast phenotype, responsible for fibrosis (Kisseleva & Brenner, 2008). In particular, stimulation of H_4R on eosinophils triggers cellular changes required for chemotaxis, actin polymerization, changes in cellular shape and expression of adhesion molecules (Ling *et al.*, 2004). The treatment with JNJ was effective in reducing the chemotaxis of eosinophils, confirming the hypothesis that this process is mediated by histamine H_4R. Considering the role of the H_4R in the reduction of the asthmatic response revealed by the studies reported here as well as the literature data on the modulation of H_4R antagonists on mast cell, eosinophil, T and dendritic cell functions, it is highly warranted to evaluate the future therapeutic use of H_4R antagonists in asthma. Many research groups have utilized the guinea pig model with success (reviewed by Riley et al., 2013). Smith and Broadley (2008) investigated the functional airways responses to inhaled 5'-AMP in actively sensitized, conscious guinea-pigs following administration of selective adenosine receptor antagonists and demonstrated that all four adenosine receptor subtypes play various roles in the airways response. The same group has recently stated that the sensitized guinea pig model is a suitable animal model of chronic airway inflammation, airway hyperresponsiveness and lung remodelling that can accurately predict drug effectiveness in human asthma. In OA-sensitized guinea pigs, the efficacy of the inhaled corticosteroid fluticasone propionate (FP), the phosphodiesterase 4 inhibitor roflumilast and the inducible nitric oxide synthase (iNOS) inhibitor GW274150 administered orally 24 hours and 30 min before OA exposure were compared. Fluticasone and roflumilast inhibited T cell lung influx and airway hyperresponsiveness and remodeling, while GW274150 only inhibited the inflammatory response but not remodeling (Evans *et al.*, 2012). In the clinical setting, inhaled corticosteroids and phosphodiesterase 4 inhibitors are relatively effective against most features of asthma whereas the iNOS inhibitor GW274150 was ineffective. Thus, while certain differences remain between the experimental and clinical effectiveness of antiasthma drugs, the chronic pulmonary inflammation guinea pig model does appear to be a good pre-clinical predictor of potential asthma therapeutics.

5.4. Histamine H_4R Ligands in Animal Models of Lung Inflammation and Fibrosis

Fibrosis can be considered an excessive reparative response of tissues to chronic injury and inflammation, and features excess deposition of collagen and other extracellular matrix (ECM) components in the interstitium. ECM accumulation disrupts the normal histological architecture of an organ, eventually leading to its dysfunction (Paz & Shoenfeld, 2009; Kisseleva & Brenner, 2008). In spite of different aetiology and target organs of fibrotic disorders, one pathological hallmark is the presence of myofibroblasts, which are activated by collagen-secreting fibroblasts originating from stromal precursor cells that are induced to proliferate and differentiate by pro-fibrotic factors released in the local inflammatory micro-environment, such as TGF-β and angiotensin II (Kisseleva & Brenner, 2008; Wynn 2007). In particular, fibrosis takes place when the synthesis of new collagen by myofibroblasts exceeds its degradation rate, leading to accumulation of collagen over time (Wynn, 2008).

Pulmonary fibrosis is the end stage of a wide range of chronic lung inflammatory diseases leading to progressive parenchymal destruction. Histopathologically it is characterized by alveolar and capillary loss due to pneumocyte and endothelial apoptosis, accumulation of myofibroblasts and an excess of interstitial collagen and abnormal remodelling of lung parenchyma (Hardie *et al.*, 2009). This process results in progressive airway stiffening and thickening of the air-blood membrane, which makes breathing difficult and eventually leads to respiratory failure. Idiopathic pulmonary fibrosis (IPF), the most common fibrotic disease of the lungs, has a particularly poor prognosis and represents a therapeutic challenge for pneumologists. In fact, the classical anti-inflammatory drugs are nearly ineffective in improving its clinical course, being unable to prevent or delay the onset of respiratory failure. Therefore, the current therapeutic approaches to IPF are oriented towards novel substances which could override the limitations of the existing anti-inflammatory drugs, such as molecules targeting TGF-β signalling, the upstream activation pathway of myofibroblasts (Gharaee-Kermani *et al.*, 2009). In this context, cyclooxygenase (COX)-inhibiting nitric oxide (NO) donors (CINODs), designed to inhibit COX-1 and COX-2 while releasing NO, have shown a significantly higher efficacy than classic anti-inflammatory drugs in reducing lung inflammation and preventing collagen accumulation in bleomycin-induced lung fibrosis in male C57BL/6 mice (Pini *et al.*, 2012), agreeing with previous studies on the cirrhotic liver (Casini *et al.*, 1997; Failli *et al.*, 2000). In the latter situation, NO supplementation by NO donors decreased liver cirrhosis by inhibiting the profibrotic activation of hepatic stellate cells induced by reactive oxygen species (ROS) or platelet-derived growth factor (Casini *et al.*, 1997; Failli *et al.*, 2000, Svegliati-Baroni *et al.*, 2001). Moreover, experimental data in cultured rat mesangial cells suggests that NO could have direct antifibrotic effects in kidney cells by down-regulating the expression of fibrosis-related genes (Wani *et al.*, 2007). In this context, a novel antifibrotic peptide with relaxin-like activity, which exerts an inhibitory effect on TGF-β1-induced collagen deposition in human dermal fibroblasts and enhanced matrix metalloproteinase (MMP)-2 expression, caused a significant reduction in lung inflammation and injury and ameliorated adverse airway remodelling and peribronchial fibrosis (Pini *et al.*, 2010).

Based on the above evidence and on the results of Kohyama and coworkers (2010), which demonstrated that the profibrotic effect of histamine on human foetal lung fibroblasts is a consequence of H_4R activation and fibronectin-induced lung fibroblast migration is blocked by JNJ, we investigated whether H_4R ligands could have a therapeutic effect in the same mouse model of bleomycin-induced lung fibrosis (Kaminski *et al.*, 2000; Moeller *et al.*, 2008).

The results indicate that H_4R antagonists have significant anti-inflammatory and anti-fibrotic properties, where H_4R antagonists consistently decreased inflammatory and oxidative stress parameters and reduced the relative number of goblet cells, the thickness of smooth layer, the level of TGF-β and collagen deposition. All of these parameters were also accompanied by a decrease in airway resistance to inflation (PAO) (Figure 5. 2) (Lucarini *et al.*, 2013).

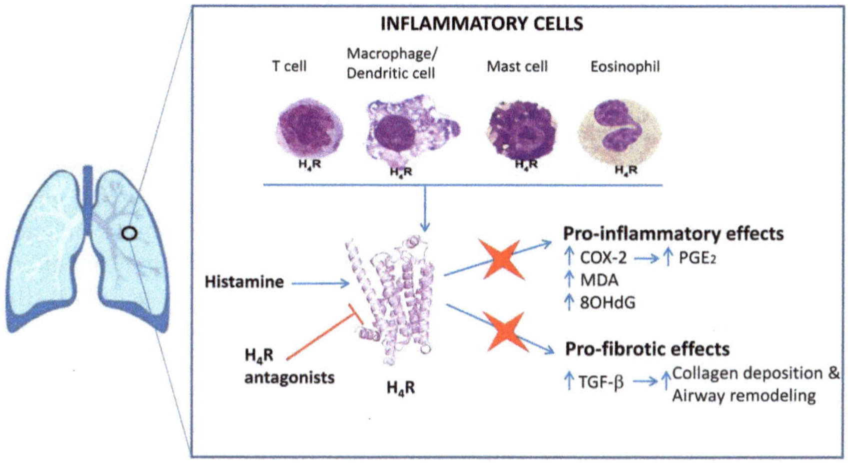

Figure 5.2 A schematic description of the protective action of H_4 receptor (H_4R) antagonists on histamine-induced pro-inflammatory and pro-fibrotic effects. COX-2: cyclooxygenase-2; MDA: malonyldialdehyde; 8OHdG: 8-hydroxy-2'-deoxyguanosine; TGF-β: transforming growth factor β.

As previously reported, histamine H_4R are functionally expressed in cells of the innate immune system and histamine is a robust chemotactic factor for these cells (Damaj *et al.*, 2007). Furthermore, H_4R antagonists inhibit the migration of immune cells, decreasing the production of the pro-fibrotic cytokine TGF-β, which has been recently identified as a target for the development of novel antifibrotic agents.

In conclusion, the dual effect of H_4R antagonists on inflammatory and fibrotic mediator production points to their therapeutic potential for the treatment of Th2-mediated diseases, including pulmonary fibrosis.

5.5. Conventional Antihistamines as Effective Treatment for Asthma in New Combination with Other Mast Cell Inhibitors

Current clinical guidelines do not support treatment of asthma with antihistamines. However, this conclusion has not been reassessed recently. The recommendation rests on early clinical data with the first generation of H_1 antagonists. This report advocates the opinion that it is now timely to revisit the use of antihistamines for the treatment of asthma. The strategy should not be through the use antihistamines as single medications, however, but as adjuvant therapy in combination with drugs that block one or two of the other major mast cell mediators. This has great potential to become new, effective treatments of asthma.

Accordingly, as reviewed in other parts of this volume, there is strong evidence from different experimental models to support this specific combination therapy. Whereas singular inhibition of one mediator class such as histamine or cysteinyl-leukotrienes (CysLT) provides small or sometimes insignificant protective effects against mast cell-dependent bronchoconstriction, a combination of inhibitors against two or more classes of mediators generally produces significant effects (e.g. Hay *et al.*, 1987; Wikström-Jonsson, 1994; Sundström *et al.* 2003). This is illustrated in Figure 5.3 where the antagonism of histamine, CysLTs and prostaglandins was studied in a step-wise fashion in allergic bronchoconstriction evoked in the peripheral airways of the guinea pig (Ressmeyer *et al.*, 2006).

These results demonstrate that antagonism of one mediator class has only a small effect on the response, antagonism of two classes has mostly greater but incomplete effects, but intervention with all three pathways confers complete abolishment of the reaction. The absolute degree of this effect with the different interventions in different systems are variable, but the overall concept that triple mediator antagonism essentially eliminates the responsiveness to antigen-challenge has been confirmed in other guinea pig models and on isolated human bronchi (Dahlén *et al.*, 1983; Hay *et al.*, 1987; Björck & Dahlén 1993; Wikström-Jonsson, 1994; Sundström *et al.* 2003;). This is not surprising, as

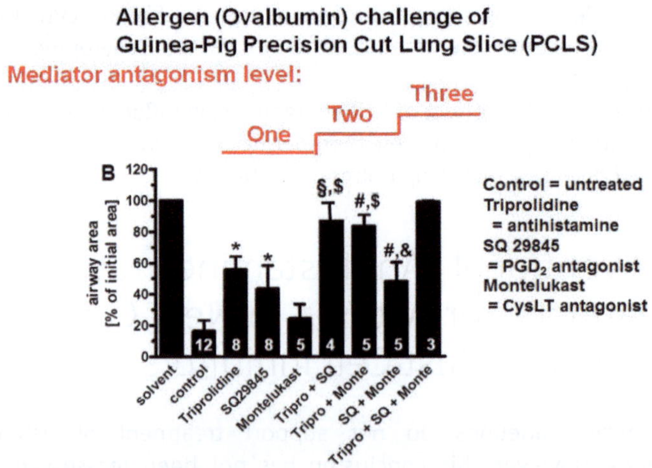

Figure 5.3 Video-microscopy of antigen challenge of guinea pig peripheral airways, 100% represents completely open airways (solvent and triple antagonism), control is antigen without antagonist. Modified from Ressmeyer et al, 2006.

the antigen-induced activation of mast cells results in the release of histamine, prostaglandin (PG) D_2 and the CysLTs (Levi-Schaffer *et al.*, 1987), which are all effective bronchoconstrictors.

Furthermore, following the original work by Adams and Lichtenstein (1979), several studies in isolated human bronchi have shown that the combination of a CysLT antagonist and an antihistamine have profound inhibitory effects on the Schultz-Dale contraction induced by challenge with a specific antigen or anti-IgE. In a series of publications during the 1980-90's, this has been extensively documented both in human bronchi from non-asthmatic subjects and occasionally in airway preparations from subjects with asthma (e.g. Dahlén *et al.*, 1983; Hay *et al.*, 1987; Björck & Dahlén, 1993). Together, these studies in human tissues confirm the value of the basic strategy of combined antagonism of mast cell mediators outlined in the predictive guinea-pig models.

However, the critical standard for establishing proof-of-concept and predicting clinical efficacy in asthma is the allergen bronchoprovocation setting. In fact, there are no currently used clinical treatments of asthma that do not display effectiveness in this human *in vivo* model (Boulet *et al.*, 2007). In the first comprehensive study of the combined effects of antihistamines and antileukotrienes (Roquet *et al.*, 1997), it was reported that the antihistamine loratadine by itself caused significant inhibition of both early and late phase allergen-induced bronchoconstriction (Figure 5.4). Although the effect of the

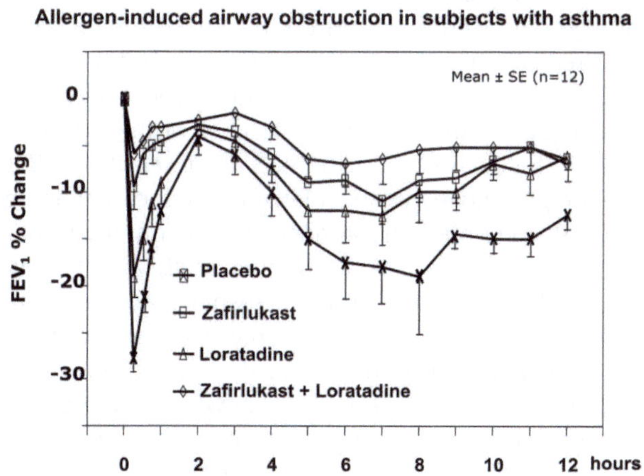

Figure 5.4 Allergen bronchoprovocation of subjects with allergic asthma in a cross-over trial where each subject was provoked four times after one week double-blind pre-treatment with placebo, the H_1R antagonist loratadine, the $CysLT_1$ antagonist zafirlukast or loratadine plus zafirlukast. At least four weeks of washout between the sessions. Airway response monitored as drop in FEV1 from prechallenge baseline. Modified from Roquet et al, 1997.

antihistamine was smaller than that of the antileukotriene zafirlukast on the early phase reaction, the effect on the late phase reaction was significant and closely similar to that of the antileukotriene. Moreover, when both drugs were combined, there was a substantial ~75% inhibition in both phases of allergen-induced bronchoconstriction (Figure 5.4).

More recently, these effects have been replicated using another leukotriene receptor antagonist (montelukast) in combination either with the metabolite of loratadine, desloratadine (Davis *et al.*, 2009) or azelastine (Richter *et al.*, 2008). In addition, Davis and coworkers found that this combination significantly blunted the allergen-elicited sputum eosinophila (Davis *et al.*, 2009).

In a 10 week placebo-controlled clinical trial with cross-over design, performed in 117 subjects with asthma, it has been further established that the combination of loratadine and montelukast had greater clinical effect than montelukast alone (Reicin *et al.*, 2000). A shorter 6 week cross-over study in 406 patients with asthma mostly agreed with this observation and also showed that the effect of the combination therapy was comparable to treatment with an ordinary dose of inhaled beclomethasone (Lu *et al.*, 2009). Considering both the clinical trials well as the extensive experimental data collected in the allergen bronchoprovocation studies, it is in our opinion support for the use of a combination of an H_1 anti-histamine drug with an antileukotriene for the treatment of mast cell-dependent asthma.

Furthermore, this implies that a combination treatment should not be restricted to allergen-induced reactions but also to asthma triggered by other stimuli such as exercise, inhalation of cold air or mannitol, as well as aspirin in aspirin/NSAID-intolerant asthma (AIA), as mast cell activation is the final pathway for inducing the bronchoconstriction in all of these responses (Sladek & Szczeklik 1993; O'Sullivan *et al.*, 1996; O'Sullivan *et al.*. 1998a; Brannan *et al.*, 2003). While the protective effects of leukotriene receptor antagonists have been extensively documented in bronchoconstriction induced by exercise (Manning *et al.*, 1990; Leff *et al.*, 1998; Dahlén *et al.*, 2002) or aspirin in AIA (Christie *et al.*, 1991; Dahlén *et al.*, 1993), the response to cumulative challenge with mannitol was found to be inhibited only with respect to the recovery phase, not the sensitivity to this cumulative challenge (Brannan *et al.*, 2001). For bronchoconstriction induced by exercise, however, the protective effect of a single intervention with antihistamines appears to have minimal effects both in children and adults (Dahlén *et al.*, 2002; Peroni *et al.*, 2002). There is also substantial release of PGD_2 after exercise (O'Sullivan *et al.*, 1998b) and mannitol challenges (Brannan *et al.*, 2003) and that TP receptor antagonists, which block bronchoconstriction to PGD_2, demonstrate protective effects on the airway response to exercise challenge (Magnussen *et al.*, 1992), which highlights the importance of this major mast cell mediator in the asthmatic responses. In line with the findings in the animal models, complete attenuation of mast cell-dependent bronchoconstriction requires antagonism of all three major classes of bronchoconstrictors. As indicated, the relative importance of histamine, CysLTs and PGD_2 varies depending upon the characteristics of the challenge and possibly also between individuals, but we postulate that when all three mediators are blocked, the airway response will be almost completely abolished (Figure 5.5).

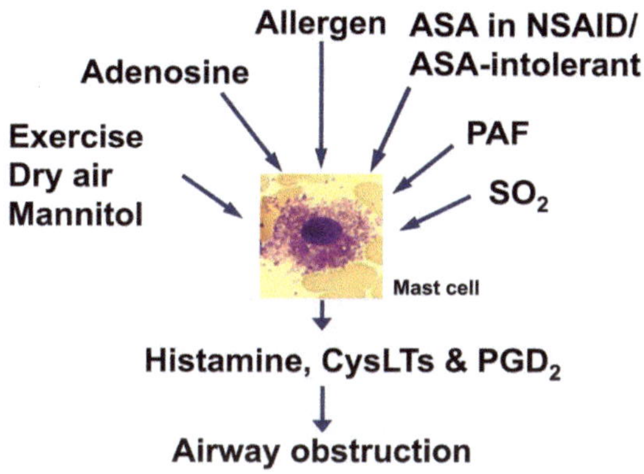

Figure 5.5 Mast cell activation as the final common path for a number of trigger-factors in asthma.

This hypothesis again rests on the large databank of results from various experimental models, including isolated human bronchi (Dahlén *et al.*, 1983: Hay *et al.*, 1987; Björck *et al.*, 1993).

Finally, whereas the currently available data discussed above supports H_1 antihistamines as part of the treatment of asthma, there is as of yet no bronchoprovocation trials or treatment studies with H_4 antagonist that provide definitive data on the use of that particular class of antagonists in asthma. As discussed in other parts of this volume, there is experimental data supporting future trials addressing H_4R as another target in combination therapy. In light of the interesting profile of activities of H_4R antagonists in experimental models, it is hypothesized that this class of drugs may add further clinical benefit by causing a broader inhibition of histamine. Likewise, in order to establish the introduction of the triple mast cell mediator concept that this communication advocates, it is necessary to perform well designed, long-term treatment trials to support changes to the current clinical guidelines. The concept of combined mediator antagonists has received comparatively little attention because there are no patents protecting the development of such therapy, and thus provides minor commercial incentive for its development. Meanwhile, many patients, especially those with occasional or seasonal troubles, are looking for simple and easy to use oral treatments for their asthma.

In conclusion, it is clear that histamine contributes to the pathophysiology of asthma. Although current clinical guidelines do not support treatment of asthma with antihistamines. This affirmation is however based on early clinical data with first generation of H_1 antagonists and it is current opinion to revisit the use of this class of drugs for the treatment of asthma. The strategy should, however, not be to use antihistamines as single medications, but rather as adjuvant therapy in combination with other drugs acting on other classes of mast cell mediators. The interesting profile of preclinical activities of histamine H_4 receptor antagonists raises the potential that a broader inhibition of histamine might add clinical benefit and introduce a new effective treatment of asthma.

Acknowledgement

The work was supported by COST action BM0806 and grants from Ente Cassa di Risparmio of Florence (Florence, Italy), the Swedish Heart-Lung Foundation, MRC, Vinnova (CiDAT), The Stockholm County Council Research Funds (ALF), the Asthma and Allergy Research Foundation, Karolinska Institutet.

References

Adams, G. K., Lichtenstein, L. (1979). In vitro studies of antigen-induced bronchospasm: effect of antihistamine and SRS-A antagonist on response of sensitized guinea pig and human airways to antigen. *J Immunol, 122*, 555-562.

Bani, D., Ballati, L., Masini, E. et al. (1997). Relaxin counteracts asthma-like reaction induced by inhaled antigen in sensitized guinea pigs. *Endocrinology, 138*, 1909-1915.

Bani, D., Giannini, L., Ciampa, A. et al. (2006). Epigallocatechin-3-gallate reduces allergen-induced asthma-like reaction in sensitized guinea pigs. *J Pharmacol Exp Ther, 317*, 1002-1011.

Barnes, P.J., Page, C. (2001). Mediators of asthma: a new series. *Pulm Pharmacol Ther, 14*, 1-2.

Bellelli, A., Morpurgo, L., Mondovì, B. et al. (2000). The oxidation and reduction reactions of bovine serum amine oxidase.A kinetic study. *Eur J Biochem, 267*, 3264-3269.

Billah, M. M., Cooper, N., Minnicozzi, M. et al. (2002). Pharmacology of N-(3,5-dichloro-1-oxido-4-pyridinyl)-8-methoxy-2-(trifluoromethyl)-5-quinoline carboxamide (SCH 351591), a novel, orally active phosphodiesterase 4 inhibitor. *J Pharmacol Exp Ther, 302*, 127-137.

Björck, T., Dahlén, S.-E. (1993). Leukotrienes and histamine mediate IgE-dependent contractions of human bronchi: Pharmacological evidence obtained with tissues from asthmatic and non-asthmatic subjects. *Pulm Pharm Ther, 6*, 87-96.

Bochenek, G., Nizankowska, E., Gielicz, A. et al. (2004). Plasma 9alpha,11beta-PGF2, a PGD2 metabolite, as a sensitive marker of mast cell activation by allergen in bronchial asthma. *Thorax, 59*, 459-464.

Boulet, L. P., Gauvreau, G., Boulay, M. E. et al. (2007). The allergen bronchoprovocation model: an important tool for the investigation of new asthma anti-inflammatory therapies. *Allergy, 62*, 1101-1110.

Bousquet, J., Jeffery, P. K., Busse, W. W. et al. (2000). Asthma.From Bronchoconstriction to Airways Inflammation and Remodeling. *Am J Respir Crit Care Med, 161*, 1720–1745

Bradding, P., Holgate, S. T. (1996). The mast cell as a source of cytokines in asthma. *Ann N Y Acad Sci, 796*, 272-281.

Brannan, J. D., Anderson, S. D., Gomes, K. et al. (2001). Fexofenadine decreases sensitivity to and montelukast improves recovery from inhaled mannitol. *Am J Respir Crit Care Med., 163*(6), 1420-5.

Brannan, J. D., Gulliksson, M., Anderson, S. D. et al. (2003). Evidence of mast cell activation and leukotriene release after mannitol inhalation. *Eur Respir J, 22*, 491-6.

Brightling, C. E., Bradding, P., Symon, F. A. et al. (2002). Mast-cell infiltration of airway smooth muscle in asthma. *N Engl J Med, 346*, 1699-1705.

Busse, W. W., Lemanske, R. F. (2001). Asthma. *N Engl J Med, 344*, 350–362

Calhoun, W. J., Reed, H. E., Moest, D. R. et al. (1992). Enhanced superoxide production by alveolar macrophages and air-space cells, airway inflammation, and alveolar macrophage density changes after segmental antigen bronchoprovocation in allergic subjects. *Am Rev Respir Dis, 145*, 317-325

Casini, A., Ceni, E., Salzano, R. et al. (1997). Neutrophil-derived superoxide anion induces lipid peroxidation and stimulates collagen synthesis in human hepatic stellate cells: role of nitric oxide. *Hepatology, 25*, 361-367.

Christie, P. E., Smith, C. M., Lee, T. H. (1991). The potent and selective sulfidopeptide leukotriene antagonist, SK&F 104353, inhibits aspirin-induced asthma. *Am Rev Respir Dis, 144*(4), 957-8.

Cirino, G., Flower, R. J., Browning, J. L. et al. (1987). Recombinant human lipocortin 1 inhibits thromboxane release from guinea-pig isolated perfused lung. *Nature, 328*, 270-272.

Cogé, F., Guénin, S. P., Rique, H. et al. (2001). Structure and expression of the human histamine H_4-receptor gene. *Biochem Biophys Res Commun, 284*, 301–309.

Comhair, S. A., Lewis, M. J., Bhathena, P .R. et al. (1999). Increased glutathione and glutathione peroxidase in lungs of individuals with chronic beryllium disease. *Am J Respir Crit Care Med, 159*, 1824-1829.

Cowden, J. M., Riley, J. P., Ma, J. Y. (2010). Histamine H_4 receptor antagonism diminishes existing airway inflammation and dysfunction via modulation of Th2 cytokines. *Respir Res, 11*, 86

D'Acquisto, F., Perretti, M., Flower, R. J. (2008). Annexin-A1: a pivotal regulator of the innate and adaptive immune systems. *Br J Pharmacol, 155*, 152-169.

Dahlén, S.-E., Hansson, G., Hedqvist, P. et al. (1983). Allergen challenge of lung tissue from asthmatics elicits bronchial contraction that correlates with the release of leukotrienesC_4, D_4 and E_4. *Proc Natl Acad Sci USA, 80*, 1712-1716.

Dahlén, B., Kumlin, M., Margolskee, D. J. et al. (1993). The leukotriene-antagonist MK-0679 blocks aspirin-induced airway obstruction induced by bronchial provocation with Lysine-aspirin in aspirin-sensitive asthmatics. *Eur Resp J, 6*, 1018-1026.

Dahlén, B., Roquet, A., Inman, M. D. et al. (2002). Influence of zafirlukast and loratadine on exercise-induced bronchoconstriction.*J. Allergy Clin Immunol, 109*, 789-793.

Damaj, B. B., Becerra, C. B., Esber, H. J. et al. (2007). Functional expression of H_4 histamine receptor in human natural killer cells, monocytes and dendritic cells. *J Immunol, 179*, 7907-7915.

Davis, B. E., Illamperuma, C., Gauvreau, G. M. et al. (2009). Single-dose desloratadine and montelukast and allergen-induced late airway responses. *Eur Respir J., 33*(6), 1302-8.

Drazen, J. M., Israel, E., O'Byrne, P. M. (1999). Treatment of asthma with drugs modifying the leukotriene pathway.*N Engl J Med, 340*(3):197-206. Review. Erratum in: *N Engl J Med 1999 Feb 25;340*(8):663. *N Engl J Med, 341*, 1632.

Dunford, P. J., O'Donnell, N., Riley, J. P. et al. (2006). The histamine H_4 receptor mediates allergic airway inflammation by regulating the activation of CD4+ T cells. *J Immunol, 176*, 7062-7070.

Evans, R. L., Nials, A. T., Knowles, R. G. et al. (2012). A comparison of antiasthma drugs between acute and chronic ovalbumin-challenged guinea-pig models of asthma. Pulm Pharmacol Ther, *25*, 453-464

Failli, P., Fazzini, A., Ruocco, C. et al. (2000). Lack of nitric oxide- and guanosine 3':5'-cyclic monophosphate-dependent regulation of alpha-thrombin-induced calcium transient in endothelial cells of spontaneously hypertensive rat hearts. *Br J Pharmacol, 130*, 1468-1476.

Galli, S. J. (1993). New concepts about the mast cell. *N Engl J Med, 328*:257-265.

Gantner, F., Sakai, K., Tusche, M. W. et al. (2002). Histamine H_4 and H_2 receptors control histamine-induced interleukin-16 release from human CD8(+) T cells. *J Pharmacol Exp Ther, 303*, 300-307

Gharaee-Kermani, M., Hu, B., Phan, S. H. et al. (2009). Recent advances in molecular targets and treatment of idiopathic pulmonary fibrosis: focus on TGF beta signaling and the myofibroblast. *Curr Med Chem, 16*, 1400-1417.

Godot, V., Arock, M., Garcia, G. et al. (2007). H_4 histamine receptor mediates optimal migration of mast cell precursors to CXCL12. *J Allergy Clin Immunol, 120*, 827-834.

Hardie, W. D., Glasser, S. W., Hagood, J. S. (2009). Emerging concepts in the pathogenesis of lung fibrosis. *Am J Pathol, 175*, 3-16.

Hay, D. W., Muccitelli, R. M., Tucker, S. S. et al. (1987). Pharmacologic profile of SK&F 104353: a novel, potent and selective peptidoleukotriene receptor antagonist in guinea pig and human airways. *J Pharmacol Exp Ther, 243*(2), 474-81

Hill, S. J., Ganellin, C. R., Timmerman, H. et al. (1997). International Union of Pharmacology. XIII. Classification of histamine receptors. *Pharmacol Rev, 49*, 253-78.

Holgate, S. T. (2012). Innate and adaptive immune response in asthma. *Nat Med, 18*, 673-683.

Jarjour, N. N., Calhoun, W. J. (1994). Enhanced production of oxygen radicals in asthma. *J Lab Clin Med, 123*, 131-136.

Johansson S. G. O., Buhl (2006). Anti-IgE Therapy for Asthma. In: *Pharmacotherapy of Asthma*, Ed.

Kaminski, N., Allard, J. D., Pittet, J. F. et al. (2000). Global analysis of gene expression in pulmonary fibrosis reveals distinct programs regulating lung inflammation and fibrosis. *Proc Natl Acad Sci USA, 97*, 1778-1783.

Kemp, J. P. (2003) Recent advances in the management of asthma using leukotriene modifiers. *Am J Respir Med, 2*, 139-156.

Kiss, R., Keserü. G. M. (2012). Histamine H_4 receptor ligands and their potential therapeutic applications: an update. *Expert Opin Ther Patents, 22*(3), 205-221

Kisseleva, T., Brenner, D. A. (2008). Fibrogenesis of parenchymal organs. *Proc Am Thorac Soc, 5*, 338-342.

Klinman, J. P. &Mu, D. (1994). Quinoenzymes in biology. *Annu Rev Biochem, 63*, 299-344.

Kohyama, T., Yamauchi, Y., Takizawa, H. et al. (2010). Histamine stimulates human lung fibroblast migration. *Mol Cell Biochem, 337*, 77-81.

Leff, J. A., Busse, W. W., Pearlman, D. et al. (1998). Montelukast, a leukotriene-receptor antagonist, for the treatment of mild asthma and exercise-induced bronchoconstriction. *N Engl J Med, 339*(3), 147-52.

Levi-Schaffer, F., Austen, K. F., Caulfield, J. P. et al. (1987). Co-culture of human lung-derived mast cells with mouse 3T3 fibroblasts: morphology and IgE-mediated release of histamine, prostaglandin D_2, and leukotrienes. *J Immunol, 139*(2), 494-500.

Ling, P., Ngo, K., Nguyen, S. et al. (2004). Histamine H_4 receptor mediates eosinophil chemotaxis with cell shape change and adhesion molecule upregulation. *Br J Pharmacol, 142*, 161-171.

Liu, C., Ma, X. J., Jiang, X. et al. (2001). Cloning and pharmacological characterization of a fourth histamine receptor (H_4) expressed in bone marrow. *Mol Pharmacol, 59*, 420–426.

Lu, S., Liu, N., Dass, S. B. et al. (2009). A randomized study comparing the effect of loratadine added to montelukast with montelukast, loratadine, and beclomethasone monotherapies in patients with chronic asthma. *J Asthma, 46*(5), 465-9.

Lucarini, L., Pini, A., Lanzi, C. et al. (2013) Effects of a selective inhibitor of poly(ADP-ribose)polymerase (PARP) on mast cell activation, bronchoconstriction and lung fibrosis. *42st Annual Meeting EHRS, 8-11 May 2013, Lodz, Poland.*

Magnussen, H., Boerger, S., Templin, K. et al. (1992). Effects of a thromboxane-receptor antagonist, BAY u 3405, on prostaglandin D_2 - and exercise-induced bronchoconstriction. *J Allergy Clin Immunol, 89*(6), 1119-26.

Manning, P. J., Watson, R. M., Margolskee, D. J. et al. (1990). Inhibition of exercise-induced bronchoconstriction by MK-571, a potent leukotriene D4-receptor antagonist. *N Engl J Med, 323*(25), 1736-9.

Masini, E., Bani, D., Bigazzi, M. et al. (1994). Effects of relaxin on mast cells.In vitro and in vivo studies in rats and guinea pigs. *J Clin Invest, 94*, 1974-1980.

Masini, E., Vannacci, A., Marzocca, C. et al. (2002). A plant histaminase modulates cardiac anaphylactic response in guinea pig. *Biochem Biophys Res Commun, 296*, 840-846.

Masini, E., Vannacci, A., Giannini, L. et al. (2004). Effect of a plant histaminase on asthmalike reaction induced by inhaled antigen in sensitized guinea pig. *Eur J Pharmacol, 502*, 253-264.

Masini, E., Bani, D., Vannacci, A. et al. (2005). Reduction of antigen-induced respiratory abnormalities and airway inflammation in sensitized guinea pigs by a superoxide dismutase mimetic. *Free RadicBiol Med, 39*, 520-531.

Masini, E., Giannini, L., Nistri, S. et al. (2008). Ceramide: a key signaling molecule in a Guinea pig model of allergic asthmatic response and airway inflammation. *J Pharmacol Exp Ther, 324*, 548-557.

Miyamoto, K., Iwase, M., Nyui, M. et al. (2006) Histamine type I receptor deficiency reduces air way inflammation in a murine asthma model. *Int Arch Allergy Immunol, 140*, 215-222.

Moeller, A., Ask, K., Warburton, D. et al. (2008). The bleomycin animal model: a useful tool to investigate treatment options for idiopathic pulmonary fibrosis? *Int J Biochem Cell Biol, 40*, 362-382.

Mondovì, B. (1985). *Structure and function of amine oxidases.* 1st Ed., CRCV Press, Boca Raton.

Mondovi, B., Fogel, W. A., Federico, R. et al. (2013). Effects of Amine Oxidases in Allergic and Histamine-Mediated Conditions.*Recent Pat Inflamm Allergy Drug Discov, 2012 Aug 29.*

Mondovi, B., Fogel, W. A., Federico, R. et al. (2013). Effects of Amine Oxidases in Allergic and Histamine-Mediated Conditions. *Recent Pat Inflamm Allergy Drug Discov, 7*, 20-34.

Morgan, R. K., McAllister, B., Cross, L. et al. (2007). Histamine 4 receptor activation induces recruitment of FoxP3+ T cells and inhibits allergic asthma in a murine model. *J Immunol, 178*, 8081-8089.

Morishita, Y., Hirayama, Y., Miyayasu, K. et al. (2005). FK506 aerosol locally inhibits antigen-induced airway inflammation in Guinea pigs. *Int Arch Allergy Immunol, 136*, 372-378.

Morse, K. L., Behan, J., Laz, T. M. et al. (2001). Cloning and characterization of a novel human histamine receptor. *J Pharmacol Exp Ther, 296*, 1058–1066.

Myers, T. R. (2000). Pediatric asthma epidemiology: incidence, morbidity, and mortality. *Respir Care Clin N Am, 6*, 1-14.

Nabe, T., Yamauchi, K., Shinjo, Y. et al. (2005). Delayed-type asthmatic response induced by repeated intratracheal exposure to toluene-2,4-diisocyanate in guinea pigs. *Int Arch Allergy Immunol, 137*, 115-124.

Nakamura, T., Itadani, H., Hidaka, Y. et al. (2000). Molecular cloning and characterization of a new human histamine receptor, HH_4R. *Biochem Biophys Res Commun, 279*, 615–620.

Ndengele, M. M., Muscoli, C., Wang, Z. Q. et al. (2005). Superoxide potentiates NF-kappaB activation and modulates endotoxin-induced cytokine production in alveolar macrophages. *Shock, 23*, 186-193.

Nguyen, T., Shapiro, D. A., George, S. R. et al. (2001). Discovery of a novel member of the histamine receptor family. *Mol Pharmacol, 59*, 427–433.

Oda, T., Morikawa, N., Saito, Y. et al. (2000). Molecular cloning and characterization of a novel type of histamine receptor preferentially expressed in leukocytes. *J Biol Chem, 275*, 36781–36786.

Ohtsu, H., Tanaka, S., Terui, T. et al. (2001). Mice lacking histidine decarboxylase exhibit abnormal mast cells. *FEBS Lett, 502*, 53-56.

O'Sullivan, S., Kumlin, M., Dahlén, B. et al. (1996). Increased urinary excretion of the prostaglandin D_2 metabolite 9 alfa,11 beta-prostaglandin F_2 following aspirin challenge supports mast cell activation in aspirin-induced airway obstruction. *J Allergy ClinImmunol, 98*, 421-432.

O'Sullivan, S., Roquet, A., Dahlén, B. et al. (1998a). Urinary excretion of inflammatory mediators during allergen-induced early and late phase asthmatic reactions. *Clin Exp Allergy, 28*, 1332-1339.

O'Sullivan, S., Roquet, A., Dahlén, B. et al. (1998b). Evidence for mast cell activation during exercise-induced bronchoconstriction. *Eur Respir J, 12*, 345-350.

O'Sullivan, S. (1999). On the role of PGD2 metabolites as markers of mast cell activation in asthma. *Acta Physiol Scand Suppl, 644*, 1-74.

Paz, Z., Shoenfeld, Y. (2010). Antifibrosis: to reverse the irreversible. *Clin Rev Allergy Immunol, 38*, 276-286.

Peebles, R. S., Hartert, T. V. (2002). Highlights from the annual scientific assembly: patient-centered approaches to asthma management: strategies for treatment and management of asthma. *South Med J, 95*, 775-779.

Peroni, D. G., Piacentini, G. L., Pietrobelli, A. et al. (2002). The combination of single-dose montelukast and loratadine on exercise-induced bronchospasm in children. *EurRespir J, 20*(1), 104-7.

Pesci, A., Foresi, A., Bertorelli, G. et al. (1993). Histochemical characteristics and degranulation of mast cells in epithelium and lamina propria of bronchial biopsies from asthmatic and normal subjects. *Am Rev Respir Dis, 147*, 684-689.

Pini, A., Shemesh, R., Samuel, C. S. et al. (2010). Prevention of bleomycin-induced pulmonary fibrosis by a novel antifibrotic peptide with relaxin-like activity. *J Pharmacol Exp Ther, 335*, 589-599.

Pini, A., Viappiani, S., Bolla, M. et al. (2012). Prevention of bleomycin-induced lung fibrosis in mice by a novel approach of parallel inhibition of cyclooxygenase and nitric-oxide donation using NCX 466, a prototype cyclooxygenase inhibitor and nitric-oxide donor. *J Pharmacol Exp Ther, 341*, 493-499.

Reicin, A., White, R., Weinstein, S. F. et al. (2000). Montelukast, a leukotriene receptor antagonist, in combination with loratadine, a histamine receptor antagonist, in the treatment of chronic asthma. *Arch Intern Med, 160*(16), 2481-8.

Ressmeyer, A-R., Larsson, A-K., Vollmer, E. et al. (2006). Antigen and mediator responses in guinea pig precision-cut lung slices: Comparison with human tissues. *Eur Respir J*, 28, 603–611.

Richter, K., Grönke, L., Janicki, S. et al. (2008). Effect of azelastine, montelukast, and their combination on allergen-induced bronchoconstriction in asthma. *Pulm Pharmacol Ther, 21*(1), 61-6.

Riley, J. P., Fuchs, B., Sjöberg, L. et al. (2013) Mast cell mediators cause an early allergic bronchoconstriction in guinea-pigs in vivo: A model of relevance to asthma. *Clin Sci (London), 125*, 533-542.

Roquet, A., Dahlén, B., Kumlin, M. et al. (1997). Combined antagonism of leukotrienes and histamine produces predominant inhibition of allergen-induced early and late phase airway obstruction in asthmatics. *Amer J Resp Crit Care Med, 155*, 1856-1863.

Rossi, G.L., Olivieri, D. (1997). Does the mast cell still have a key role in asthma? *Chest, 112*, 523-529.

Salcedo, C., Pontes, C., Merlos, M. (2013). Is the H_4 receptor a new drug target for allergies and asthma? *Front Biosci (Elite Ed), 5* (Jan 1), 178-87.

Savall, B. M. (2013) Discovery of JNJ 39758979 and SAR of Related 2-Aminopyrimidine Histamine H_4 Antagonists. *Abstract for EHRS meeting in Lodz 8-11 May 2013.*

Schuiling, M., Zuidhof, A. B., Meurs, H. et al. (1999). Role of tachykinin NK2-receptor activation in the allergen-induced late asthmatic reaction, airway hyperreactivity and airway inflammatory cell influx in conscious, unrestrained guinea-pigs. *Br J Pharmacol, 127*, 1030-1038.

Sladek, K., Szczeklik, A. (1993). Cysteinyl leukotrienes overproduction and mast cell activation in aspirin-provoked bronchospasm in asthma. *Eur Respir J, 6*(3), 391-9.

Smith, N., Broadley, K. J. (2008) Adenosine receptor subtypes in the airways responses to 5'-adenosine monophosphate inhalation of sensitized guinea-pigs. *Clin Exp Allergy, 38*, 1536-47.

Somma, T., Cinci, L., Formicola, G. et al. (2013) A selective H_4 receptor antagonist prevents antigen-induced airway inflammation and bronchoconstriction in guinea pig: involvement of lipocortin-1. *Br. J. Pharmacol, 170*, 200-213.

Sundström, E., Långbom, L., Ryrfeldt, A. et al. (2003). Interactions among three classes of mediators explain antigen-induced bronchoconstriction in the isolated perfused and ventilated guinea pig lung. *J Pharmacol Exp Ther, 307*, 408-418.

Svegliati-Baroni, G., Saccomanno, S., van Goor, H. et al. (2001). Involvement of reactive oxygen species and nitric oxide radicals in activation and proliferation of rat hepatic stellate cells. *Liver, 21*, 1-12.

Thurmond, R. L., Desai, P. J., Dunford, P. J. et al. (2004). A potent and selective histamine H_4 receptor antagonist with anti-inflammatory properties. *J Pharmacol Exp Ther, 309*, 404-413.

Thurmond, R. L., Gelfand, E. W., Dunford, P. J. (2008). The role of histamine H_1 and H_4 receptors in allergic inflammation: the search for new antihistamines. *Nat Rev Drug Discov, 7*, 41-53.

Wani, J., Carl, M., Henger, A. et al. (2007). Nitric oxide modulates expression of extracellular matrix genes linked to fibrosis in kidney mesangial cells. *Biol Chem, 388*, 497-506.

Wardlaw, A. J., Dunnette, S., Gleich, G. J. et al. (1988). Eosinophils and mast cells in bronchoalveolar lavage in subjects with mild asthma.Relationship to bronchial hyperreactivity. *Am Rev Respir Dis, 137*, 62-69.

Wikström-Jonsson, E., Dahlén, S.-E. (1994). Interactions between leukotrienes and histamine in the anaphylactic contraction of guinea pig lung parenchyma. *J Pharmacol Exp Ther, 271*, 615-623.

Williams, C. M., Galli, S. J. (2000). Mast cells can amplify airway reactivity and features of chronic inflammation in an asthma model in mice. *J Exp Med, 192*, 455-462.

Wynn, T. A. (2007). Common and unique mechanisms regulate fibrosis in various fibroproliferative diseases. *J Clin Invest, 117*, 524-529.

Wynn, T. A. (2008). Cellular and molecular mechanisms of fibrosis. *J Pathol, 214*, 199-210.

Xie, Q. M., Chen, J. Q., Shen, W .H. et al. (2002). Effects of cyclosporin A by aerosol on airway hyperresponsiveness and inflammation in guinea pigs. *Acta Pharmacol Sin, 23*, 243-247.

Zampeli, E. &Tiligada, E. (2009). The role of histamine H_4 receptor in immune and inflammatory disorders. *Br J Pharmacol, 157*, 24-33.

Zhang, M., Thurmond, R. L., Dunford, P. J. (2007). The histamine H_4 receptor: a novel modulator of inflammatory and immune disorders. *Pharmacol Ther, 113*, 594-606.

Zhu, Y., Michalovich, D., Wu, H. et al. (2001). Cloning, expression, and pharmacological characterization of a novel human histamine receptor. *Mol Pharmacol, 59*, 434–441.

Chapter 6

Histamine in Atopic Disorders: Atopic Dermatitis and Pruritus

W. Bäumer[1,2], F. Glatzer[3], K. Roßbach[1], H. Ohtsu[4], M. Seike[5], S. Mommert[3], T. Werfel[3], R. Gutzmer[3]

[1]Department of Pharmacology, Toxicology and Pharmacy, University of Veterinary Medicine Hannover, 30559 Hannover, Germany

[2]Department of Molecular Biomedical Sciences, College of Veterinary Medicine, North Carolina State University, Raleigh, USA, e mail: wbaeumer@ncsu.edu

[3]Division of Immunodermatology and Allergy Research, Hannover Medical School, Germany

[4]Applied Quantum Medical Engineering, Department of Engineering, Tohoku University, Japan

[5]Department of Food and Nutrition Science, Sagami Women's Junior College, Japan

6.1. Role of Histamine and Histamine H_4 Receptor in Inflammatory Skin Diseases in Man

6.1.1. Introduction

The skin is the primary interface between the environment and the inside of an organism and protects the body against harmful agents, pathogens, allergens, physical trauma, UV radiation and excessive electrolyte and water loss. Different diseases interfere with this function, including the extremely common chronic inflammatory skin diseases. Chronic inflammatory skin diseases are characterized by chronic inflammation of the skin, often with severe pruritus. The most diseases include psoriasis and eczema, including atopic dermatitis (AD) and allergic contact dermatitis (ACD). Numerous different inflammatory cell populations are involved in the pathogenesis of inflammatory skin diseases, including T cells, antigen presenting cells (APC), granulocytes and keratinocytes. The inflamed skin in these diseases becomes dry, red, itchy and scaly, resulting in decreased quality of life. At present, the responsible mechanisms that are involved in inflammation and pruritus in these disorders and how they lead to pathogenesis are not completely understood.

A major mediator of inflammation and allergic reactions is histamine, which plays an important role in acute and chronic inflammation as well as hypersensitivity reactions. High amounts of histamine are released during allergic and inflammatory disorders, where increased levels of histamine have been reported in lesions of psoriasis and eczematous skin diseases (Stander & Steinhoff, 2002) with possible peak values of 10^{-5} to 10^{-3} molar during immediate hypersensitivity reactions following mast cell degranulation (Simons & Simons, 1994). Histamine may hold many therapeutic capabilities, but therapeutic blockade of the H_1 and H_2 histamine receptors are not efficient to relieve pruritus in AD or psoriasis (Akdis et al., 2006). A possible explanation for this observation is that histamine might mediate effects via other receptors, such as the H_4 receptor (Thurmond et al., 2008). The H_4 receptor is the most recently discovered histamine receptor and there is evidence that it is preferred histamine target in the immune system (Zampeli & Tiligada, 2009). This makes it an attractive target for developing novel therapeutics of inflammatory skin diseases (Tiligada et al., 2009).

This article summarizes the findings regarding the differential expression and function of H_4 receptor on cell types relevant for human chronic inflammatory skin diseases and pruritus. The role of histamine in animal models of atopic dermatitis is also discussed.

6.1.2. Histamine H_4 Receptor Expression and Function on Cell Types Involved in Inflammatory Skin Diseases

The expression of H_4 receptor is described for various cells types relevant for inflammatory skin diseases and for some cell populations, functional studies of the H_4 receptor also exist.

6.1.2.1. T Cells

T cells play a key role as immune effector cells in AD (Novak et al., 2003), and various studies have been conducted to characterize their role in this disease. Gene expression studies have shown the expression of the H_4 receptor on mRNA and protein level in T cells, where H_4 receptor expression on CD4+ T cells can be upregulated by an IL-4-dominated micromilieu. Furthermore, it was demonstrated that H_4 receptor expression on Th2 cells is higher compared with Th1 cells and stimulation of H_4 receptor with specific agonists results in an induction of transcription factor AP-1 in Th2 but not in Th1 cells (Gutzmer et al, 2009). Stimulation of the H_4 receptor results in upreglation of IL-31 on the mRNA level in Th2 cells (Gutzmer et al., 2009). Moreover, H_4 receptor might be involved in the induction of proinflammatory cytokines in activated monocytes and macrophages by enhanced IL-31 release (Kasraie et al., 2010). Stimulation of the H_4 receptor also inhibits the antigen-specific human T cell responses by decreased IFN-γ and IL-5 expression (Sugata et al., 2007). Th17 cells, another important subtype of T cells, also express the H_4 receptor. Th17 cells are increased in peripheral blood of AD patients and play a potential role in AD, and the number of this cell type is associated with the severity of AD (Koga et al., 2008). The H_4 receptor was detected on human memory T cells, polarized into Th17 cells by IL-1β and IL-23, and on IL-17-positive cells in lesions of inflamed skin. Furthermore, it was demonstrated that a stimulation of H_4 receptor results in an upregulation of IL-17 mRNA and secreted protein and an induction of the transcription factor AP-1 (Mommert et al., 2012).

Ou et al. (2004) observed that the blood of patients with AD showed high numbers of CD4+CD25+ regulatory T cells (T_{regs}) in combination with increased induction of the transcription factor forkhead box P3 (FOX P3). Another study by Schnopp et al. (2007) demonstrated the presence of T_{regs} in the skin of AD patients (Ou et al., 2004, Schnopp et al., 2007).

CD8+ T cells also express H_4 receptor at the mRNA and protein levels and stimulation of the H_4 receptor in CD8+ T cells results in a release of the T cell chemoattractant IL-16 (Gantner et al., 2002). Invariant natural killer T cells (iNKT cells) have been found in lesional skin of patients suffering ACD, AD and other types of inflammation (Simon et al., 2009). Activation of these cells causes the expression of IL-4 and IFN-γ, but it is unclear under which conditions they trigger a Th1 or Th2 response.

6.1.2.2. Antigen Presenting Cells

Antigen presenting cells (APCs), which include monocytes and dendritic cells (DCs), are one of the major components in the initiation of allergic inflammation. A major role of these cells in allergic inflammation is antigen uptake and presentation followed by the secretion of cytokines and chemokines to generate a cytokine network. In early investigations, it was demonstrated that human monocytes and monocyte-derived dendritic cells (MoDCs) express the H_4 receptor on the mRNA and protein level and that the receptor is involved in control of cytokine and chemokine production. H_4 receptor stimulation in monocytes and MoDCs result in a suppression of IL-12, IL-27 and CCL2, leading to calcium influx and chemotaxis (Gutzmer et al., 2005, Damaj et al., 2007, Dijkstra et al., 2007, Gschwandtner et al., 2012).

Further studies focused on several subtypes of DC which play an important role in inflammatory skin disorders, such as inflammatory dendritic epidermal cells (IDECs), plasmacytoid DCs (pDCs), myeloid DCs (mDCs) and Langerhans cells (LCs). The H_4 receptor on murine and human LCs is expressed on the mRNA and protein levels and stimulation of the H_4 receptor results in decreased expression of CCL2 in human LCs as well as enhanced the migration of LCs from the epidermis in *ex vivo* migration assays using human epidermis (Gschwandtner et al., 2010). Recent findings showed the expression of H_4 receptor on different subtypes of DCs demonstrated different functional activities. IDEC represents a DC subtype, which was found in lesions of AD. Stimulation of the H_4 receptor expressed on IDECs result in a decreased expression of IL-12 and CCL2 (Dijkstra et al., 2008). Inflammatory DCs characterized by 6-Sulfo LacNAc groups on the cell surface (SLAN-DCs), a major population of DCs in human blood, expresses the H_4 receptor on mRNA and protein level and shows high proinflammatory capacity through T cell response induction and the release of proinflammatory cytokines. SLAN-DCs are recruited in inflamed skin, where subsequent stimulation of the H_4 receptor results in a strong decrease of lipopolysaccharide-induced TNF-α and IL-12 production. Through this mechanism, histamine reduces the proinflammatory capacity of these cells (Gschwandtner et al., 2011b). Expression of the H_4 receptor has also been detected on pDCs and mDCs. Interestingly, pDCs from patients with psoriasis express more H_4 receptor compared with healthy or AD donors. Preliminary functional studies on the role of H_4 receptor in pDC revealed decreased expression of the pro-inflammatory cytokines TNF-α, Interferon-α and the chemokine CXCL8. Moreover, an active migration of pDCs in response to stimulation with a H_4 receptor agonist was observed (Gschwandtner et al., 2011a). For mDCs, the stimulation of H_2 and H_4 receptors lead to reduced expression of the Th1-associated CXCL10. This suggests that the H_4 receptor has major anti-inflammatory effects in APCs by downregulating proinflammatory cytokines such as IL-12, TNF-α and IL-27 and the chemokines CCL2, CXCL8 and CXCL-10.

6.1.2.3. Granulocytes

An increased number of circulating eosinophils in AD patients and the detection of activated eosinophils in lesional skin possibly provide evidence for a role of granulocytes in the pathology of inflammatory skin diseases (Simon et al., 2004, Werfel, 2009). Human eosinophils and neutrophils express the H_4 receptor and histamine acts as chemoattractant for eosinophils. (Clark et al., 1975, Clark et al., 1977, Oda et al., 2000, Morse et al., 2001). Subsequent studies identified the H_4 receptor as the inducer of chemotaxis and showed that histamine stimulation resulted in calcium mobilization, the induction of actin polymerization, the increase in adhesion molecules CD11b/CD18 and CD54 and the alteration in cell shape (O'Reilly et al., 2002, Buckland et al., 2003, Ling et al., 2004). H_4 receptor activation also leads to a migration of eosinophils from blood into inflamed tissues (Barnard et al., 2008). In contrast, for human basophils and neutrophils no chemoattractant effect via H_4 receptor was observed.

6.1.2.4. Natural Killer Cells

Natural killer cells (NK cells) are a subpopulation of lymphocytes in peripheral blood (15 %) that act as early effector cells in the innate immune system and are characterized by CD3- and CD56+ expression. In atopic eczema, NK cells are located near DCs in lesional skin and it is suspected that NK cells modulate the function of DCs in AD (Buentke et al., 2002). Expression studies revealed H_1 and H_4 receptor expression on the protein level for human NK cells with induction of chemotaxis by histamine. This effect is inhibited by the H_3/H_4 receptor antagonist thioperamide. NK cells are lacking the H_3 receptor, so this effect has been attributed to thioperamide binding to the H_4 receptor (Damaj et al., 2007).

6.1.2.5. Skin cells (Keratinocytes, Fibroblasts, Mast Cells)

Keratinocytes form the outermost surface of the body and are directly in contact with the environment. These cells are responsible for the recruitment of other cell types to the site of inflammation and regulate the first step in the production of antimicrobial peptides as well as proinflammatory cytokines and chemokines. They play an important role in inflammatory skin diseases, such as AD (Wittmann and Werfel, 2006). Early studies showed keratinocytes express the H_1 receptor and described the various immunomodulatory effects of histamine on these cells (Kohda et al., 2002, Kanda & Watanabe, 2003, Giustizieri et al., 2004, Kanda & Watanabe, 2004). Histamine modulates the differentiation from epidermal keratinocytes and impairs the skin barrier function via H_1 receptor (Gschwandtner et al., 2013).

Yamaura et al. (2009) initially reported the expression of H_4 receptor in keratinocytes through immunohistochemical studies on human epidermal tissue. The H_4 receptor appeared to be upregulated during differentiation of keratinocytes in the upper layer of epidermis versus keratinocytes in basal layer (Yamaura et al., 2009a). The expression of H_4 receptor on human primary keratinocytes has also been demonstrated. Furthermore, outer root sheath keratinocytes from patients with AD show significantly increased expression of the H_4 receptor compared with keratinocytes of healthy donors. The stimulation of the H_4 receptor results in an enhanced proliferation of foreskin and outer root sheath keratinocytes in different *in vitro* cell proliferation assays and scratch assays. This effect was blocked by pre-incubation with the H_4 receptor specific antagonist JNJ7777120 (Glatzer et al., 2013).

Dermal fibroblasts also express the H_4 receptor on mRNA and protein levels. The expression of this receptor can be upregulated by stimulation of fibroblasts with lipopolysaccharide, indomethacin or dexamethasone (Ikawa et al., 2008). The involvement of fibroblasts in skin diseases like AD has not been well understood until recently. Leung et al. (1982) showed morphology changes and cell mediated cytotoxicity against skin fibroblasts in AD skin (Leung et al., 1982). Another study showed a possible gene expression profile associated with AD in fibroblasts based on proteomic analysis. These findings suggest a possible role for fibroblasts in the pathogenesis of AD (Park et al., 2006).

Human mast cells express the H_4 receptor on the mRNA and protein levels and stimulation of this receptor in mast cells results in enhanced CXCL12-mediated recruitment of precursor mast cells into the dermis (Hofstra et al., 2003, Lippert et al., 2004, Godot et al., 2007). A recent study demonstrated enhanced production of IL-6 via ERK and phosphoinositide 3-kinase (PI3K) activation following stimulation of the H_4 receptor in murine mast cells (Desai & Thurmond, 2011). Based on these investigations, it can be hypothesized that the H_4 receptor on mast cells plays an important role for the accumulation of these cells in allergic tissue. Otherwise, H_4 receptor stimulation has no relevant influence on the mast cell degranulation and subsequent mediator release from these cells (Hofstra et al., 2003).

6.1.3. Molecular Mechanisms in Regulation of H_4 Receptor Expression

Recent studies have reported that alterations in the H_4 receptor gene are associated with numerous diseases, including chronic inflammatory skin disorders. The H_4 receptor is significantly upregulated in the course of different skin diseases (Gschwandtner et al., 2011a, Gschwandtner et al., 2011b), but the possible genetic or epigenetic reasons behind these observations are unknown. Yu et al. (2010) detected three different single nucleotide polymorphisms (SNPs) within the H_4 receptor gene significantly associated with AD, indicating that the H_4 receptor plays a role in the development of this disease. (Yu et al., 2010b). The same group demonstrated a correlation between the presence of increased copies (Copy number variation) of the H_4 receptor gene and the risk of developing systemic lupus erythematosus (Yu et al., 2010a). An association between genetic variations of H_4 receptor and infectious asthma has also been described (Simon et al., 2012).

6.1.4. Conclusions

Together, all of the above studies provide evidence for a pathogenetic and immunomodulatory role of the H_4 receptor in chronic inflammatory skin diseases and pruritus. The H_4 receptor modulates the function of relevant

Table 6.1

Expression and function of the human H4 receptor in relevant cell types for inflammatory skin diseases (AP-1 activating protein 1, DC dendritic cell, IDEC inflammatory dendritic epidermal cell, IFN-γ Interferon gamma, IL Interleukin, SLAN-DC 6-Sulfo LacNac-expressing dendritic cell, Th T helper cell, TNF-α tumor necrosis factor alpha, Treg regulatory T cell).

Cell subpopulation	**Expression H4 receptor**	**Function H4 receptor**	**Reference**
Antigen presenting cells			
Monocytes	mRNA and protein level	↑ Calciuminflux ↓ CCL2 ↓ IL-12 ↓ IL-27, ↓ IL-23 ↓ IP10	(Damaj et al., 2007), (Dijkstra et al., 2007), (Gschwandtner et al., 2012a) unpublished data
Monocyte-derived DC	mRNA and Protein level	↑ Chemotaxis ↓ IL-12 Induction of AP-1	(Gutzmer et al., 2005)
IDEC	protein level upregulation by IFN-γ	↓ CCL2 ↓ IL-12	(Dijkstra et al., 2008)
SLAN-DC	mRNA and protein level	↓ IL-12 ↓ TNF-α	(Gschwandtner et al., 2011b)
Plasmacytoid DC	mRNA and protein level	↑ Chemotaxis ↓ TNF-α ↓ IFN-γ ↓ CXCL8	(Gschwandtner et al., 2011a)
Myeloid DC	mRNA level	↓ IP10	unpublished data
Langerhans cells	protein level	↑ Chemotaxis ↓ CCL2	(Gschwandtner et al., 2010)
T cells			
CD4+ Th 1/2	mRNA and protein level	↑ IL-31 ↓ IFN-γ ↓ IL-5 Induction of AP-1	(Gutzmer et al., 2009b), (Sugata et al., 2007)
CD4+ Th17	mRNA and protein level	↑ IL-17 Induction of AP-1	(Mommert et al., 2012)
CD4+ Treg	Not described	4-methylhistamine recruited T_{reg}	(Morgan et al., 2007)
CD8+	mRNA level	↑ IL-16	(Gantner et al., 2002)
Invariant natural killer T cells	mRNA and protein level	Not described	Unpublished data
Other			
Natural killer cells	mRNA and protein level	↑ Chemotaxis	(Damaj et al., 2007)
Keratinocytes	mRNA and protein level upregulation in differentiated cells and in atopic dermatitis	↑ Proliferation	(Glatzer et al., 2013)
Fibroblasts	mRNA and protein level upregulation by LPS and indomethacin on mRNA upregulation by dexamethasone on protein	Not described	(Ikawa et al., 2008)
Mast cells	mRNA and protein level	↑ Recruitment of CXCL12 expressing precursors, ↑ chemotaxis Calcium mobilization	(Thurmond et al., 2008), (Hofstra et al., 2003), (Lippert et al., 2004), (Godot et al., 2007)

cell populations by influencing chemotaxis and cytokine production, among other mechanisms. By influencing chemotaxis and cytokine production, the H_4 receptor may influence Th cell polarization linking innate and adaptive immune pathways. This makes the H_4 receptor a desirable therapeutic target in chronic inflammatory skin diseases. Table 1 summarizes the different cell populations expressing the H_4 receptor and the possible described functions of this receptor on these cells.

6.2. Role of Histamine in Animal Models of Atopic Dermatitis

6.2.1. Introduction

Histamine is a ubiquitous chemical messenger that displays numerous functions mediated through four known pharmacologically distinct receptors. The H_1 receptor is expressed in peripheral nerves, keratinocytes and endothelial cells (Baumer and Rossbach, 2010). The characteristic features of H_1 receptor activation in the skin are exemplified by itching transactions (Hagermark et al., 1979) and increased vascular permeability (Sercombe et al., 1986). As many of these functions contribute to allergic responses, H_1 receptor antagonists have been successfully used as drugs for treating certain forms of allergies (Simons & Simons, 2011). The H_2 receptor is expressed in keratinocytes, melanocytes, macrophages and lymphocytes (Akdis & Simons, 2006), but its exact function in the skin remains unclear. The H_3 receptor is predominantly expressed in the central nervous system (Sander et al., 2008, Baumer & Rossbach, 2010) but the exact physiological roles it plays in the skin remain to be explored. The H_3 receptor is also expressed in mast cells and sympathetic and parasympathetic nerves and may regulate histamine, serotonin, acetylcholine and other neurotransmitters (Sander et al., 2008). Recently, a fourth histamine receptor was identified as H_4 (Nakamura et al., 2000), where H_4 receptor expression has been observed in eosinophils, T cells, dendritic cells, mast cells and primary sensory neurons (de Esch et al., 2005). The H_4 receptor mediates chemotactic activity of histamine for mast cells and eosinophils (Zampeli & Tiligada, 2009).

Assessing the effect of histamine in the development of eczematous lesions *in vivo* has been difficult, as most observations involve the use of histamine receptor antagonists. When using such agents, problems such as adverse effects, short half-life and selectivity are unavoidable. However, histamine-deficient HDC(-/-) mice, produced by disrupting the histidine decarboxylase (HDC) gene, were studied to circumvent these problems (Ohtsu et al., 2001).

6.2.2. Histamine Facilitates Development of Eczematous Lesions in Contact Dermatitis: Role of Histamine H_1 Receptor

The role of histamine in scratching behaviour (see chapter 6.3) and the extent of chronic eczema was investigated by using HDC(-/-) mice of the 129/Sv inbred strain (Seike et al., 2005a). Chronic contact dermatitis was induced with daily application of diphenylcyclopropenone (DCP) on a hind paw of HDC(+/+) and HDC(-/-) mice for 2 months. Histological examination of the skin sample revealed that the mice displayed inflammatory cell infiltration, hyperplastic epidermis and newly spreading neuronal processes, but the magnitude of these changes were more significant in HDC(+/+) mice.

Further studies using the C57B6 strain were conducted to support the previous evidence, to observe the effects in our new experimental strain and to provide further information about the role of each histamine receptor (Seike et al., 2005b). Development of eczematous lesions in contact dermatitis was reduced in HDC(-/-) mice versus HDC(+/+) mice, as observed in the previous experiment with the 129/Sv strain. The H_1 agonist histamine trifluoromethyl toluidide (HTMT) promoted development of eczematous lesion in HDC(-/-) mice and correspondingly, an H_1 receptor antagonist (loratadine) reduced development of eczematous lesions in HDC(+/+) mice. However, an H_2 agonist (dimaprit) and H_2 antagonist (cimetidine) were ineffective in HDC(-/-) and HDC(+/+) mice, respectively. These results suggest that histamine facilitates the development of eczematous lesions in a murine model of contact dermatitis mainly via H_1 receptors, which was confirmed with another H_1 receptor antagonist, olopatadine hydrochloride. Mice administered with olopatadine develop mild erythema with slight hair loss, whereas mice without olopatadine develop scaly erythema with marked hair loss and erosions. Skin repeatedly challenged with hapten displays marked epidermal hyperplasia and infiltration of inflammatory cells, including mast cells in the dermis, and olopatadine suppresses these histological changes (Hamada et al., 2006).

6.2.3. Histamine H_4 Receptor: a New Player in the Game

Soon after discovery of the fourth histamine receptor, it became apparent that this receptor might mediate central immune and inflammation-mediated processes like immune cell migration and modulation of cytokine secretion (Zampeli & Tiligada, 2009a, Gutzmer et al., 2011). These actions might also have an impact on the development of allergic skin diseases (Gutzmer et al., 2011).

Effects of H_4 receptor antagonists have been tested in several murine models of allergic dermatitis, in the skin of laboratory beagles and in a canine model of atopic dermatitis. In some studies acute models of allergic contact dermatitis were used and in one additional study acute lesions were compared to chronic lesions induced by repetitive topical administration of the relevant hapten.

The selective H_4 receptor antagonist JNJ7777120 did not reduce the ear swelling induced by the haptens dinitrochlorobenzene (DNCB) and toluenediisocyanate (TDI), which differ in their induction of a Th1- and Th2-dominated inflammatory response (Baumer et al., 2004, Rossbach et al., 2009b). Seike et al. used trinitrochlorobenzene as hapten to elicit ACD in mice, a hapten similar to dinitrochlorobenzene. Even suprapharmacological doses of the H_4 receptor antagonist JNJ7777120 did not reduce the acute allergic response. By chronification of the lesions, however, a reduction of inflammatory response accompanied by a diminished mast cell and eosinophilic infiltration was observed in this study published 2010 (Seike et al., 2010).

More recently, reduction of the inflammatory response induced by topical administration of the hapten fluorescein isothiocyanate (FITC) was also observed in the acute phase of allergic dermatitis by treatment with JNJ7777120 (Cowden et al., 2010). One characteristic of the FITC model is a distinct eosinophilia, which is less pronounced in other models of hapten-induced contact dermatitis (such as by DNCB or TDI). In this respect, it is interesting that the dual H_3/H_4 receptor antagonist thioperamide was effective in reducing inflammation in a modified model of picryl chloride-induced allergic dermatitis, in which blood eosinophilia was induced by cyclophosphamide. A combination of thioperamide with the H_1 receptor antagonist pyrilamine even enhanced the anti-inflammatory effect (Hirasawa et al., 2009).

Taken together, these results indicate that H_4 receptor blockade reduces inflammation in chronic ACD and acute ACD with predominant Th2 milieu, but there appears to be only a minor role for H_4 receptor antagonism in lesions of acute ACD with predominant Th1 milieu.

A very recent study revealed that the H_1 receptor antagonist olopatadine and H_4 receptor antagonist JNJ7777120 improved scratching behavior and skin inflammation in a model of chronic allergic dermatitis established in a NC/Nga mouse model (Ohsawa & Hirasawa, 2012), where mice have a genetic defect in barrier function that provides a model of AD closer to the human disease than the aforementioned contact dermatitis models (Jin et al., 2009). However, atopic-like lesions were triggered by frequent topical administration of the hapten picryl chloride. Thus, this model might be viewed as an intermediate between an allergic contact dermatitis model and an atopic dermatitis model.

Interestingly, in a recent study, mice were sensitized epicutaneously against ovalbumine (OVA), which resembles another widely accepted atopy mouse model (Jin et al., 2009). Wildtype (BALB/c) and H_4 receptor knockout mice were epicutaneously sensitized with OVA according to a standard protocol with slight modifications (Spergel et al., 1998). The clinical development of dermatitis was markedly attenuated in H_4 receptor knockout mice and was accompanied by a significantly diminished influx of inflammatory cells and reduced epidermal hyperproliferation, as revealed by histological examination. Interestingly, the serum of knockout mice showed a significantly reduced amount of OVA-IgE. In spleen and skin draining lymph nodes, a significantly decreased number of CD4+ T cells and F4/80+ macrophages were observed. The INF-γ production of *ex vivo*-stimulated splenocytes was also significantly reduced in knockout mice (Rossbach et al., 2012).

In less severe skin lesions, the reduced cell influx into the skin as well as the reduced number of splenocytes and lymph node cells clearly indicate an anti-inflammatory role of the H_4 receptor in this chronic model of atopic dermatitis.

6.2.4. Histamine in Dogs

Dogs suffering from atopic dermatitis share several similarities (albeit with some differences) compared to human AD. This has caused debate as to whether atopic dogs may be a suitable model for human AD (Marsella & Girolomoni, 2009, Marsella et al., 2011). A role of histamine has been disputed for decades and enhanced concentration of histamine has been reported in lesional skin of dogs (Marsella & Olivry, 2001). There also seems to be higher "reagibility" of mast cells in atopic dogs (Marsella & Olivry, 2001). However, H_1 receptor as well as H_2 receptor antagonists show only moderate effects on lesions and pruritus in atopic dogs (DeBoer & Griffin, 2001, Olivry et al., 2010). Nevertheless, the role of histamine was re-evaluated using the first findings, where the H_4 receptor might have immunomodulatory functions. In 2008, Jiang et al reported the first pharmacological characterization of the canine H_4 receptor (Jiang et al., 2008) and robust expression of the H_4 receptor was demonstrated on the mRNA level in canine skin (Eisenschenk et al., 2011). Our group explored the role of H_4 receptor in canine skin by intradermal injection of histamine and putative H_4 receptor agonists to laboratory beagles with the intention to study a classical weal and flare reaction and pruritus. Histamine as well as the H_4 receptor agonists clobenpropit and VUF4830 induced a weal and flare reaction after intradermal injection. Unexpectedly, no pruritus was induced by either substance (Rossbach et al., 2009a). However, when the beagles were pre-treated topically with the H_4 receptor antagonist JNJ7777120 before histamine application, the weal and flare reaction was reduced by approximately 30%, further indicating a role of the H_4 receptor in this reaction (Rossbach et al., 2009a).

The prevention of skin lesions by H_1 and H_4 receptor antagonists was also tested in an alternative canine model of atopic dermatitis (Pucheu-Haston et al., 2008). Six atopic Maltese-beagle crossbred dogs experimentally sensitized to *Dermatophagoides farinae* (Df) were enrolled into blinded placebo and active controlled experiments. H_4 receptor antagonists (JNJ7777120 or JNJ28307474) were applied topically before allergen challenge and JNJ28307474 was also given orally after allergen challenge. A triamcinolone acetonide solution applied topically was used as a positive control, and skin lesions that developed after the application of the Df allergen were graded at the site of allergen application. Twenty four hours after the challenge, placebo treated animals and animals treated with topical and oral JNJ28307474 or topical JNJ7777120 showed a comparable lesion score, whereas the triamcinolone solution prevented all dogs from having any lesions (Baumer et al., 2011). The systemic administration of cetirizine and hydroxyzine was also tested in the same model. A prophylactic administration of hydroxyzine and cetirizine in doses, which significantly reduce the histamine or anti-canine, IgE-induced weal and flare reaction, also did not reduce the median inflammatory score of the placebo treatment (Baumer et al., 2011). These data provide evidence that the preventive administration of H_1 receptor as well as H_4 receptor antagonists has no impact on the development of acute inflammatory skin lesions induced by the topical administration of a relevant allergen. It is thus intended to study effects of H_4 receptor antagonists in chronic skin lesions.

6.3. Role of Histamine Receptors in Pruritus Transmission

6.3.1. Introduction

Pruritus has been defined as an unpleasant sensation that triggers a desire to scratch (Ikoma et al., 2011). A wide range of substances have been implicated for the induction of itch (Steinhoff et al., 2006), but histamine remains as the best-known endogenous agent and serves as a classical inducer of itch under experimental settings (Magerl et al., 1990, Schmelz et al., 1997). In a mouse model of chronic contact dermatitis induced by daily application of diphenylcyclopropenone (DCP) for 2 months, HDC(+/+) mice showed significantly increased inflammatory cell infiltration, hyperplastic epidermis and new spreading of neuronal processes than the HDC(-/-) mice (Seike et al., 2005a). Scratching behaviour was induced in HDC(+/+) mice but was barely observed in HDC(-/-) mice, suggesting that histamine production is important for the itch sensation. Aside from the H_2 receptor, all histamine receptors are involved in the transmission of histamine-induced itch (Bell et al., 2004, Rossbach et al., 2011). The first indications that the H_4 receptor is involved in pruritus were made by Bell et al (2004), where it was demonstrated that intracutaneously administered clobenpropit (H_3 receptor antagonist/H_4 receptor agonist) as well as imetit (H_3/H_4 receptor agonist) induced scratching behaviour in mice. Furthermore, the clobenprobit-induced itch was antagonised by systemic administration of the H_3/H_4 receptor -antagonist thioperamide. Moreover, intradermal injection of 4-methylhistamine, a H_4 receptor agonist that displays no affinity at the H_3 receptor, induced itch in mice, confirming the findings by Bell et al. (2004) and clearly pointing out the role of the H_4 receptor in histamine-induced itch transmission (Dunford et al., 2007). Pruritic properties of intradermal administered thioperamide and clobenpropit indicated that the H_3 receptor is also involved in histamine-induced itch response (Hossen et al., 2003, Sugimoto et al., 2004). However, since systemically administered thioperamide dose dependently reduced histamine- or clobenpropit-induced pruritus (Hossen et al., 2003, Bell et al., 2004, Sugimoto et al., 2004), the precise role of the H_3 receptor remained unclear. It has recently been shown that the H_3 receptor is also involved in histamine-induced itch. Intradermal injection of the selective H_3 receptor inverse agonist pitolisant dose-dependently induced scratching behaviour in mice. Interestingly, this itch response could be blocked by pre-treatment with H_1 or H_4 receptor antagonists (Rossbach et al., 2011).

Histamine, 4-methylhistamine and thioperamide induce scratching behaviour in mice independently of mast cells or other haematopoietic cells (Hossen et al., 2003, Dunford et al., 2007), indicating a direct effect on sensory nerves. Additionally, stimulation of the H_4 receptor on Th2 cells leads to an increased release of IL-31 (Sonkoly et al., 2006, Gutzmer et al., 2009). This newly discovered cytokine is an important itch inducer and is mainly produced by activated T cells. The IL-31 receptor complex (IL-31 receptor A and the oncostatin-M receptor) is found in the skin on sensory C-fibers and keratinocytes as well as in the dorsal root ganglia (DRG), where it probably contributes to the transmission of itch signal (Raap et al., 2008). Thus, IL-31 might represent a mediator contributing to pruritus induced by H_4 receptor stimulation (Gutzmer et al., 2009).

In humans, itch induced by histamine is transmitted via specific mechanoinsensitive C fibers. These "itch" fibers are preferentially activated by pruritogens like histamine and respond to histamine application with a time course of excitation that reflects the sensation of itch (Schmelz et al., 1997). Besides the histaminergic pathway, electrophysiological studies suggest the existence of a second peripheral pathway for the transmission of itch (Schmelz, 2010). Cell bodies of sensory nerves are located in the DRG and expression of the H_1, H_3 and H_4 receptors have been observed on a subpopulation (about 15%) of DRG neurons (Kashiba et al., 1999, Cannon et al., 2007, Strakhova et al., 2009, Rossbach et al., 2011). Stimulation with H_1 or H_4 receptor agonists as well as inhibition of the H_3 receptor increases intracellular $[Ca^{2+}]_{free}$ levels in these neurons (Kim et al., 2004, Han et al., 2006, Shim et al., 2007, Rossbach et al., 2011) and histamine requires the activation of TRPV1 to excite sensory neurons (Kim et al., 2004, Shim et al., 2007, Kajihara et al., 2010, Rossbach et al., 2011). Moreover, mice pretreated with a TRPV1 blocker (or mice lacking TRPV1) showed significantly reduced scratching behaviour in response to histamine application.

Shim et al. (2007) further demonstrated that histamine excites sensory neurons by activating TRPV1 via phospholipase A_2 and lipoxygenase stimulation. In addition, Han et al. (2006) showed that phospholipase Cβ3 mediates the scratching response induced by activation of the H_1 receptor on C fiber neurons (Han et al., 2006). These results strongly suggest that histamine requires the activation of TRPV1 to excite sensory neurons via H_1 and H_4 receptors and cause itching. The precise mechanisms underlying the mediation of itch via the H_3 receptor are still unclear. The H_3 receptor modifies the release of histamine and other neurotransmitters not only in the CNS but also in peripheral tissues (Arrang et al., 1983, Ohkubo et al., 1995, Nemmar et al., 1999, Blandizzi et al., 2000). The H_3 receptor may modulate the release of histamine directly from DRG neurons or possibly regulate the release of other neurotransmitters such as substance P, which in turn could activate surrounding cells to release histamine. Substance P has found to be involved in the mediation of histamine-induced itch, where H_4 receptor antagonism inhibits substance P-induced pruritus and intradermal injection of a tachykinin NK1 antagonist decreases the pruritus induced by the H_3 receptor antagonist/H_4 receptor agonist clobenpropit (Hossen et al., 2006, Yamaura et al., 2009). A decreased threshold or even an enhanced neurotransmitter release in response to H_3 receptor inverse agonism might activate H_1 receptor and H_4 receptor on a subset of sensory neurons, which in turn could result in the excitation of itch-mediating histamine-sensitive sensory nerves, triggering the itch response (Rossbach et al. 2011). However, it can not be excluded that skin cells other than mast cells, such as keratinocytes, are required for a possible enhanced release of histamine via blockade of the H_3 receptor.

Sensory nerves in the skin transmit the itch signal to the DRG and from there it reaches the spinal cord. From the lamina I, a superficial layer within the dorsal horn of the spinal cord, the signal is projected to the thalamus. The expression of c-Fos was specifically upregulated in lamina I of the spinal dorsal horn following repeated DCP application in mice (Seike et al., 2005a). Increased expression of c-Fos and substance P in this region were downregulated by olopatadine (Hamada et al., 2006), suggesting that scratching behaviour in chronic contact dermatitis in mice are mainly mediated via histamine and the afferent pathways of sensation to the central nervous system are mediated through lamina I of the spinal dorsal horn. Histamine-induced itch is transmitted via a distinct neuronal pathway consisting of specialized mechano-insensitive primary afferent fibers and mechano-insensitive dorsal horn spinothalamic projection neurons (Schmelz et al., 1997, Andrew and Craig, 2001, Schmelz, 2010). These histamine-sensitive spinothalamic tract neurons project mainly to the ventral, posterior and inferior nucleus as well as the ventral periphery of the ventral, posterior and lateral nucleus of the lateral thalamus, whereas nociceptive spinothalamic tract neurons project mainly to the nucleus submedius of the medial thalamus (Ikoma et al., 2011). However, the histamine-responsive fibres are also excited by at least one algogen, namely capsaicin (ligand of TRPV1) and are thus not itch-specific (Schmelz, 2010). The impact of the different histamine receptor subtypes on the spinal itch transmission to our knowledge has not been elucidated yet. Intrathecal injection of histamine induces a behavioral response consisting of biting and licking with occasional hindlimb scratching in mice, which seems to be dependent on the injected histamine doses mediated by H_1, NK1 and NMDA receptors (Sakurada et al., 2002, Watanabe et al., 2008, Mizoguchi et al., 2011). Recently, the expression of the H_4 receptor in the spinal cord has been shown but its functional role has not yet been clarified (Strakhova et al., 2009). While histamine has been a well-known mediator of pruritus for over 100 years, the understanding of itch signalling via histamine receptors is far from being completely understood.

6.3.2. Role of Histamine in Scratching Behaviour in Murine Allergic Dermatitis Models

Pruritus is a major symptom of allergic skin diseases like atopic dermatitis but often difficult to control. There is little information on the role of histamine in scratching behaviour and sensory transmission of atopic dermatitis and chronic eczema. Fukamachi et al. (2011) showed an increased expression of semaphorin 3A in normal human epidermal keratinocytes by histamine. Semaphorin 3A plays an inhibitory role for C-fiber elongation in upper layer of epidermis and decreased expression of semaphorin 3A has been found in lesional skin of AD (Fukamachi et al., 2011). In murine keratinocytes, the expression of semaphorin 3A mRNA is reduced by incubation with histamine and this reduction can be reversed by an H_1 receptor antagonist (olopatadine) but not by an H_4 receptor antagonist (JNJ7777120) (Ohsawa and Hirasawa, 2012). The efficacy of classical H_1 receptor antihistamines under clinical conditions is limited to a few diseases (like urticaria and insect bite reactions) and chronic pruritus as seen in atopic dermatitis patients does not respond to H_1 receptor blockade (Akdis et al., 2006b, Zuberbier et al., 2009). The discovery of the H_4 receptor has rekindled the interest in histamine receptors as antipruritic targets. The H_4 receptor seems to be a promising target for the treatment of pruritc skin diseases; H_4 receptor antagonists showed similar or even superior effects compared to traditional H_1 receptor antihistamines in the attenuation of experimental pruritus (Dunford et al., 2007, Rossbach et al., 2009b, Cowden et al., 2010). In two mouse models of allergen-mediated pruritus induced by repetitive administration of strong sensitizers (2,4-dinitrochlorobenzene or toluene-2,4-diisocyanate, respectively), a combination of the H_1 receptor antagonist cetirizine and the H_4 receptor antagonist JNJ7777120 led to a reduction of scratching bouts to up to 90%. According to Dunford et al. 2007, the antiprurinergic potential of the H_4 receptor antagonist JNJ7777120 because of any sedative properties can be excluded, even though JNJ7777120 crosses the blood brain barrier. Moreover, in a mouse model of chronic dermatitis induced in NC/Nga mice by repeated challenge with the hapten picryl chloride, the combination of the H_1 receptor antagonist olopatadine and JNJ7777120 showed an antipruritic efficacy as potent as that of prednisolone (Ohsawa & Hirasawa, 2012). Thus, a combination of H_4 and H_1 receptor antagonism might be a new strategy to treat pruritus related to allergic diseases like atopic dermatitis. Since the H_3 receptor serves as an inhibitory receptor which possibly increases the threshold for histamine-induced itch, the addition of an H_3 receptor agonist might be even more beneficial. To date, there is only one study that analysed the involvement of H_3 receptor in allergen-induced pruritus, but this study revealed that the H_3 receptor inverse agonist pitolisant had no further effect on allergen-induced itch in a murine model of allergic contact dermatitis (Rossbach et al., 2011).

6.4. Summary

Taken together the aforementioned studies provide evidence for an immunmodulatory role of the H_4 receptor in allergic inflammatory skin diseases and pruritus. However, not all results confirm a pro-inflammatory role for H_4 receptor. In particular the modulation of cytokine secretion in antigen presenting cells favour an anti-inflammatory role mediated by the H_4 receptor. Nevertheless, pro-inflammatory outcomes in mouse models of allergic dermatitis treated with H_4 receptor antagonists or performed with H_4 receptor knockout-mice have not yet been reported. In most settings, a clear anti-inflammatory effect has been demonstrated by blocking the H_4 receptor. Also as far as pruritus is concerned, a blockade of H_4 receptor might be a new option for the treatment of pruritus associated with allergic skin diseases.

References

Akdis C.A., Akdis M., Bieber T., et al. (2006) Diagnosis and treatment of atopic dermatitis in children and adults: European Academy of Allergology and Clinical Immunology/American Academy of Allergy, Asthma and Immunology/PRACTALL Consensus Report. *J Allergy Clin Immunol, 118*, 152-169.

Akdis C.A., Simons F.E. (2006) Histamine receptors are hot in immunopharmacology. *Eur J Pharmacol, 533*, 69-76.

Andrew D., Craig A.D. (2001) Spinothalamic lamina I neurons selectively sensitive to histamine: a central neural pathway for itch. *Nat Neurosci, 4*, 72-77.

Arrang J.M., Garbarg M., Schwartz J.C. (1983) Auto-inhibition of brain histamine release mediated by a novel class (H3) of histamine receptor. *Nature, 302*, 832-837.

Barnard R., Barnard A., Salmon G., et al. (2008) Histamine-induced actin polymerization in human eosinophils: an imaging approach for histamine H4 receptor. *Cytometry A, 73*, 299-304.

Baumer W., Rossbach K. (2010) [Histamine as an immunomodulator]. *J Dtsch Dermatol Ges 8*, 495-504.

Baumer W., Seegers U., Braun M., et al. (2004) TARC and RANTES, but not CTACK, are induced in two models of allergic contact dermatitis. Effects of cilomilast and diflorasone diacetate on T-cell-attracting chemokines. *Br J Dermatol, 151*, 823-830.

Baumer W., Stahl J., Sander K., et al. (2011) Lack of preventing effect of systemically and topically administered histamine H(1) or H(4) receptor antagonists in a dog model of acute atopic dermatitis. *Exp Dermatol, 20*, 577-581.

Bell J.K., McQueen D.S., Rees J.L. (2004) Involvement of histamine H4 and H_1 receptors in scratching induced by histamine receptor agonists in Balb C mice. *Br J Pharmacol, 142*, 374-380.

Blandizzi C., Tognetti M., Colucci R., et al. (2000) Histamine H(3) receptors mediate inhibition of noradrenaline release from intestinal sympathetic nerves. *Br J Pharmacol, 129*, 1387-1396.

Buckland K.F., Williams T. J., Conroy D.M. (2003) Histamine induces cytoskeletal changes in human eosinophils via the H(4) receptor. *Br J Pharmacol, 140*, 1117-1127.

Buentke E., Heffler L.C., Wilson J.L., et al. (2002) Natural killer and dendritic cell contact in lesional atopic dermatitis skin--Malassezia-influenced cell interaction. *J Invest Dermatol, 119*, 850-857.

Cannon K.E., Chazot P.L., Hann V., et al. (2007) Immunohistochemical localization of histamine H3 receptors in rodent skin, dorsal root ganglia, superior cervical ganglia, and spinal cord: potential antinociceptive targets. *Pain, 129*, 76-92.

Clark R.A., Gallin J.I., Kaplan A.P., (1975) The selective eosinophil chemotactic activity of histamine. *J Exp Med, 142*, 1462-1476.

Clark R.A., Sandler J.A., Gallin J.I., et al. (1977) Histamine modulation of eosinophil migration. *J Immunol, 118*, 137-145.

Cowden J.M., Zhang M., Dunford P.J., et al. (2010) The histamine H4 receptor mediates inflammation and pruritus in Th2-dependent dermal inflammation. *J. Invest. Dermatol, 130*, 1023-1033.

Damaj B.B., Becerra C.B., Esber H.J., et al. (2007) Functional expression of H4 histamine receptor in human natural killer cells, monocytes, and dendritic cells. *J Immunol, 179*, 7907-7915.

de Esch I.J., Thurmond R.L., Jongejan A., et al. (2005) The histamine H4 receptor as a new therapeutic target for inflammation. Trends in pharmacological sciences 26:462-469.

DeBoer D.J., Griffin C.E. (2001) The ACVD task force on canine atopic dermatitis (XXI): antihistamine pharmacotherapy. *Vet. Immunol Immunopathol, 81*, 323-329.

Desai P., Thurmond R.L. (2011) Histamine H(4) receptor activation enhances LPS-induced IL-6 production in mast cells via ERK and PI3K activation. *Eur J Immunol, 41*, 1764-1773.

Dijkstra D., Leurs R., Chazot P., et al. (2007) Histamine downregulates monocyte CCL2 production through the histamine H4 receptor. *J Allergy Clin Immunol, 120*, 300-307.

Dijkstra D., Stark H., Chazot P.L., et al. (2008) Human inflammatory dendritic epidermal cells express a functional histamine H4 receptor. *J Invest Dermatol, 128*, 1696-1703.

Dunford P.J., Williams K.N., Desai P.J., et al. (2007) Histamine H4 receptor antagonists are superior to traditional antihistamines in the attenuation of experimental pruritus. *J Allergy Clin Immunol, 119*, 176-183.

Eisenschenk M.N., Torres S.M., Oliveira S., et al. (2011) The expression of histamine H_4 receptor mRNA in the skin and other tissues of normal dogs. *Vet Dermatol, 22*, 396-400.

Fukamachi S., Bito T., Shiraishi N., et al. (2011) Modulation of semaphorin 3A expression by calcium concentration and histamine in human keratinocytes and fibroblasts. *J Dermatol Sci, 61*, 118-123.

Gantner F., Sakai K., Tusche M.W., et al. (2002) Histamine h(4) and h(2) receptors control histamine-induced interleukin-16 release from human CD8(+) T cells. *J Pharmacol Exp Ther, 303*, 300-307.

Giustizieri M.L., Albanesi C., Fluhr J., et al. (2004) H1 histamine receptor mediates inflammatory responses in human keratinocytes. *J Allergy Clin Immunol, 114*, 1176-1182.

Glatzer F., Gschwandtner M., Ehling S., et al. (2013) Histamine induces proliferation in keratinocytes from atopic dermatitis patients, J Allergy Clin Immunol (in press)

Godot V., Arock M., Garcia G., et al. (2007) H4 histamine receptor mediates optimal migration of mast cell precursors to CXCL12. *J Allergy Clin Immunol, 120*, 827-834.

Gschwandtner M., Bunk H., Kother B., et al. (2012) Histamine down-regulates IL-27 production in antigen-presenting cells. *J Leukoc Biol, 92*, 21-29.

Gschwandtner M., Mildner M., Mlitz V., et al. (2013) Histamine suppresses epidermal keratinocyte differentiation and impairs skin barrier function in a human skin model. *Allergy* 68, 37-47.

Gschwandtner M., Mommert S., Kother B., et al. (2011a) The histamine H4 receptor is highly expressed on plasmacytoid dendritic cells in psoriasis and histamine regulates their cytokine production and migration. *J Invest Dermatol, 131*, 1668-1676.

Gschwandtner M., Rossbach K., Dijkstra D., et al. (2010) Murine and human Langerhans cells express a functional histamine H4 receptor: modulation of cell migration and function. *Allergy, 65*, 840-849.

Gschwandtner M., Schakel K., Werfel T., et al. (2011b) Histamine H(4) receptor activation on human slan-dendritic cells down-regulates their pro-inflammatory capacity. *Immunology, 132*, 49-56.

Gutzmer R., Diestel C., Mommert S., et al. (2005) Histamine H4 receptor stimulation suppresses IL-12p70 production and mediates chemotaxis in human monocyte-derived dendritic cells. *J Immunol, 174*, 5224-5232.

Gutzmer R., Gschwandtner M., Rossbach K., et al. (2011) Pathogenetic and therapeutic implications of the histamine H4 receptor in inflammatory skin diseases and pruritus. *Front Biosci (Schol Ed), 3*, 985-994.

Gutzmer R., Mommert S., Gschwandtner M., et al. (2009) The histamine H_4 receptor is functionally expressed on T(H)2 cells. *J. Allergy Clin. Immunol, 123*, 619-625.

Hagermark O., Strandberg K., Gronneberg R. (1979) Effects of histamine receptor antagonists on histamine-induced responses in human skin. *Acta dermato-venereologica, 59*, 297-300.

Hamada R., Seike M., Kamijima R., et al. (2006) Neuronal conditions of spinal cord in dermatitis are improved by olopatadine. *Eur. J Pharmacol, 547*, 45-51.

Han S.K., Mancino V., Simon M.I. (2006) Phospholipase Cbeta 3 mediates the scratching response activated by the histamine H1 receptor on C-fiber nociceptive neurons. *Neuron, 52*, 691-703.

Hirasawa N., Ohsawa Y., Katoh G., et al. (2009) Modification of the picryl chloride-induced allergic dermatitis model in mouse ear lobes by 12-O-tetradecanoylphorbol 13-acetate, and analysis of the role of histamine in the modified model. *Int Arch Allergy Immunol, 148*, 279-288.

Hofstra C.L., Desai P.J., Thurmond R.L., et al. (2003) Histamine H4 receptor mediates chemotaxis and calcium mobilization of mast cells. *J Pharmacol Exp Ther, 305*, 1212-1221.

Hossen M.A., Inoue T., Shinmei Y., et al. (2006) Role of substance P on histamine H(3) antagonist-induced scratching behavior in mice. *J. Pharmacol. Sci, 100*, 297-302.

Hossen M.A., Sugimoto Y., Kayasuga R., et al. (2003) Involvement of histamine H_3 receptors in scratching behaviour in mast cell-deficient mice. *Br J Dermatol, 149*, 17-22.

Ikawa Y., Shiba K., Ohki E., et al. (2008) Comparative study of histamine H4 receptor expression in human dermal fibroblasts. *J Toxicol Sci, 33*, 503-508.

Ikoma A., Cevikbas F., Kempkes C., et al. (2011) Anatomy and neurophysiology of pruritus. *Semin Cutan Med Surg, 30*, 64-70.

Jiang W., Lim H.D., Zhang M., et al. (2008) Cloning and pharmacological characterization of the dog histamine H4 receptor. *Eur J Pharmacol, 592*, 26-32.

Jin H., He R., Oyoshi M., et al. (2009) Animal models of atopic dermatitis. *J Invest Dermatol, 129*, 31-40.

Kajihara Y., Murakami M., Imagawa T., et al. (2010) Histamine potentiates acid-induced responses mediating transient receptor potential V1 in mouse primary sensory neurons. *Neurosci, 166*, 292-304.

Kanda N., Watanabe S. (2003) Histamine enhances the production of nerve growth factor in human keratinocytes. *J Invest Dermatol, 121*, 570-577.

Kanda N., Watanabe S. (2004) Histamine enhances the production of granulocyte-macrophage colony-stimulating factor via protein kinase Calpha and extracellular signal-regulated kinase in human keratinocytes. *J Invest Dermatol, 122*, 863-872.

Kashiba H., Fukui H., Morikawa Y., et al. (1999) Gene expression of histamine H1 receptor in guinea pig primary sensory neurons: a relationship between H1 receptor mRNA-expressing neurons and peptidergic neurons. *Brain Res Mol Brain Res, 66*, 24-34.

Kasraie S., Niebuhr M., Werfel T. (2010) Interleukin (IL)-31 induces pro-inflammatory cytokines in human monocytes and macrophages following stimulation with staphylococcal exotoxins. *Allergy, 65,* 712-721.

Kim B.M., Lee S.H., Shim W.S., et al. (2004) Histamine-induced Ca(2+) influx via the PLA(2)/lipoxygenase/TRPV1 pathway in rat sensory neurons. *Neurosci Lett, 361,* 159-162.

Koga C., Kabashima K., Shiraishi N., et al. (2008) Possible pathogenic role of Th17 cells for atopic dermatitis. *J Invest Dermatol, 128,* 2625-2630.

Kohda F., Koga T., Uchi H., et al. (2002) Histamine-induced IL-6 and IL-8 production are differentially modulated by IFN-gamma and IL-4 in human keratinocytes. *J Dermatol Sci, 28,* 34-41.

Leung D.Y., Parkman R., Feller J., et al. (1982) Cell-mediated cytotoxicity against skin fibroblasts in atopic dermatitis. *J Immunol, 128,* 1736-1741.

Ling P., Ngo K., Nguyen S., et al. (2004) Histamine H4 receptor mediates eosinophil chemotaxis with cell shape change and adhesion molecule upregulation. *Br J Pharmacol, 142,* 161-171.

Lippert U., Artuc M., Grutzkau A., et al. (2004) Human skin mast cells express H2 and H4, but not H3 receptors. *J Invest Dermatol, 123,* 116-123.

Magerl W., Westerman R.A., Mohner B., et al. (1990) Properties of transdermal histamine iontophoresis: differential effects of season, gender, and body region. *J Invest Dermatol, 94,* 347-352.

Marsella R., Girolomoni G. (2009) Canine models of atopic dermatitis: a useful tool with untapped potential. *J Invest Dermatol, 129,* 2351-2357.

Marsella R., Olivry T. (2001) The ACVD task force on canine atopic dermatitis (VII): mediators of cutaneous inflammation. *Vet. Immunol. Immunopathol, 81,* 205-213.

Marsella R., Olivry T., Carlotti D.N. (2011) Current evidence of skin barrier dysfunction in human and canine atopic dermatitis. *Vet Dermatol, 22,* 239-248.

Mizoguchi H., Komatsu T., Iwata Y., et al. (2011) Partial involvement of NMDA receptors and glial cells in the nociceptive behaviors induced by intrathecally administered histamine. *Neurosci Lett, 495,* 83-87.

Mommert S., Gschwandtner M., Koether B., et al. (2012) Human memory Th17 cells express a functional histamine H4 receptor. *Am J Pathol, 180*, 177-185.

Morgan R. K., McAllister B., Cross L. et al. (2007) Histamine 4 receptor activation induces recruitment of FoxP3+ T cells and inhibits allergic asthma in a murine model. *J Immunol, 178*, 8081-8089.

Morse K.L., Behan J., Laz T.M., et al. (2001) Cloning and characterization of a novel human histamine receptor. *J Pharmacol Exp Ther, 296*, 1058-1066.

Nakamura T., Itadani H., Hidaka Y., et al. (2000) Molecular cloning and characterization of a new human histamine receptor, HH4R. *Biochem Biophys Res Commun, 279*, 615-620.

Nemmar A., Delaunois A., Beckers J.F., et al. (1999) Modulatory effect of imetit, a histamine H3 receptor agonist, on C-fibers, cholinergic fibers and mast cells in rabbit lungs in vitro. *Eur. J Pharmacol, 371*, 23-30.

Novak N., Bieber T., Leung D.Y. (2003) Immune mechanisms leading to atopic dermatitis. *J Allergy Clin Immunol, 112*, S128-139.

O'Reilly M., Alpert R., Jenkinson S., et al. (2002) Identification of a histamine H4 receptor on human eosinophils--role in eosinophil chemotaxis. *J Recept Signal Transduct Res, 22*, 431-448.

Oda T., Morikawa N., Saito Y., et al. (2000) Molecular cloning and characterization of a novel type of histamine receptor preferentially expressed in leukocytes. *J Biol Chem, 275*, 36781-36786.

Ohkubo T., Shibata M., Inoue M., et al. (1995) Regulation of substance P release mediated via prejunctional histamine H3 receptors. *Eur. J. Pharmacol, 273*, 83-88.

Ohsawa Y., Hirasawa N. (2012) The antagonism of histamine H1 and H4 receptors ameliorates chronic allergic dermatitis via anti-pruritic and anti-inflammatory effects in NC/Nga mice. *Allergy, 67*, 1014-1022.

Ohtsu H., Tanaka S., Terui T., et al. (2001) Mice lacking histidine decarboxylase exhibit abnormal mast cells. *FEBS Lett, 502*, 53-56.

Olivry T., Foster A.P., Mueller R.S., et al. (2010) Interventions for atopic dermatitis in dogs: a systematic review of randomized controlled trials. *Vet Dermatol, 21*, 4-22.

Ou L.S., Goleva E., Hall C., et al. (2004) T regulatory cells in atopic dermatitis and subversion of their activity by superantigens. *J Allergy Clin Immunol, 113*, 756-763.

Park Y.D., Lyou Y.J., Lee K.J., et al. (2006) Towards profiling the gene expression of fibroblasts from atopic dermatitis patients: human 8K complementary DNA microarray. *Clin Exp Allergy, 36*, 649-657.

Pucheu-Haston C.M., Jackson H.A., Olivry T., et al. (2008) Epicutaneous sensitization with Dermatophagoides farinae induces generalized allergic dermatitis and elevated mite-specific immunoglobulin E levels in a canine model of atopic dermatitis. *Clin Exp Allergy, 38*, 667-679.

Raap U., Wichmann K., Bruder M., et al. (2008) Correlation of IL-31 serum levels with severity of atopic dermatitis. *J Allergy Clin Immunol, 122*, 421-423.

Rossbach K., Hartwig C., Kietzmann M., et al. (2012) Role of histamine H4 receptor in a murine model of atopic dermatitis. *J Vet Pharmacol Therap, 35* (Suppl. 3), 138.

Rossbach K., Nassenstein C., Gschwandtner M., et al. (2011) Histamine H1, H3 and H4 receptors are involved in pruritus. *Neuroscience, 190*, 89-102.

Rossbach K., Stark H., Sander K., et al. (2009a) The histamine H receptor as a new target for treatment of canine inflammatory skin diseases. *Vet Dermatol, 20*, 555-561.

Rossbach K., Wendorff S., Sander K., et al. (2009b) Histamine H4 receptor antagonism reduces hapten-induced scratching behaviour but not inflammation. *Exp. Dermatol, 18*, 57-63.

Sakurada S., Orito T., Sakurada C., et al. (2002) Possible involvement of tachykinin NK(1) and NMDA receptors in histamine-induced hyperalgesia in mice. *Eur. J. Pharmacol, 434*, 29-34.

Sander K., Kottke T., Stark H. (2008) Histamine H3 receptor antagonists go to clinics. *Biol. Pharm. Bull, 31*, 2163-2181.

Schmelz M. (2010) Itch and pain. *Neurosci Biobehav Rev, 34*, 171-176.

Schmelz M., Schmidt R., Bickel A., et al. (1997) Specific C-receptors for itch in human skin. *J Neurosci, 17*, 8003-8008.

Schnopp C., Rad R., Weidinger A., et al. (2007) Fox-P3-positive regulatory T cells are present in the skin of generalized atopic eczema patients and are not particularly affected by medium-dose UVA1 therapy. *Photodermatol Photoimmunol Photomed, 23,* 81-85.

Seike M., Furuya K., Omura M., et al. (2010) Histamine H(4) receptor antagonist ameliorates chronic allergic contact dermatitis induced by repeated challenge. *Allergy, 65,* 319-326.

Seike M., Ikeda M., Kodama H., et al. (2005a) Inhibition of scratching behaviour caused by contact dermatitis in histidine decarboxylase gene knockout mice. *Exp Dermatol, 14,* 169-175.

Seike M., Takata T., Ikeda M., et al. (2005b) Histamine helps development of eczematous lesions in experimental contact dermatitis in mice. *Arch. Dermatol. Res, 297,* 68-74.

Sercombe R., Verrecchia C., Philipson V., et al. (1986) Histamine-induced constriction and dilatation of rabbit middle cerebral arteries in vitro: role of the endothelium. *Blood Vessels, 23,* 137-153.

Shim W.S., Tak M.H., Lee M.H., et al. (2007) TRPV1 mediates histamine-induced itching via the activation of phospholipase A2 and 12-lipoxygenase. *J Neurosci, 27,* 2331-2337.

Simon D., Braathen L. R., Simon H.U. (2004) Eosinophils and atopic dermatitis. *Allergy, 59,* 561-570.

Simon D., Kozlowski E., Simon H. (2009) Natural killer T cells expressing IFN-gamma and IL-4 in lesional skin of atopic eczema. *Allergy, 64,* 1681-1684.

Simon T., Semsei A.F., Ungvari I., et al. (2012) Asthma Endophenotypes and Polymorphisms in the Histamine Receptor HRH4 Gene. *Int Arch Allergy Immunol, 159,* 109-120.

Simons F.E., Simons K.J. (1994) The pharmacology and use of H1-receptor-antagonist drugs. *N Engl J Med, 330,* 1663-1670.

Simons F.E., Simons K.J. (2011) Histamine and H1-antihistamines: celebrating a century of progress. *J Allergy Clin Immunol, 128,* 1139-1150 e1134.

Sonkoly E., Muller A., Lauerma A.I., et al. (2006) IL-31: a new link between T cells and pruritus in atopic skin inflammation. *J Allergy Clin Immunol, 117,* 411-417.

Spergel J.M., Mizoguchi E., Brewer J.P., et al. (1998) Epicutaneous sensitization with protein antigen induces localized allergic dermatitis and hyperresponsiveness to methacholine after single exposure to aerosolized antigen in mice. *J Clin Invest, 101*,1614-1622.

Stander S., Steinhoff M. (2002) Pathophysiology of pruritus in atopic dermatitis: an overview. *Exp Dermatol, 11*, 12-24.

Steinhoff M., Bienenstock J., Schmelz M., et al. (2006) Neurophysiological, neuroimmunological, and neuroendocrine basis of pruritus. *J Invest Dermatol, 126*, 1705-1718.

Strakhova M.I., Nikkel A.L., Manelli A.M., et al. (2009) Localization of histamine H4 receptors in the central nervous system of human and rat. *Brain Res., 1250*, 41-48.

Sugata Y., Okano M., Fujiwara T., et al. (2007) Histamine H4 receptor agonists have more activities than H4 agonism in antigen-specific human T-cell responses. *Immunology, 121*, 266-275.

Sugimoto Y., Iba Y., Nakamura Y., et al. (2004) Pruritus-associated response mediated by cutaneous histamine H3 receptors. *Clin Exp Allergy, 34*, 456-459.

Thurmond R.L., Gelfand E.W., Dunford P.J. (2008) The role of histamine H1 and H4 receptors in allergic inflammation: the search for new antihistamines. *Nat Rev Drug Discov, 7*, 41-53.

Tiligada E., Zampeli E., Sander K., et al. (2009) Histamine H3 and H4 receptors as novel drug targets. *Expert Opin Investig Drugs, 18*, 1519-1531.

Watanabe C., Orito T., Watanabe H., et al. (2008) Intrathecal high-dose histamine induces spinally-mediated nociceptive behavioral responses through a polyamine site of NMDA receptors. *Eur. J. Pharmacol, 581*, 54-63.

Werfel T. (2009) The role of leukocytes, keratinocytes, and allergen-specific IgE in the development of atopic dermatitis. *J Invest Dermatol, 129*, 1878-1891.

Wittmann M., Werfel T. (2006) Interaction of keratinocytes with infiltrating lymphocytes in allergic eczematous skin diseases. *Curr Opin Allergy Clin Immunol, 6*, 329-334.

Yamaura K., Oda M., Suwa E., et al. (2009) Expression of histamine H4 receptor in human epidermal tissues and attenuation of experimental pruritus using H4 receptor antagonist. *J Toxicol Sci, 34,* 427-431.

Yu B., Shao Y., Li P., et al. (2010a) Copy number variations of the human histamine H4 receptor gene are associated with systemic lupus erythematosus. *Br J Dermatol, 163,* 935-940.

Yu B., Shao Y., Zhang J., et al. (2010b) Polymorphisms in human histamine receptor H4 gene are associated with atopic dermatitis. *Br J Dermatol, 162,* 1038-1043.

Zampeli E., Tiligada E. (2009) The role of histamine H4 receptor in immune and inflammatory disorders. *Br J Pharmacol, 157,* 24-33.

Zuberbier T., Asero R., Bindslev-Jensen C., et al. (2009) EAACI/GA(2)LEN/EDF/WAO guideline: management of urticaria. *Allergy, 64,* 1427-1443.

Chapter 7

Non-professional Histamine Producing Cells, Immune Responses and Autoimmunity

Yrjö T. Konttinen[1], Henrik Husu[1], Xia Han[1], M. Beatrice Passani[2], Clara Ballerini[3], Vasily Stegaev[1], Tarvo Sillat[1], Zygmunt Mackiewicz[4]

[1]Department of Medicine, University of Helsinki and Helsinki University Central Hospital, Helsinki, Finland, e-mail: yrjo.konttinen@helsinki.fi

[2]Dipartimento di Farmacologia Preclinica e Clinica, Universita' di Firenze, Firenze, Italy

[3]Dipartimento di Scienze Neurologiche, Universita' di Firenze, Firenze, Italy

[4]State Research Institute Center of Innovative Medicine, Vilnius, Lithuania

Abstract

Histamine is a monovalent cationic biological amine synthesized by professional and non-professional histamine producing cells, which form two distinct cell categories. Professional cells produce, store and burst release their histamine from storage granules so that locally and temporarily for short periods high micromolar histamine concentrations are achieved. Non-professional cells generally produce and release histamine into the cytoplasm continuously without intermediated storage, to release it along its concentration gradient via histamine channels. These channels might play an important role in balancing, maintaining and regulating low-level histamine fluctuations, which tend to occur due to variation in the histamine content of the food, histamine production by microflora, histamine release from professional cells and enzymatic degradation and reuptake of histamine.

Effects of the professionally-produced high histamine levels are mediated via two low-affinity histamine receptors, H_1R and H_2R. In contrast, the non-professionally-produced, low nascent histamine levels are only able to activate two novel high-affinity histamine receptors, H_3R and H_4R. The former are active during short-term emergency "on-state" whereas the latter are active during a basic or long term homeostatic "off-state". Micromolar histamine/H_1R/H_2R amplifies expulsion of irritants (e.g. pollen, helminths) by adding to a direct irritant-induced reaction an amplifying IgE-dependent immunological component. Nanomolar histamine/H_3R/H_4R seems to participate in dendritic cell-mediated antigen (Ag) presentation to T-cells in induction of self-tolerance, sensitization phase in allergy and delayed lymphocyte-mediated immune reactions. From this point of view, it is interesting that H_4R is best known for being expressed by myelopoietic cells, such as dendritic cells and lymphocytes, but also polymorphonuclear neutrophils, basophils, eosinophils and mast cells (MC).

The role of histamine has been mainly recognized during exacerbation of asthma, allergic rhinitis and conjunctivitis, urticaria and anaphylaxis as well as gastric acid secretion after a meal. Antihistamines are generally helpful during the acute stages of various allergic states associated with burst release of histamine. However, it seems that the basis of these conditions lies in a steadier underlying immune activity, e.g. Ag presentation by dendritic cells to T-cells and T-cell help to B-cell-dependent IgE immunoglobulin synthesis against T-cell-dependent Ag. Normally this system participates in the maintenance of health (immune tolerance and immune responses against pathogens), but in patients it can participate in the initiation and perpetuation of diseases and contributes to the recurrence of the symptoms and signs upon renewed exposure to the symptom triggers (allergens in allergies and auto-Ag in autoimmune diseases). Therefore, antihistamines or H_1R-antagonists and H_2R blockers provide limited therapeutic effectiveness in such conditions, but these conditions may in part be regulated by low histamine and high- affinity histamine receptors, in particular H_4R. Histamine H_4R modulation by small molecules administered perorally or locally might provide effective and affordable new remedies for various autoinflammatory, allergic and autoimmune diseases.

7.1. Introduction

Despite intense research efforts in the past, new observations on its fundamental function have recently been made regarding two intimately interrelated concepts, its non-professional synthesis and its novel high-affinity receptors.

The so-called "professional" histamine producing cells produce histamine rapidly in copious amounts and store it in storage granules, from which histamine and other granular contents can be rapidly released by exocytosis in a process that is generally regulated but sometimes apparently stochastic. "Non-professional" histamine producing cells generates histamine at a 100- to 1000-fold lower pace and release it directly to the cell cytoplasm without an intermediate storage stage. Thus, during the "off-state" they first produce histamine in the cytoplasm and then passively release it to the extracellular space along the concentration gradient, apparently at the same pace as it is produced. This histamine release is mediated by "equilibrative uniporter" histamine channels.

Apart from rapid, temporary and robust local histamine release or slow, low level constitutive histamine release, histamine uptake via these "equilibrative uniporters" is possible as well.

Extracellular burst-released micromolar histamine exerts it effects on histamine target cells via classical, low-affinity H_1R and H_2R receptors. In contrast, basal non-professional histamine exerts its effects directly only via high-affinity H_3R and H_4R receptors.

During MC activation and the high histamine "on-state", the classical histamine receptors are rapidly desensitized based on phenomena like receptor phosphorylation, G-protein uncoupling, receptor internalization and receptor gene transcription. At the same time, the numerous non-professional histamine producing cells assume an alternative histamine N-methyl transferase (HNMT)-depedent function as high-capacity histamine sinks, removing histamine effectively from the extracellular space for cytoplasmic degradation. These novel cells now work, not to maintain but to return back to the homeostatic nanomolar histamine concentrations. However, it is not clear how the histamine-sensitive, high-affinity histamine receptors cope with the excessive overstimulation during the local and temporary high histamine on-states.

7.2. Histamine

In human cells, histamine is synthesized from the essential amino acid (AA) L-histidine and this synthesis is dependent on L-histidine decarboxylase (HDC). Histamine release occurs either via exocytosis or passively via histamine channels. Exocytotic release is regulated and occur only after stimulation by allergens, substance P as well as other tachykinins, lectin-like structures, physical forces (mechanical stimulation) and various foodstuffs such as citrus fruit, strawberries and nuts (for a more complete list, see Maintz & Novak 1999). In addition, other food itemssuch as such as various fishes (mackerel, herring, sardine, tuna etc.), cheeses (Emmental, Swiss, Cheddar, Camembert) and alcohol (in particular red wine) also contain histamine. Human bodyis an ecosystem, which contains 10 times as many bacteria as it contains human cells, so the relase of histamine synthesized by intestinal bacterial flora can greatly contribute to physiological histamine concentrations. The latter source is notable because it suggests that HDC knockout mice are not "histamine-free".

7.2.1. Histamine Synthesis

It was traditionally thought that MC, basophils, enterochromaffin-like (ECL) cells and histaminergic neurons loaded with histamine-containing secretory granules were also responsible for histamine synthesis in the human body. It has since been noted that histamine is an important immunomodulatory agent already at low nanomolar concentrations, and that leukocytes other than MC and basophils can also produce histamine, albeit at a much smaller scale. Furthermore, dendritic cells (DC), T-cells, monocyte/macrophages and neutrophils also produce histamine (Tanaka & Ichikawa 2011; Kubo & Nakano 1999; Szeberényi et al. 2001). Even non-immune cells such as fetal liver cells, ductal epithelial cells of the salivary glands, sperm cells, epithelial cells of the mammary glands and skeletal muscle cells can synthesize histamine. This suggests that histamine has numerous previously undiscovered physiological and pathological effects, apparently mediated by the recently discovered H_3R and H_4R receptors. Accordingly, it has been proposed that histamine can maintain homeostasis of salivary glands, stimulate neonatal hematopoiesis, regulate reproductive organs and improve adaptation of muscle cells to exercise (Tanaka & Ichikawa 2011; Niijima-Yaoita et al. 2012). Thus, the cells can be divided to two different categories: professional and non-professional histamine producing cells (Table7.1), which essentially utilize two different sets of receptors, secretory and intake pathways to regulate very different sets of bodily events.

The molecular basis of professional large scale histamine synthesis and non-professional small scale histamine synthesis as well as the differences between each process has been investigated, where at least two functionally relevant isoforms of HDC have been reported (Ichikawa et al. 2010; Tanaka & Ichikawa 2011). The full-length, low-activity 74-kDa HDC is in professional histamine producing cells proteolytically cleaved to a 53-kDa isoform, which is 100-1000-fold more active than its full-length precursor. In non-professional histamine producing cells the full-length post-translational 74 kDa isoform with a low enzymatic activity produces histamine to the cell cytoplasm (Tanaka & Ichikawa 2011; Kubo & Nakano 1999). The activity of this low activity HDC in non-professional, low scale histamine producing cells is not particularly stable in this transformation, but is known to be regulated at the transciptional and protein levels by various cytokines, including IL-1, IL-3, IL-12, IL-18, GM-CSF and TNF-α (Jutel et al. 2009). HDC activity changes in various infections, inflammation and the rejection of transplanted organs.

7.2.2. Histamine Release

Professional histamine producing cells store histamine in cytoplasmic secretory granules. Histamine in these granules is stored by ionic linkage to the carboxyl groups of proteins and heparin. A murine connective tissue MC contains 10-30 pg histamine per cell compared to < 0.2 pg histamine per cell in a mucosal MC. If 110 pg/ml of histamine is equivalent to 1 nM, it can be calculated that ~18500 connective tissue MCs could release enough histamine to 1 mL to reach 50 μM. Given that 10^4-10^5 MCs can be isolated from one gram of human tissues,

Table 7.1
Comparison of non-professional and professional histamine (HA) producing cells.

Property	**Non-professional cells**	**Professional-cells**
Rate of HA synthesis	Slow	100-1000 times faster
HA synthesizing enzyme	74-kD HDC	53-kD HDC
Handling of synthesized HA	Released to the cytoplasm	Stored in secretory granules
Requirement for release	Concentration gradient, need to overcome the resting cell membrane potential	Stimulus required[1]
Release of HA	Passively along the concentration gradient via equilibrative uniporters[2]	Actively via exocytosis
HA concentrations attained	Nanomolar	Micromolar
Concentration range	Probably very narrow	First very high, then rapidly falling off
HA concentration modulation in the active concentration range	HA from food and microbes	Burst release from professional cells
Spatial range	Universally present in the body	Usually local, at the hot spot
Time range	Long-term	Short-term
Active during	"Off-state" (most of the time)	"On-state" (short bursts)
Receptors utilized	High affinity H_3R, H_4R	Low affinity H_1R, H_2R (H_3R, H_4R)
Handling of extracellular HA excess	Degradation by DAO[3],uptake via equilibrative uniporters[1]	Degradation by DAO[3], uptake via equilibrative uniporters[1]
Handling of HA taken up	Degradation by HNMT[4]	Reuptake to granules by VMAT[5]

[1]Mechanical, HA releasing foods, lectins, substance P and other tachykinins, allergens (acting on IgE-sensitized cells)
[2]Organic cation transporter 2 (OCT2, SLC22A2), organic cation transporter 3 (OCT3, SLC22A3), plasma membrane-associated monoamine transporter (PMAT, SLC29A3, equilibrative nucleoside transporter 3/ENT3) are potential candidates
[3]Diamine oxidase
[4]Histamine N-methyl transferase
[5]Vesicular monoamine transporter 2

MCs can have the capacity to produce micromolar histamine concentrations in tissues. However, even in optimally timed plasma samples from patients with anaphylaxis, histamine levels may be within normal limits (Simons et al. 2008). At present, there is lack of technological *in vivo* histamine sensors which could continuously follow locally histamine concentrations in human tissues. Histamine burst by exocytosis release can be triggered bysuch as neuropeptides, some food items, helminths, mechanical stimulation, wounding or allergic reactions. It can lead to micromolar histamine concentrations, as estimated through the pK_i values of H_1R (4.2) and H_2R (4.3), of up to 50-100 μM (Lim et al. 2005), although the local and temporary nature of professional histamine release and its subsequent rapid degradation and dilution are reflected in the relatively low 0.6 μM concentration measured in bronchoalveolar lavage fluid after Ag challenge (Calhoun et al. 1998).

Non-professional histamine producing cells produce and secrete 100- to 1000-fold lower concentrations of histamine and their histamine release (or uptake) is mainly driven along the concentration gradient between the intra- and extracellular space (Jutel et al. 2009; Tanaka & Ichikawa 2011; Kubo & Nakano 1999). Effective concentrations at this low histamine level are nanomolar, again mainly estimated through the pK_i values of H_3R (8.0) and H_4R (8.2).

7.2.3. Histamine Receptors in General

Histamine affects cells via four histamine receptors which all belong to the large G-protein-coupled receptor (GPCR) family. These receptors contain extracellular, transmembrane and intracellular domains where the ligand-binding structures are located on the extracellular part of the receptor molecule. The intracellular domain of the receptor binds various G-proteins which are dependent on the receptor and cell type. Thus, G-proteins mediate the extracellular signals affecting the extracellular part of the receptor intracellularly via downstream activation and/or inhibition of various intracellular signal transduction cascades. This ultimately leads to various rapid events such as release of intracellular calcium stores and the phosphorylation ordephosphorylation of proteins as well as additional slow changes in gene transcription events.

7.2.3.1. Histamine H_1 receptors

Although the first antihistamines, were regularly used by the 1940s, it took until 1966 before Ash *et al.* introduced the term H_1R to describe a subgroup of histamine sensitive receptors which could be blocked by the first generation antihistamines or H_1R-antagonists (Ash & Schild 1966). Since H_1R-antagonists only inhibited some of the histamine effects, such as contraction of the smooth muscle but did not affect some others, like production of gastric acid, led the authors to the conclusion that the effects of histamine must be mediated by at least two different receptors. On the other hand, many of the so called antihistamines also inhibit muscarinic, α-adrenergic and serotonin receptors and modulate ion channels, which is reflected insuch as adverse effects such as dry mouth and other anticholinergic effects, orthostatic hypotension, increased appetite and arrhythmias, like torsades de pointes.

H_1R is best known for its localization on the vascular endothelial and smooth muscle cells, whereas the best known localization of H_2R is on the parietal cells of the stomach. High concentrations of histamine are already known to be responsible for vasodilation, increased vascular permeability, the expression of adhesion molecules on vascular endothelial cells, contraction of the bronchial smooth muscle, stimulation of gastric acid production, neurotransmission in the central nervous system (CNS) and even modulation of immune-inflammatory responses via esosinophils, basophils, neutrophils/macrophages, MC, DC, T- and B-lymphocytes (Akdis & Blase 2003; Simons 1999; Haas et al. 2008; Huang & Thurmond 2008 ; Feng et al. 2013). Many of these already known histamine effects are rather acute and are apparently mediated by the classical H_1R and H_2R, where the requisite, high histamine concentrations are only achievable transiently following regulated, professional burst release. High-affinity H_3R and H_4R react to basal histamine concentrations, which are only 1/1000 – 1/10000 of the concentrations effectively activating H_1R and H_2R.

H_1R mRNA is encoded by HRH1 gene in chromosome 3p25. The H_1R protein is composed of 487 amino acids (de Backer et al. 1998). H_1R is coupled via $G_{q/11}$ protein to activation of phospholipase C (PLC). This produces 1,4,5-inositoltrisphosphate (IP_3) and 1,2-diacylglycerol (DAG), which cause calcium (Ca^{2+}) mobilization from intracellular stores and activation of Ca^{2+} and/or DAG-dependent protein kinase C molecules. H_1R activation can also activate phospholipase A_2 (which releases arachidonic acid from membrane phospholipids), phospholipase D (which releases phosphatidic acid from phospholipids) and transcription factor Nuclear Factor kappa K (NFκB) (Leurs et al. 2002). Some of the H_1R on the cell surface membrane shows intrinsic (constitutive) activity without any histamine ligand binding, but the physiological consequences of this observation have not been clarified yet. Increased expression of H_1R has been noticed in several diseases, such as allergic rhinitis, rheumatoid arthritis (RA) and in some areas of the brain as a result of stroke.

Histamine promotes inflammatory responses and cytokines and some of these effects are mediated via H_1R (Dy & Schneider 2004). H_1R-antagonists, have shown effectiveness in the treatment of allergic rhinitis, conjunctivitis and skin reactions caused by exposure to allergens. CNS-related side effects, in particular sedation, have limited the use of the lipophilic first generation H_1R-antagonists. More modern second generation H_1R-antagonists are more hydrophilic molecules so their penetration through the blood-brain-barrier (BBB) and sedative effects are much smaller. Sedative side effects have been used in the treatment of allergic diseases associated with sleep disturbances. Cardiomyocytes express H_1R and therefore H_1R-antagonists could affect heart muscle and function, nevertheless no cardiotoxic effects have been described by the second generation H_1R-antagonists (Simons 2004).

By and large, H_1R activation seems to increase Ag presentation and costimulation to induce Th1/M_1 type of immune responses, associated with e.g. IL-12 and IFN-γ production, at the same time when it is blocking humoral immunity and IgE production (Simons 2004).

7.2.3.2. Histamine H_2 Receptors

In the 1980s, the introduction of H_2R-antagonists into the clinic revolutionized the treatment of gastric and duodenal ulcer and gastro-esophageal reflux disease (GERD). Because H_2R-antagonists strongly inhibited the production of gastric acid ("no acid, no ulcer"), the earlier, mainly surgical treatments, like pyloroplasty or Billroth I or II type gastric resections, without or with vagotomy, were rapidly abandoned.

The H_2R is coded by HRH2 gene in 5q35.2 and the corresponding protein is composed of 359 amino acids (Traiffort et al. 1995) (Entrez Gene, HRH2). The H_2R is coupled to $G_{\alpha s}$-type G-protein. Binding of agonist to H_2R stimulates adenylate cyclase, cAMP, protein kinase A (PKA) and cAMP response element-binding protein (CREB), modifying both phosphorylation and more slowly gene transcription. Activation of H_2R stimulates c-Fos, c-Jun (which together form the activating protein-1 (AP-1) transcription factor), protein kinase C and p70S6kinase (Dy & Schneider 2004; Jutel et al. 2009).

H_2R is found on parietal cells, but also on epithelial cells, endothelial cells, smooth muscles cells surrounding blood vessels, bronchiole, stomach and bowel, neurons, cardiomyocytes and cells of the immune system (neutrophils, eosinophils, monocyte/macrophages, DC, T- and B-lymphocytes). In the CNS, H_2R is a postsynaptic receptor often occurring in the same locations as H_1R, which may have synergistic effects between them (Haas et al. 2008). H_2R knockout mice display impairment of cognitive functions and changed nociception. In contrast to H_1R activation, H_2R activation diminishes eosinophil and neutrophil chemotaxis (Simons 2004). H_2R-agonism diminishes production of IL-12 by monocyte-derived DCs (MoDC), where the cytokine polarizes T_h0 to Th1 cells, but this diminished IL-12 production can be prevented by H_4R agonists (Gutzmer et al. 2005). Thus, H_2R-activation seems to downregulate Th1/M_1-type cell-mediated immuno-inflammatory responses and support Th2-inducing DCs and humoral immunity (Simons 2004).

Physiological effects mediated by H_2R include relaxation of the smooth muscle layer of blood vessels, increased vascular permeability, increased production of mucus and bronchodilation, and positive chrono- and inotropic effects on the heart (Simons 2004). CNS-effects are uncommon, perhaps in part because H_2R-antagonists are not lipophilic and do therefore not easily penetrate the BBB.

In 1975, Clark *et al.* discovered that low concentrations of histamine are chemotactic to eosinophils, whereas high histamine concentrations inhibit eosinophil chemotaxis, allowing them to participate in allergic inflammatory processes (Clark et al. 1975). After the discovery of H_4R, which has a very high affinity for histamine, it has been shown that low, 10-100 nM histamine concentrations cause changes in the form of the cell, upregulation of adhesion molecules (CD11b) and polymerization of the actin cytoskeleton via H_4R (Buckland et al. 2003; Ling et al. 2004; Barnard et al. 2008). The low, nascent histamine concentrations that are generated and modulated by non-professional histamine synthesis have been shown to cause maturation of the secretory granules of MC, migration of MC and immunomodulation of DC, monocytes/ macrophages, Kuppfer cells and basophils (Ichikawa & Sugimoto 2010).

The high-affinity receptors can therefore be proposed to mediate the effects of low concentrations of histamine, such as far away from the acutely inflamed hot spots (or around it), where they can downregulate inflammation, at the same time as the histamine concentration gradient acts as a chemotactic stimulus attracting leukocytes to the site of ongoing inflammation. Perhaps even more importantly, the high-affinity receptors can be proposed to mediate important processes in between the histamine burst releases when influenced by the non-professionally produced, low nanomolar nascent histamine fluctuating concentrations or modulated by food-derived or microbially-produced histamine. It is not yet known how high micromolar histamine concentrations affect the highly sensitive histamine sensors H_3R and H_4R, which are already stimulated by nanomolar histamine concentrations and/or to a large extent spontaneously active.

Histamine can also bind to non-conventional proteins, such as some enzymes of the cytochrome P450 superfamily, histamine transporters and proteins secreted by some insects, like nitrophorins and lipocalins (Dy & Schneider 2004). Lipocalins are highly specific histamine binding proteins, which are secreted by different types of ticks such as the hard tick (Ixodidae) and probably help the ticks to resist tick-bite-induced, locally-elicited host responses (Paesen et al. 1999).

7.2.3.3. Histamine H_3 Receptors

H_3R (Lovenberg et al. 1999) is encoded by HRH3 gene in 20q13.33 (Tardivel-Lacombe et al. 2001). H_3R protein is composed of 445 amino acids, but over 20 different isoforms are known (Cogé et al. 2001; Wellendorph et al. 2002; Wiedemann et al. 2002; Hancock et al. 2003), which suggests tissue, cell and context-dependent fine tuning in ligand binding and signal transduction.

One prominent H_3R location is on the histaminergic neurons in the CNS and in peripheral nervous tissues. In the CNS, H_3R acts as a presynaptic autoreceptor, where its activation inhibits neuronal histamine release (Haas et al. 2008; Passani & Blandina 2011). H_3R also occurs as a presynaptic receptor on other non-histaminergic neurons, where they are referred to as a heteroreceptor. In these cells, H_3R activation also inhibits the release of the corresponding neurotransmitters, such as acetylcholine, gamma-aminobutyric acid (GABA) and glutamate. All histaminergic CNS neurons, about 64,000 in the human body, originate from the tuberomamillary nucleus, from which their axons pass to most other parts of the brain. Additionally, histamine released from the vagal complex of the *nucleus tractus solitarii* can probably modulate immune responses via H_3R (Haas et al. 2008).

H_3R is coupled to the $G_{i/o}$ protein (Cogé et al. 2001; Bakker 2004; Leurs et al. 2005), which inhibits adenylate cyclase, but H_3R activation can also activate mitogen-activated protein kinase (MAPK), increase intracellular calcium and activate phospholipase A_2.

Histamine in the CNS is associated with regulation of the sleep-wake state, ability to concentrate, learning and memory as well as other phenomena. It has therefore been natural to assume that H_3R-modulation could affect diseases which impair these functions, such as memory disorders in Alzheimer's disease, narcolepsy and attention deficit/hyperactivity disorder (ADHD). Additional hopes have been raised of its potential in the treatment neuropsychiatric diseases, such as schizophrenia.

H_3Rs are also found in the peripheral autonomic and somatosensory nerve terminals, which in turn can affect parasympathetic, sympathetic and sensory functions, such as pain and itch (4).

No drugs targeting mainly H_3R are at clinical use yet, but effects of H_3R inverse-agonists are studied in narcolepsy, sleep disturbances in Parkinson's disease, ADHD, schizophrenia and Alzheimer's disease as Phase II-III clinical studies (Passani & Blandina 2011). It is worthwhile to mention that betahistin, which is used in Ménière's disease, is a weak H_1R agonist, but a much more potent H_3R inverse agonist. It was originally thought that H_1R agonism relaxes pre-capillary sphincter muscles in small arterioles, thus improving the capillary circulation in cochlea (Ihler et al. 2012). Maybe the inverse agonist effect of H_3R autoreceptors on neurotransmitter release and afferent sensory signaling in the vestibular nucleus and cochlea are more important in diminishing vertigo in Ménière's disease (Desmadryl et al. 2012).

7.2.3.4. Histamine H_4 Receptors

H_4R (Oda et al. 2000; Liu et al. 2001; Morse et al. 2001; Nguyen et al. 2001; Zhu et al. 2001; Nakamura et al. 2002) is encoded by HRH4 gene 18q11.2 (Cogé et al. 2001). The full length H_4R is composed of 390 amino acids, but two non-signaling receptor isoforms have been identified, H_4(302 aa)R and H_4(67 aa)R (van Rijn et al., 2008). Co-expression of these different isoforms may participate in downregulation of the full-length H_4R on cell membrane, probably via receptor oligomerization. The amino acid sequence of H_4R has a 37 % homology with that of H_3R (58 % in the transmembrane region) (van Rijn et al. 2008). H_4R is coupled to the pertussis-toxin sensitive inhibitory $G_{i/o}$ protein (Gantner et al. 2002). The identification of JNJ-7777120 and other potent and selective H_4R

antagonists (Thurmond et al. 2008) made it possible to elucidate the roles of the H_4R in a variety of allergic and inflammatory processes.

Intracellular signal transduction of $G_{i/o}$-coupled H_4R occurs via inhibition of adenylate cyclase and lowered cAMP (Oda et al. 2000; Nakamura et al. 2000), increased phosphorylation of mitogen-activated protein kinase (MAPK) (Morse et al. 2001) and activation of PLC followed by increased cytosolic calcium and formation of DAG (Oda et al. 2000; Nguyen et al. 2001; Zhu et al. 2001). In a superficial way, the signal transduction cascades utilized by the novel H_3R and H_4R have resemblance, but this does not mean overlapping functions, because the cells and the cellular processes these receptors are coupled show wide variation. Due the common denominator of histamine for all histamine receptors known at present, combined with their generally prominent intrinsic activity, the system may maintain a rather complicated balance composed of interactions at the pre-, receptor- and post-receptor levels in cellular networks of various kinds (Nguyen et al. 2001; Xu et al. 2008; Hishinuma et al. 2010; Shi et al. 2012). This makes it difficult to extrapolate from simple *in vitro* experiments to human (patho)physiology.

H_4R has a relatively high intrinsic constitutive activity. In spite of this half-maximally active state, H_4R reacts to histamine concentrations as low as 5-10 nM (Thurmond et al. 2008; Jablonowski et al. 2003; Akdis & Simons 2006).

Expression of H_4R has been well described in haematopoietic cells. Both myeloid and lymphoid cells express H_4R, in particular eosinophils (Oda et al. 2000; Buckland et al. 2003; Ling et al. 2004; Barnard et al. 2008), MC (Hofstra et al. 2003; Lippert et al. 2004), DC (Zhu et al. 2001; Damaj et al. 2007; Geng et al. 2012), Langerhans cells (LC, dendritic cells in the epithelium of the skin and mucosa) (Gschwandtner et al. 2010), $CD4^+$ (Zhu et al. 2001; Sugata et al. 2007; Gutzmer et al. 2009) and $CD8^+$ T-cells (Zhu et al. 2001; Gantner et al. 2002), $CD16^+$ natural killer (NK) cells and CD1d-lipid complex binding invariant natural killer T-cells (iNKT) (Damaj et al. 2007; Leite-de-Moraes et al. 2009). H_4R but not H_3R is expressed on neutrophils (Oda et al. 2000; Zhu et al. 2001) and in experimental studies H_4R-agonism seems to mobilize neutrophils from the bone marrow (Takeshita et al. 2003, Takeshita et al. 2004). On the other hand, H_4R-agonism diminished monocyte-mediated production and secretion of CCL2 chemokine, also known as monocyte chemotactic protein-1 (Dijkstra et al. 2007). Low H_4R mRNA levels have been reported in B cells but the effects of receptor activation on B cells are not known yet (Zhu et al. 2001), although B_{reg} chemotaxis and IL-10 and/or TGF-β production to alleviate arthritis (Mauri et al. 2003), systemic lupus eryhtematosus (SLE) (Lenert et al. 2005), inflammatory bowel disease (IBD) (Mizoguchi et al. 2002) and experimental allergic encephalopathy (EAE) (Mann et al. 2007) by these cells is one possibility.

Except for immune cells, H_4R has been demonstrated thus far in skin fibroblasts and keratinocytes, fibroblast-like type B and macrophage-like type A synovial lining cells and chondrocytes (Yamaura et al. 2013) as well as acinar and ductal epithelial cells in human salivary glands (Stegaev *et al.* 2012). Low and somewhat inconsistent expression of H_4R mRNA has been reported in osteoclasts, but no functional effects have been described (Biosse-Duplan et al. 2009). Expression of H_4R in neuronal cells has been a matter of debate, but more recent studies report H_4R in neurons of the nasal mucosa (Nakaya et al. 2004), primary vestibular neurons in rats (Desmadryl et al. 2012), neurons of the anterior horn of the spinal gray matter in mice (suggesting expression in motoneurons) (Lethbridge et al. 2010), in dorsal root ganglia in mice (suggesting expression in primary sensory neurons) (Kajihara et al. 2010), and in CNS in humans and rodents (Zhu et al. 2001; Connelly et al. 2009; Strakhova et al. 2009; Shan et al. 2012). H_4R-agonist 4-methylhistamine (4-MeHA) caused hyperpolarization of neurons of the somatosensory cortex in mice (Connelly et al. 2009). H_4R, together with H_1R and H_3R, has been reported in dorsal root ganglion of skin-specific sensory neurons in mice (Rossbach et al. 2011). H_4R stimulation led to an increase in cytosolic calcium. In experimental studies, H_4R-antagonist has an anti-pruritogenic effect, which could be due to a direct inhibition of the histamine-sensitive free nerve terminals in the skin (Thurmond et al. 2008). H_4R mRNA, again together with H_1R and H_3R mRNA, has been shown in human enteric neurons in the Meissner's submucosal neural network, where neurons were excited by H_4R stimulation (Breunig et al. 2007).

7.3. Histamine Transport

At physiological pH, histamine is a monocationic biogenic amine. Newly synthesized histamine is under homeostatic conditions released from the non-professional histamine producing cells either via OCT2 and 3, also known as solute carriers SLC22A2 and SLC22A3 respectively or PMAT, also known as solute carrier SLC29A3 or equilibrative nucleoside transporter 3, (ENT3).

After burst release of histamine, both professional and non-professional cells have an outside-in histamine gradient. Both utilize OCT2/3 (SLC22A2/SLC22A3) or PMAT (SLC29A3 or ENT3). PMAT transports other monoamines like serotonin and dopamine more effectively, with low affinity and high capacity. Equilibrative uniporters are not ATP-dependent active transporters, able to work against a concentration gradient, but let histamine and some other monovalent cations/biogenic amines pass along their concentration gradient. Histamine degranulation produces high histamine concentrations, perhaps up to 50 or even 100 µM. Part of the excessive histamine load is extracellularly degraded by diamineoxidase (DAO), which in general is considered to be responsible for 15-30% of total histamine degradation. Locally released histamine is rapidly diluted by diffusion and carried away by body fluids.

In cultured human HSG cells, 100 nM of radio-labelled histamine is enough to stimulate OCT3-mediated cellular uptake of histamine. On the other hand, after loading of these cells with 100 nM histamine, these cells release histamine from the cytoplasm to culture medium (Stegaev et al., 2013). This shows that OCT3 is functioning at rather low histamine concentrations, driven rather by the concentration gradient (and the negative resting cell membrane potential) than its high histamine affinity.

Extracellular *de novo* synthesized or dietary histamine can also be taken up by cells, which enables intracellular degradation (Duan & Wang 2010; Tanaka & Ichikawa 2011).

Both HDC isoforms are located on the endoplasmic reticulum, but only the 53-kDa HDC localizes close to intracellular granules, probably to facilitate their storage of the newly synthesized histamine via vesicular monoamine transporter 2 (VMAT-2) -mediated internalization (Tanaka & Ichikawa 2011).

7.4. Histamine Degradation

After burst release of histamine from both professional and non-professional cells, a steep outside-in downhill histamine gradient is formed. Both cell types utilize then the aforementioned histamine channels for cellular uptake, but differ in their handling of intracellular histamine. In professional histamine producing cells, some of the intracellular histamine is subjected to granular re-uptake via VMAT2 for recirculation. Apart from histamine synthesis and release by host and commensal microbial cells, as well as the histamine content of food, histamine concentrations in the body are dependent on the mostly extracellular DAO-mediated and intracellular HNMT-mediated histamine degradation processes mentioned above, which degrade histamine into imidazole acetate and N-methylhistamine, respectively. After cellular synthesis, DAO is initially located in vesicles located close to cell membranes, which upon stimulation release DAO into the extracellular space (Maintz & Novak 2007; Jutel et al. 2009). In mammals, DAO has been localized to the small intestine, ascending colon, placenta and kidneys. In renal tubular cells, DAO specifically localizes to the central clear matrix of the peroxisomes. Extracellular release of DAO and its activity in the extracellular space can be inhibited by various drugs, substances and alcohol. DAO-mediated extracellular degradation of histamine decreased in this way can increase histamine concentration and may cause histamine intolerance, depending on the individual and the context-dependent histamine response threshold (Maintz & Novak 2007). However, it is apparent that in the low nanomolar range levels of histamine concentration is continuously subjected to changes in the human body and sometimes peak to cause symptoms. In that respect, the basic non-professional histamine synthesis and intracellular degradation may exert important balancing ("buffering") effects on the cellular histamine micromilieu.

7.5. Constitutive Receptor Activity

H_4R has a high intrinsic constitutive activity meaning that it can exist in an active signaling state without a ligand bound to it. Normally ligand-free G protein-coupled receptor is in an inactive R state and assumes an active R* conformation (R*) first after binding its specific ligand, but intrinsically active receptors have assumed a similar activated H_4R* state in the absence of a receptor-bound ligand (Milligan et al. 1995; Leff 1995). The constitutive or basal H_4R activity can therefore be described by the ratio between R* and R, and this basal activity is considered to be approximately 50% of the total H_4R activity. In spite of this half-maximally active state H_4R reacts already to histamine concentrations as low as 5-10 nM (Thurmond et al. 2008; Jablonowski et al. 2003; Akdis & Simons 2006). This relatively high constitutive activity may suggest that there is a low and potentially harmful threshold value below which the cells cannot go without damage, and to avoid such deprivation of histamine effects, a quite large pool of the receptors is maintained in an active transducing state even in the absence of any ligands.

Another important consequence of the constitutive receptor activities is that various receptor ligands can act as agonists, neutral antagonists or inverse agonists. Binding of natural or synthetic agonists favour the active conformation and subsequent signal transduction. Binding of a neutral antagonist to the receptor does not favour active or inactive receptor conformation alone and therefore does not by itself affect constitutive signalling. However, a neutral antagonist competes with an agonist for receptor binding, and therefore inhibits agonist-mediated receptor activation and signal transduction. Binding of inverse agonists (which earlier were also called antagonists) stabilizes and thus promotes the inactive receptor conformation, which leads to a shift of spontaneously active receptors to the inactive receptor pool and thus inhibits the constitutive receptor activity. This way, an inverse agonist diminishes the spontaneous receptor activity below the regular constitutive signalling level. All four histamine receptors have been shown to be generally constitutively active, but this is probably particularly important for H_3R and H_4R (Smit et al. 1995; Bakker et al. 2000; Morisset et al. 2000; Oda et al. 2000)., However, the importance of the balance between the H_1R-H_4R constitutive signalling pathways and the "off-state" and "on-state" function of various cells remains an interesting but open question.

7.6. Recognized Histamine-associated Diseases

Most known histamine-associated diseases in man seem to relate to enforced expulsion of micro- and macrostructures from the human body, such as pollen (~10-1000 µm) or parasitic worms (largest of which is broad tapeworm, which grow usually 5-6 meters long, being then almost as long as the average human small intestine)(Pulendran & Artis, 2012). Pollen, eggs or larvae can intrude the human body via air (pollen via conjunctivae, nasal mucosa, airways), by mouth to the bowel and other internal organs (eggs ot tape worms etc.) or through the skin from where they can also reach the bowel and other internal organs (e.g. larvae in schistosomiasis). Many times they can be expelled, in part via burst released histamine triggered by lectin-like, MC activating structures (Imamura et al. 1996; Pramod et al. 2007), which is followed by tearing, sneezing, coughing or expulsion from the bowel. If the condition develops further, these reactions develop, like wound healing, to a robust type 2 rather than type 1 immune responses. This engages DC2, Th2 and M2 but for the initation of the type 2 response innate MCs and basophils and their cytokines such as IL-5 and IL-13 (later produced by Th2 cells together with IL-4) play important roles (Gordon 2003). Parasites participate actively in the induction of the host response by producing excretory/secretory products (proteinases, chitins) and egg Ag, which via danger receptors and T- and B cells receptors (TCR/BCR) interact with DC2, MCs, basophils, eosinophils, goblet cells, enterocytes, M2, Th2 and B cells (Wilson et al., 2007). In more chronic stages, the intruder can in part be isolated to granulomas, fibrosis and calcified cysts. One promising strategy in the treatment of these diseases is vaccination aiming to strengthen type 1 responses or immunotherapy aiming to induce Treg and suppression/tolerance, in which H_4R may plays a role.

Before the host response caused by direct irritation and naïve inflammatory responses e.g. of the airways, such as sneezing or coughing, can be strengthened by IgE-enhanced, MC- and basophil-mediated immediate-type I immune responses, the body must be sensitized. This engages DC-T-cell interactions, which are in part regulated by nanomolar histamine and H_3R and H_4R. If the body in spite of the initial non-immunological, irritative and naïve "get rid of the intruder"-type host responses cannot get rid of the intruder, its components are taken up and apart by the professional Ag-processing and -presenting DCs. Foreign Ag (linearized epitopes) are then in secondary lymphatic tissues presented in the context of major histocompatibility complex class II (MHC II) to TCR. For the T-cell to be activated by this Ag presentation, it needs also costimulatory signals, which are produced as a response to exogenous or endogenous (tissue damage) danger signals. Also B-cells present these Ag in the context of their surface MHC II to T-cells. If the Ag presenting B-cell recognizes an epitope of the same Ag in its soluble form by its surface immunoglobulin (*viz.* BCR, which was used to internalize the Ag in the first place), it becomes a plasma cell producing soluble antibodies. This is a joint result of the T-cell help, which provides co-stimulatory signals for the B-cell activation, and the Ag presentation to the B-cell via BCRs. In type 2 IL-4 and IL-13 driven response the immunoglobulin isotype switches from surface IgM and surface IgD to soluble IgE or IgG_4. This might in part be related to the route of entrance of the irritatant immunological stimulus and its character. Sensitization and production of IgE help to expel the irritant even more effectively because the weaker and direct host responses are now enforced by an effective immunological activation of MCs and basophils. If this IgE production is local and the exposure only temporary, it could serve a useful purpose. However, if the IgE production is excessive, IgE spreads to more distant sites and the condition transforms to a disease-like and disturbing state known as allergy.

Thus, DC-lymphocyte interactions, sensitization and delayed T-cell-dependent immunological reactions related to H_3R and H_4R preceed the development of immediate, allergic repellent host responses related to H_1R and H_2R. It has long been known that this sequence is nicely depicted in reactions to mosquito bites (Mellanby 1946). Non-exposed and non-sensitized individuals do not produce any macroscopic skin response (stage I). After a number of mosquito bites the Ag dose is enough to lead to a delayed response apparent after one to several days after the bite (stage II). If the exposure to mosquito bites continues, an immediate response starts to develop to the bite in some 15 minutes and remains symptomatic for 1-2 hours (stage III). If the exposure to the salivary proteins released by the mosquito during its blood meal continues, the delayed-type response disappears and only the immediate-type response continues (stage IV); possibly as a result of development of inducible regulatory T-cells (iTreg) and blocking IgG_4. No more IgE antibodies are produced but some IgE still passively maintains the immediate type I reaction mode, a similar phenomenon as is seen in the Prausnitz-Küstner test. Upon further exposure, Ag-specific IgE diminishes and this leads to complete tolerance (stage V) so that in Lapps (Sami people) mosquito bites do not cause any response at all.

7.7. Immune Defence and Autoimmune Diseases

Self-tolerance fails severely in autoimmune diseases, which leads to cell and tissue pathology, patient symptoms and other clinical signs and findings ("horror autotoxicus").

According the clonal selection theory of immunity, central tolerance of the bone marrow-derived $CD4^-$, $CD8^-$ and TCR^- progenitor cells is established during their differentiation in the thymus. In cortical thymus, T-cell receptor gene rearrangement occurs followed by positive selection (clonal amplification) of properly finely-tuned lymphocytes, showing moderate self (MHC I or II) reactivity. T-cells showing overly weak reactivity against self-MHC molecules are subjected to negative selection via death by neglect, as they respond too weakly for their function even to Ag-loaded MHC complexes. Similarly, T-cells reacting too strongly against self-MHC antigens undergo activation-induced apoptosis, because they would be able to cause rejection of autologous tissue.

From the cortical thymus, $CD4^+$, $CD8^+$ double-positive T-cells pass to the thymic medulla, where they are now confronted with various self-Ag loaded to MHC, produced ectopically *in situ* as a result of molecules like AIRE (autoimmune regulator): very weakly reactive T cell clones die out of neglect, weakly self-reactive T cell clones can become conventional effector T cells (T_{eff}), moderately self-reactive and costimulated T cell clones with a quite wide TCR diversity become stable natural regulatory T cells (nT_{reg}; $CD4^+$, $CD25^+$, $Foxp3^+$ nT_{reg}) and strongly self-reactive T cell clones undergo activation induced apoptosis. Roughly 5% of the total T-cells subjected to clonal selection are then released as naïve T cells into the periphery. What is often forgotten is that the thymus undergoes involution after its peak development in puberty. Body is then largely dependent on the 10^8 T naïve cell clones produced, each clone consisting of 1000-10000 cells, depending on whether the total number of T cells is estimated to 10^{11} or 10^{12} (Arstila et al. 1999; Arstila et al. 2000). Interestingly, most of these T-cells never meet their specific antigenic determinant, as memory T-cells only make up < 1 % of the total cell diversity.

Due to various harmful events, such as trauma leading to a release of potential auto-Ag, it is also necessary to establish peripheral tolerance. This seems to be in part based on the danger signals. In the absence of danger signal-induced costimulatory factors, presentation of Ag alone leads to apoptosis of the lymphocyte (dying away), anergy of the lymphocyte (weaning away) or formation of $CD4^+$, $FoxP3^+$ inducible regulatory T cells (iT_{reg}, active resistance), preferably in the presence of a low dose of a high affinity antigenic determinant together with CD28 co-stimulation. iT_{reg} cells are important for example for tolerance against food Ag and commensal flora, where the development of these populations require a suitable cytokine milieu, in particular TGF-β (Yuan et al. 2012).

Anti-idiotypic responses also provide a means to control immune activation. All TCR and BCR contain highly variable but always clone-specificidiotopes, which are the structures responsible for the binding of epitopes (antigenic determinants). An oligoclonal, Ag-driven response defines a set of idiotopes, which is collectively referred to as an idiotype. For down-regulation of the immune response, anti-idiotypic TCRs and antibodies, representing "internal images" of the Ag, are produced. They bind to the antigen-specific TCR and BCR and can block Ag binding and responses. However, anti-idiotypic antibodies are not always restricted to blocking, as histamine burst release stimulating anti-idiotypic antibodies against IgE have been reported (Boutin et al. 1993).

The peripheral tolerance presented above is reactive, but it is also possible that some degree of proreactive autoimmunity is necessary for the maintenance of the immune balance. Self reactive lymphocyte clones could be maintained for active surveillance to recognize and respond to biomarkers, which reveal the site of the offence through tissue specific Ag, the type of the offence (cytokines, tissue damage induced inflammation, etc.) and the stress associated with it (stress proteins and the condition of the local cells) (Cohen 2007).

Because central selection is not 100% effective, autoimmunity could be based on *molecular mimicry* between microbial and self-Ag. Even in the absence of any initial cross-reactivity, a persistent infection can lead to *epitope spreading*, which reaches a point in the Ag where self-mimicking, cross-reactive epitope scripts are met. Professor Shunichi Shiozawa has introduced a model in which a persistent release of self-Ag even in the absence of infection (as a result of auto-Ag release upon exposure to UV light in SLE, for example) overstimulates the immune system so the *self-organized criticality* limit is surpassed, causing immune receptors to undergo *de novo* revisions leading to new reactivities which cover self-Ag (Tsumiyama et al. 2009). In *bystander activation* the antimicrobial response occurs in an infected and damaged tissue containing self-Ag in an environment invaded by mature and competent antigen presenting cells (APC) in a cytokine milieu favoring expression of co-stimulatory molecules. This leads to an immune attack against an innocent bystander self-Ag. Self-tolerance can be maintained by apoptosis which produces dominant tolerogenic antigenic determinants that only stimulate synthesis of anti-inflammatory cytokines. If this normal Ag processing is diverted as a result of tissue pathology, the very same Ag can instead produce *hidden (cryptic) immunogenic epitopes*, creating new self-derived molecules which lymphocytes did not meet during its education, eliciting an immune response.

7.8. Mast Cells and Autoimmunity

MCs are derived from the hematopoietic stem cells of the bone marrow, their precursors circulate in blood and they differentiate into granule-loaded "explosive" mature MCs in different tissues (Abraham & St. John 2010). In human tissues, tryptase-negative precursors or proliferation of MCs seem to be rare (Čeponis et al. 1998). Some tissue specificity in the final maturation is suggested by the dichotomy of MCs in rodents to mucosal MCs positive for mast cell tryptase and connective tissue MCs also positive for mast cell chymase. In contrast to murine MCs, human MCs do not produce IL-4 or express CD14 and IL-3R, FcγRI or functional TLR receptors are expressed in very small amounts, but they are strong IL-5 producers (Bischoff 2007). IL-5 stimulates B cell proliferation, increases immunoglobulin secretion and activates eosinophils.

MCs are best known for their role in immediate type I hypersensitivity reactions. IgE molecules with specificity against allergens like pollen) bind to MC surface FcεRI. When these sensitized MCs encounter allergens, cross-linking of the cell surface-bound IgE molecules leads to MC degranulation and activation. MCs can be stabilized by the drug sodium cromoglycate but also naturally by inducible OX40 on the surface of activated iT_{reg} cells, which interacts with OX40L on the MC surface (Mekori & Hershko 2012).

MCs contain also other preformed granular mediators like mast cell tryptase, heparin, IL-5, GM-CSF (granulocyte-monocyte colony stimulating factor), TNF-α, IL-17A and IL-17C. These cells can synthesize various eicosanoids as well as chemo- and cytokines. In a MC-dependent mouse model of zymosan-induced peritonitis, an H_4R-antagonist blocked neutrophil infiltration (Thurmond et al. 2004). Due to their role in DC maturation, polarization, migration, Ag presentation and other immunomodulatory processes, MCs participate in many autoimmune diseases, such as multiple sclerosis, bullous pemphigoid and RA, as demonstrated by lack of arthritis in MC-deficient mouse model and in RA by the improvement of type c-kit tyrosine kinase inhibitors (Eklund & Joensuu 2003).

MCs can produce IL-12 and IFN-γ, which can polarize D0, M0 and Th0 cells to type 1 cells; human MCs contain IL-5, both preformed and de novo produced, which can contribute to the polarization of other cells to type 2 cells; IL-6 and TGF-α, which can polarize cells to type 17 cells; and finally TGF- and IL-10, which can help polarize cells to T_{reg}. The outcome of the process to a large extent is context dependent. In particular, histamine decreased the production of poly I:C-induced, type 1 polarizing IL-12 by monocyte-derived DCs upon H_4R-agonist pretreatment (Gutzmer et al. 2005). Histamine at high concentrations also diminished lipypolysaccharide (LPS)-induced IL-12 production and promoted type 2 polarisation of DC and T_h0 cells via H_1R and H_2R (Caron et al. 2001; Kitawaki et al. 2006). MC can even present Ag to $CD4^+$ T-cells (Fox et al. 1994) and cross-present Ag to $CD8^+$ T-cells (Stelekati et al. 2009).

7.9. Dendritic Cells and Autoimmunity

H_4R mediates chemotaxis of human human myeloid DCs (Gutzmer et al. 2005). DCs are professional cells responsible for Ag sampling and surveillance. They capture (for which the outspread dendritic morphology is very suitable), transport (often through losing the dendritic morphology) and proteolytically process Ags and present it to T-cells after induced maturation, such as by danger signals, certain cytokines and immune complexes. DCs are often referred to as professional APCs, due to their critical role in primary immunization and sensitization, as no other cells are able to activate naïve T-cells (Steinman & Banchereau 2007). This functionality may depend on their capacity to migrate to secondary lymphoid tissue where they effectively interact with densily packed T-cells as interdigitating DCs in the paracortical T-cell area.

Ag presentation by DCs can result in induction of tolerance or immune activation. The two main categories of human DCs are conventional or myeloid $CD11c^+$, $CD123^-$ DCs (cDC or mDC, which can express TLR1-6 and TLR8) and plasmacytoid $CD11c^-$, $CD123^+$ DCs (pDC, naïve interferon type I-producing cells, characterized by TLR7 and TLR9) (Hashizume et al. 2005). CD11c is an integrin

α_x subunit, part of the Int $\alpha_x\beta_2$ complement receptor 4 for inactivated-C3b (iC3b), whereas CD123 is interleukin-3 receptor (IL-3R).

CD11⁺ DC may have several subsets: CD16⁺ DCs in blood, CD1c⁺ (BDCA-1⁺), which strongly expresses TLR4 and produces IL-10 and IDO in response to whole *E. coli* (Kassianos et al. 2012) and CD141⁺ (BDCA-3), which strongly expresses TLR3 and effectively induces Th1 responses (Jongbloed et al. 2010).

DCs can also be classified as immature and mature, migratory (collecting Ags in the periphery as peripheral sentinels then migrating along lymphatic vessels to secondary lymphatic tissues to present them) and resident (possibly arising from blood-borne precursors or formed *in situ* in one lymphoid organ, where antigens are collected and presented in the organ before the cells undergo apoptosis) as well as non-polarized (DC_0) or polarized, where the latter is comprised of DC_1, which are produced by contact with stimuli such as bacteria including mycobacteria or poly I:C and induce a T_h0-to-Th1 shift by producing IL-12 and DC_2, which are produced by contact with stimuli such as helminths, thymic stromal lymphopoietin (also produced in epithelia) and PGE_2 and induce a T_h0-to-Th2 shift by producing IL-4 and other Th2 polarizing cytokines.

Both immature and mature DCs express H_1R, H_2R, H_3R and H_4R (Gutzmer et al. 2005; Jutel et al. 2009). DCs produced from human monocytes with GM-CSF and IL-4 belong to the non-professional histamine producing cells containing HDC and histamine, both of which increase up to 300-400% upon culture up to day 5. HDC inhibition as well as H_1R-antagonists and H_2R-antagonists decrease the DC differentiation-associated CD45 expression from day 3 by 60-80 % of control values. DPPE, inhibits histamine binding to microsomal and nuclear structures and downregulates CD40, CD86, CD33, HLA-DR and CD11c (Szeber nyi et al. 2001). Higher histamine concentrations achieved via MC degranulation exceed previously established peripheral tolerance via loss of T_{reg} suppression and lead to T-cell-mediated acute rejection of the skin graft (de Vries et al. 2009). However, this may relate to deprivation of tryptophan, another important MC product (Nowak et al. 2012).

Further support for the role of H_4R in the maturation of DCs, possibly via H_4R-agonism, has been obtained in asthma models. HDC knockout mice maintained on a histamine-free diet had reduced OVA-sensitized and challenged airway hyperreactivity, BALF eosinophilia, OVA allergen-specific IgE and late phase cytokine levels (Kozma et al. 2003). The questions are how much histamine was in these mice produced by the intestinal microbial flora and how important a role the constitutive activity of H_4R plays in this model. Histamine-binding tick protein given before OVA-Ag challenge effectively prevented murine OVA-induced allergic asthma associated bronchial hyperreactivity, peribronchial inflammation, pulmonary eosinophilia, mucus production and IL-4 and IL-5 secretion (Couillin et al. 2004). These histamine effects may be mediated via H_4R because H_4R knockout and H_4R-antagonist treated mice have decreased lung inflammation and Th2 responses (Dunford et al. 2006). H_4R-agonism favours DC_2/Th2 responses, whereas lack of histamine and H_4R-antagonism diminish Th2 responses.

$CD1a^+$, $CD207^+$ (Langerin, a C-type lectin, which is a crucial component of Birbeck granules) Langerhans cells (LC) are perhaps the best known DC, residing in the epithelia of skin and mucosa LCs are monocyte/macrophage-derived DCs but seem to replenish to a large extent by local self-renewal. They belong to the so-called migratory DCs and can capture Ag by macropinocytosis, phagocytosis or receptor-mediated endocytosis after a local insult, leading to production of LC mobilizing TNF-α and IL-1β and then transport it to local lymph nodes. Once localized, LCs seem to participate in the maintenance of self-tolerance via activation, clonal proliferation and the maturation of iT_{reg} (Idoyaga et al. 2013). Interestingly, naïve T-cells seem to enter lymph nodes via high endothelial venules (HEV), whereas memory/effector T-cells enter mainly via afferent lymphatics (Förster et al. 2012).

Dermal cDCs residing in the upper dermis are composed of two subtypes: $CD1a^{high}$, $CD14^-$, $CD206^{high}$ and $CD1a^{low}$, $CD14^+$, $CD206^{low}$, CD209 (DC-SIGN) subsets, both of which express the coagulation factor XIII (Toebak et al. 2009; Chu et al. 2011). A special subtype of DCs, inflammatory dendritic epidermal cells (IDECs) are equipped with low- and high-affinity FcεRs (type II and I) and occur in skin lesions in atopic dermatitis. Cross-lining of the high affinity FcεRI IgE-receptor leads to IL-12 and IL-18 release, and Th1 polarisation. Th2-mediated inflammation with IL-4, IL-5 and IL-13 as well as IgE-driven MC degranulation prevail during the acute phase of atopic dermatitis, whereas eosinophils, macrophages and Th1 response with IFN-γ production are characteristic for the chronic phase with thickening of the epidermis and dermal fibrosis. IFN-γ increased H_4R on human monocyte-derived IDEC (Dijkstra et al. 2008).

Histamine can via H_2R stimulate endocytosis of Ag (Amaral et al. 2007). Regarding resident murine spleen DCs, histamine decreases their Ag presentation capacity in an apparently H_4R-mediated manner, as this effect was inhibited by a H_4R-antagonist and enhanced antigen presentation was observed in H_4R-/- DCs (Simon et al. 2011). After uptake of Ag, DCs can also pass them from the endocytotic compartment to the cell cytoplasm to be processed by the proteasome for cross-presentation of the antigenic epitopes in the context of MHC I. Cross-presentation is necessary for the mobilization of $CD8^+$ killer T-cells against malignant and virally infected cells if the DC themselves are not affected and would thus not be able to present tumour or viral Ags. Via H_3R and H_4R, histamine at low concentrations can stimulate cross-presentation of soluble but not particulate endocytosed Ags (Amaral et al. 2007). MHC I-restricted cross-presentation of self-Ag leads to induction of tolerance through deletion of autoreactive $CD8^+$ T-cells (Kurts et al. 1997). Therefore, it is possible that H_4R stimulation maintains production of tolerogenic DCs (tol-DC) during a steady low histamine state, whereas adaptive immunity is raised against infections, which trigger DC-mediated production of IL-12 and IFN-α (Steinman et al. 2003). Activated DCs can rapidly recruit cells with a capacity to produce more cytokines, namely NK cells, which express H_4R (Damaj et al. 2007), and iNKT, in which H_4R is required to IL-4 and IFN-γ synthesis (Leite-de-Moraes et al. 2009).

DCs regulate the polarization of naïve $CD4^+$ T_h0 T-cells, which can be polarized to Th1 effector cells via IFN-γ and IL-12, producing IFN-γ, GM-CSF, CCR5 and CXCR3, to Th2 effector cells via IL-4, producing IL-4, IL-5, IL-13, CCR3 and CCR4, to Th17 effector cells via IL-6 and TGF-β, producing IL-6, IL-17, IL-22, CCR6 and IL23R, to T_{FH} cells via IL-6 and IL-27, together with interactions with follicular B cells or to iT_{reg} cells via IL-10 and TGF-β, producing IL-10, TGF-β and IL2R (Steinman & Banchereau 2007). Th1 and Th2 cells predominantly (but not exclusively) express H_1R and H_2R, respectively. H_4R deficient mice and mice treated with H_4R antagonist had decreased OVA-induced allergic lung inflammation and *ex vivo* induced T-cell-mediated IL-4, IL-5 and IL-13 production (Dunford et al. 2006).

H_4R expression has been shown on Th1, Th2 and Th17 cells (Mommert et al. 2011). Stimulation of H_4R on Th17 cells increased production of IL-17A and intracellular activation of activating protein-1 (AP-1) (Mommert et al. 2012). Th17 cells have been shown in psoriatic skin lesions and in acute atopic dermatitis. H_4R-agonists stimulate Th2 cells to produce IL-31, which causes itch (Mommert et al. 2011). This suggest that H_4R plays a role in the modulation of T-cells, which participate in the pathomechanisms of inflammatory skin diseases. Although we still lack detailed mechanistic studies, iNKT-cells have been associated with autoimmune diseases like primary biliary cirrhosis (PBC), psoriasis, MS, SS, SLE and RA (Simoni et al. 2013). In experimental PBC model, iNKT-cells play a key role in the initiation of the disease whereas iNKT-cells in other autoimmune diseases are supposed to play a stronger role during the chronic stages. These disease-promoting effects can depend on inappropriate activation of iNKT-cells (such as in psoriasis) but also on defects in their immunoregulatory functions (such as MS, RA).

7.10. Regulatory T-cells and Autoimmunity

There are two main types of regulatory T cells. Naturally occurring nT_{reg} are centrally-produced in thymus from naïve T-cells (T_n) by exposure to IL-2, TGF-β and Ag and include $CD4^+$, $CD25^+$ and $Foxp3^+$. Adaptive iT_{reg} are produced in the periphery in lymphocyte tissue via tolerogenic mechanisms by exposure of T_n to IL-2, TGF-β and Ag presented by tol-DC.

Differentiation of DCs to tol-DCs upon exposure to IL-10 and TGF-β (or corticosteroids) starts by expression of tolerogenic immunoglobulin-like transcripts (ILT3 and ILT4) at the same time when costimulatory molecules (CD40, CD80, CD86 and IL-12) and MHC II are down-regulated. Tol-DCs produce IL-10 and TGF-β upon Ag presentation in mucosal surfaces and induce T_h0 cells to iT_{reg} which cause oral tolerance. Production of indoleamine 2,3-dioxygenase (IDO) by tol-DC contributes to immunotolerance. H_3R/H_4R-agonist enhance cross-presentation and tolerance by deletion of autoreactive $CD8^+$ cells, whereas the presence of danger signals, such as TLR-ligands and GM-CSF, would favor formation of $CD8^+$ effector T cells.

In humans, Foxp3 is not restricted to iT_{reg}, which can be recognized as $CD127^{lo}$, $CD25^+$ and $CD4^+$ cells. IL-10-produces inducible type 1 regulatory T cells (Tr1 T_{reg} and TGF-β- Th3 T_{reg}, which both are $Foxp3^-$. Regulatory T cells in general down-regulate immune activation by producing IL-10, TGF-β and IL-35 (Burrell et al. 2012), by expressing immunosuppressive surface molecules, such as CTLA-4, and by inducing production of IDO in APC. H_4R-agonists are chemotactic to T_{reg} and can stimulate IL-10 production (Morgan et al. 2007).

Regulatory B10 cells are produced when B10-cell precursor via BCR is stimulated by an auto-Ag. Auto-Ag is internalized and its determinants are presented in a B-cell surface MHC II to TCR, whereas a B-cell surface CD40 stimulates the T-cell via CD154. This induces T-cell produces interleukin-21, which via B-cell surface IL-21R stimulates the matured B10 cell to produce IL-10 to down-regulate T-cell function.

7.11. Atopic Dermatitis and Inflammatory Skin Diseases

The natural history of atopic manifestations ("allergic march") is characterized by a sequence of atopic dermatitis and food allergy (peak prevalence at the age of 1 year), followed by asthma (peak prevalence at the age of 5-8 years) and finally allergic rhinitis (and conjunctivitis) reaching a peak plateau at the age of 12 years. Increased lesional histamine concentrations have been described in prick-tests, urticaria, allergic contact dermatitis and psoriasis (Thurmond et al. 2008; Gutzmer et al. 2011), but in spite of some older suggestive studies, the situation is not clear in atopic dermatitis (Ständer & Steinhoff 2002). In a dog model for atopic dermatitis, histamine increased only locally for 1-3 hours after challenge (Bäumer et al. 2011). H_1R-antagonists have incomplete effects and it has therefore been suggested that maybe H_4R modulation could be helpful here. H_4R has been shown in keratinocytes, dermal fibroblasts, MCs, eosinophils, dermal DCs, LCs and nerve cells in the skin, but this depends on the differentiation, localization and external stimuli (Damaj et al. 2007; Dijkstra et al. 2008; Ikawa et al. 2008; Gschwandtner et al. 2010). H_4R is higher in the upper than in the lower epidermis (Yamaura et al. 2009) and stimulation with LPS, indomethacin and dexamethasone increases H_4R mRNA and/or protein in dermal fibroblasts.

Atopic dermatitis is characterized by increased numbers of MCs in skin lesions and often dietary or pollen allergies. Increased serum IgE values (Liu et al. 2011) correlate positively with the disease activity and treatment with anti-IgE (e.g. omalizumab) is occasionally effective. IDEC are characteristic for atopic eczema and disappear upon successful treatment (Wollenberg et al. 1996; Wollenberg et al. 2001). Acute atopic dermatitis is characterized by a Th2 response (IL-4, IL-5, IL-13) but a chronic disease by Th1 milieu (IFN-γ) and lichenification, epidermal hyperplasia and dermal fibrosis caused by scratching (Dijkstra et al. 2008; Simon et al. 2004). Dermal eosinophil infiltrates are prominent and eosinophilic cationic protein ECP can be used to evaluate disease activity.

Three different single nucleotide polymorphisms (snp) have been described in HRH4 gene coding H_4R in atopic dermatitis, suggesting that this receptor may play a role (Yu et al. 2010). MC and eosinophil H_4R may mediate chemotaxis. IDEC is H_4R positive and stimulation of H_4R (but not modulation of the other histamine receptors) of IDEC led to diminished production of Th2-chemokine CCL2 (attracting monocytes, activated T-cells and NK-cells) and Th1-polarizing cytokine IL-12 (Dijkstra et al. 2008), suggesting that low level histamine, if maintained, may down-regulate the immune-inflammatory activity of these cells.

LCs in atopic dermatitis can probably act as pro-inflammatory, immunosuppressive or tolerogenic cells. LCs express functional H_4R which by low histamine concentrations or by specific H_4R-agonists down-regulates CCL2 production. This also increases their migration *ex vivo* from the skin, normally indicating migration towards secondary lymphatic organs (Gschwandtner et al. 2010; Chu et al. 2011).

Th1, Th2 and Th17 cells are immunological effector cells in human inflammatory skin diseases and they can all express H_4R (Mommert et al. 2011), which is *in vitro* upregulated by the Th2 cytokine IL-4. H_4R stimulation increases production of pruritogenic IL-31 (Gutzmer et al. 2009), in particular in atopic dermatitis.

Also chronic idiopathic urticaria is associated with increased MC numbers, in which disease IgG auto-Ab against the FcεRI can trigger MC degranulation. MCs in skin express mostly H_2R and H_4R (Lippert et al. 2004). In a FITC-induced Th2-driven inflammatory contact dermatitis in mice H_4R-antagonists had anti-inflammatory effects (Cowden et al. 2010b), perhaps via DC by diminishing their mobilization from skin to lymph nodes and thus Th2 polarization. Symptomatic effects were not seen in MC-KO mice suggesting that here H_4R-antagonists acted directly on MCs. Similar results, combined with diminished serum IgE levels, were obtained in a TNCB-driven mouse model with H_4R antagonists started before sensitization and continued during TNCB-treatment (Suwa et al. 2011). In contrast, in an acute dog (Bäumer et al. 2011) and mouse (Seike et al. 2010) models H_4R-antagonists were not helpful.

Recently, dual inhibition of H_1R/H_4R has been advocated (Matsushita et al. 2012; Ohsawa et al. 2012). Monotherapy with H_4R-antagonist diminished IL-4, IL-5 and IL-6, but increased IL-12 and IFN-γ. Increase of IL-12 and IFN-γ was not seen in animals treated with dual inhibition. Dual inhibition helped against itch and was as effective in the overall treatment as prednisolone for which it might offer an alternative in the future. In contrast to H_1R-antagonist, H_4R-antagonist had an effect both on histamine and substance P induced itch (Yamaura et al. 2009), which may be mediated directly via H_4R^+ sensory nerves (Dunford et al. 2007), where H_3R inverse agonist caused dose-related itch which was totally blocked by a combined treatment with H_1R- and H_4R-antagonists (Rossbach et al. 2011). Therefore, it was concluded H_1R and H_4R stimulation and H_3R-inverse agonist all cause itch (and calcium increase in the neuronal cells).

7.12. Asthma

Asthma is a chronic, inflammatory and often immune-based disease that mostly targets the small contractile bronchioles of the airways, characterized by a smooth muscle wall and eosinophil, MC and T-cell infiltrates (Bousquet et al. 2000). Dyspnea, cough and wheezing are typical symptoms. These features are inducible by H_1R-agonists, but in spite of this H_1R-antagonists are not effective in the treatment of asthma (Thurmond et al. 2008). Asthma may be extrinsic (atopic, with type 1 hypersensitivity to allergens) and intrinsic (non-atopic) and both Th2 lymphocytes/IgE/MC and innate helper type 2(I_h2)/IL-5/eosinophils play important roles in it.

Chemokines and cytokines are not only produced by the rapidly acting MCs, basophils and eosinophils, but also by DC, macrophages and in particular T- and B-cells, which are readily present in newly-diagnosed asthma (Laitinen et al. 1993). In untreated and newly diagnosed airway epithelium (BALF may not accurately reflect tissue events), 612 lymphocytes (110 in controls, $p<0.05$), 120 MCs (7 in controls, $p<0.001$), 75 eosinophils (0 in controls, $p<0.05$), and 15 macrophages (0 in controls, $p < 0.05$) per mm^2 were counted (Laitinen et al. 1993). The situation was possibly further lymphocyte-dominated in airway lamina propria, where the lymphocyte values were 656 *vs.* 155 ($p<0.001$), plasma cells were 285 *vs.* 27 ($p<0.001$), macrophages were 44 *vs.* 3 ($p<0.001$) and monocytes were 11 *vs* 1 (NS), whereas the MCs were 124 *vs.* 88 (NS) and eosinophils were 134 *vs.* 3 ($p<0.001$). These findings suggest that asthma may basically be a lymphocyte-based immunological disease.

As demonstrated by the cellular pathology above, asthma and other atopic diseases can be divided to two different phases, the sensitization phase and the challenge phase (Little et al. 2003). In the sensitization phase, the mucosal membrane and mucosal-associated lymphatic tissue (MALT)) are exposed to a potential aeroallergen (Bielory et al. 2013), leading to IgE production in type 2 response in pulmonary hilar lymph nodes with involvement of DCs, $CD4^+$ T-cells, $CD8^+$ T-cells, B-cells and plasma cells.

Type 2 macrophage polarization is characterized by suppression of pro-inflammatory cytokines, intracellular killing and Ag presentation, but accompanied by production of moderate levels of IL-10 (Gordon et al. 2003; Martinez et al. 2009). Membrane receptors with scavenger function are upregulated together with a variety of molecules implicated in tissue regeneration, wound healing, granuloma formation and immunity against large parasites (Martinez et al. 2009). Via inducation of arginase 1, arginine metabolism shifts from production of NO to production of L-ornithine, utilized to produce polyamines and proline important for cell growth and collagen synthesis. M2 macrophages secrete chemokines such as CCL17, CCL18, CCL22 and CCL24 which recruit Th2 cells, basophils and eosinophils (Mantovani et al. 2004). IgE antibodies produced do not fix complement and do not pass placenta and form only 0.05% of serum immunoglobulins. However, they are empowered by binding to high affinity FcεRI IgE-receptors on the surface of highly explosive MC and basophils.

H_4R has been shown in most leukocytes, which are believed to play a role in asthma (Gantner et al. 2002). Stimulation of H_4R causes MC and eosinophil chemotaxis (Hofstra et al. 2003) and increases CD11b/CD18 and ICAM-1 adhesion molecules on eosinophils (Buckland et al. 2003; Ling et al. 2004; Barnard et al. 2008). H_4R mRNA has also been shown in non-immune cultured cells originating from lung tissue, including smooth muscle cells, epithelial cells and vascular endothelial cells. Experimental data suggest that H_4R-antagonist JNJ 7777120 administered perorally diminishes T cell infiltration into lung tissue in already established airway inflammation (Cowden et al. 2010a).

Upon Ag challenge, IL-16 increases in BALF in both patients and experimental asthma models. Histamine stimulates production of IL-16 in pulmonary epithelial cells and $CD8^+$ T-cells (Cruikshank et al. 2000; Gantner et al. 2002). The stimulating effect of histamine on $CD8^+$ T-cells is mediated via H_2R and H_4R, but in asthma most IL-16 is probably derived from epithelial cells. The amount of IL-16 in BALF correlates with the $CD4^+$ T-cell infiltrates in asthma (Cruikshank et al. 2000). This agrees with the observation that IL-16 binds to the CD4-receptor and promotes recruitment of $CD4^+$ cells. In OVA-sensitized mice, intra-tracheal administration of IL-16 before OVA challenge lead to inflammatory inilftrates and diminished Th2-cytokine IL-5 in BALF (Little et al. 2003). This could be due to recruitment and induction of T_{reg} (McFadden et al. 2007).

DCs have a capacity to act in a tolerogenic, anti-inflammatory or pro-inflammatory way, but the factors regulating the type of DC response are not known in detail. Histamine *in vitro* can stimulate migration of monocyte-derived cDC via H_2R and H_4R (Gutzmer et al. 2005). The same study reported that stimulation of cDC via H_4R inhibited production of Th1-cytokine IL-12 and therefore promotes Th2-cytokine milieu. Screening of gene expression in histamine stimulated cDC demonstrated increased expression of certain Th2-related cytokines, such as TNF-α and TGF-β2 (Lundberg et al. 2011).

However, histamine stimulation also causes gene expression of various chemokines and cytokines that are related to Th1-, Th17- and T_{reg}-mediated immune responses. Histamine affected the maturation of DCs via H_1R and H_4R. The neutral H_4R-antagonist JNJ7777120 inhibited allergen-provoked T-cells *in vitro*.

Conflicting results have also been published. Inhalation of 4-MeHA, which is a double agonist of H_4R and H_2R (with approximately 100-fold preference for H_4R) inhibited OVA-induced inflammation, probably due to the T_{reg} recruitment to lungs (Morgan et al. 2007) or indirectly via effects on DCs. It is possible that this difference is simply caused by a different route of drug application, as inhaled 4-MeHA causes a local chemotactic concentration gradient to H_4R and H_2R positive leukocytes, whereas a systemic or genetic inhibition of H_4R and H_4R antagonists directly inhibit H_4R^+ cells everywhere (Cowden et al. 2010a; Dunford et al., 2006).

7.13. Allergic Rhinitis

Allergic rhinitis, either seasonal or persistent/perennial, is characterized by recurrent sneezing, increased mucus production, obstruction of the nasal passages and itching. (Skoner et al. 2001). It has a strong genetic component favouring development of Th2-IgE-MC-mediated immune responses upon challenge and sensitization to allergens. Common allergens are fecal proteins from dust mites, cat and dog dandruff and pollen. The central mechanism in the sensitization phase is DC-mediated Ag presentation to T_h0 cells favoring polarization to IL-3, IL-4, IL-5 and IL-13 producing Th2 cells and plasma cell-mediated production of IgE antibodies. Upon Ag re-challenge nasal mucosal membranes become infiltrated by MC, eosinophils and plasma cells. Also basophils have been found in nasal lavage fluid in such patients (Shiraishi et al. 2013). Local or systemic administration of H_1R-antagonists, local sympathomimetic drugs (pseudoephedrin) and local corticosteroids are used in its treatment.

H_1R is expressed on the epithelial, endothelial and neural cells in the lower nasal passages, H_2R on epithelial and glandular (mucus production) cells and both H_3R and H_4R on neural cells (Nakaya et al. 2004). It has been proposed that H_3R participates in the homeostatic maintenance of the nasal mucosa and regulates mucus production (Suzuki et al. 2008). Phase II clinical study did not find any advantage of H_1R/H_3R antagonist over H1R antagonist upon exposure to allergen in a challenge chamber (Daley-Yates et al. 2012).

Experimental animal studies showed that repeated local administration of H_4R-antagonist JNJ7777120 decreased sneezing and scratching of the nose at only 1 nM concentration (Takahashi et al. 2009). Repeated peroral administration decreased serum IgE concentrations while decreased IL-4and increased IFN-γ levels were found concurrently in nasal lavage fluid, which is notable because IL-4 and IFN-γ increase and decrease plasma cell-mediated IgE production. Results encourage to further studies of single or double action drugs.

H_4R-antagonists could possibly have symphatomimetic, pseudoefedrin-like effects via induced endogenous cathecolamine release (Chan et al. 2012). Inhibition of H_4R with antagonists or gene knockout is antipruritogenic and stimulation of H_4R with histamine causes itch, with a similar effect obtained via manipulation of H_1R (Dunford et al. 2007; Thurmond et al. 2008; Rossbach et al. 2011). A third possible mechanism of action could relate to diminished MC and eosinophil chemotaxis along with diminished DC, Th2 and plasma cell dependent IgE synthesis, which would affect both the acute and chronic phase of allergic rhinitis.

7.14. Allergic Conjunctivitis

Allergic conjunctivitis is characterized by bilateral itch, burning, redness, tearing and swelling of the conjunctiva and lids. Seasonal allergic conjunctivitis (SAC) and perennial (persistent) forms (PAC) are relatively mild Ag-driven diseases, but atopic keratoconjunctivitis (AKC) and vernal keratoconjunctivitis (VKC) can affect the cornea and vision (Bonini et al. 2009). Comorbidity between AKC, allergic rhinitis, atopic dermatitis, urticaria and asthma is common, suggesting shared disease mechanisms in form of sensitization and subsequent challenge phases. Although specific allergens cannot be shown in a large proportion of VKC-patients, inflammation is Th2-driven with increased histamine concentrations in tear fluid (Leonardi et al. 2011), which may contribute to pathological tissue remodeling and chronicity. As a result, local and peroral H_1R-antagonists, mast cell-stabilizing drugs (such as cromoglycate and nedochromil) and corticosteroid eye drops form the mainstay of treatment (Bonini et al. 2009). In difficult cases, immunosuppressive calcineurin-inhibitors and sometimes desensitization can improve tolerance to allergens.

Locally administered levocabastine (H_1R-antagonist) and JNJ7777120 (H_4R-antagonist) diminished scraping response, hyperemia and swelling in a dose-dependent manner in a 2-10 μM histamine-induced experimental mouse model, but statistical significance was only seen for the H_1R-antagonist, but the most potent effect was obtained by combined use (Nakano et al. 2009). Immunostaining revealed increased H_1R, H_2R and H_4R levels in conjunctival cells for VKC patients compared to controls (Leonardi et al. 2011), which was confirmed by messenger RNA levels that were 5-fold increased for H_2R and H_4R. H_1R was particularly increased on vascular endothelial cells and H_4R in stromal inflammatory cells. Expectations of H_4R-modulators are high but require conformation by randomized clinical trials.

7.15. Rheumatoid Arthritis

Reumatoid arthritis (RA) is a symmetrical autoimmune polyarthritis of usually small, peripheral synovial joints associated with rheumatoid factor (RF) and anti-citrullinated protein antibodies (ACPA). Its midline symmetry and involvement of the most densely innervated small peripheral joints (with widest representation in hominiculus) has evaded firm explanations, but the most strongly advocated theory refers to neurogenic inflammation (Konttinen et al. 1994). This is interesting considering the close spatial and functional relationship between small peripheral C and Aδ neuropeptide nerves and MC (Hukkanen et al. 1991) combined with the presence and role of H_3R and H_4R in the function of those nerves. Furthermore, MCs produce corticotropin-releasing hormone (Kempuraj et al. 2004) and respond to it (Cao et al. 2006), suggesting the hypothalamus-pituitary-adrenal or HPA axis might participate in the pathogenesis of RA similar to the arthritis model in Lewis rats (Sternberg et al. 1989).

The fundamental cause of RA is unknown and it has been suggested that the triggering stimulus (loss of tolerance) and the perpetuating stimulus (chronicity) are different, such as an initial triggering infection causing a mostly continuous release of cartilage-derived damage-associated molecular patterns (DAMPs) and auto-Ag. In this context, it is interesting to note that in rheumatoid synovitis, mature DCs capable of presenting antigen and activated cytokine-producing T-cells do not recur to the joint from which all cartilage has been removed in association to total joint replacement (Li et al. 2001; Li et al. 2002).

Synovitis is the best known pathological manifestation of RA, although its pathology is also evident in the bone marrow and it can be assumed that the preclinical immunological processes occur in the lymphoid organs. Synovitial stroma is characterized by the presence of MHC II^+ macrophages, specifically $CD4^+$ T-cells and plasma cells, but also by DCs, MCs, fibroblasts, endothelial cells, adipocytes and peripheral nerve fibers and terminals. Synovial lining is composed of fibroblast-like type B lining cells and macrophage-like type A lining cells. Cultured RA (fibroblast-like synoviocytes) contain mRNA coding H_1R, H_2R and H_3R but lack H_4R coding mRNA (Ikawa et al. 2005) but the presence of the H_4R protein was detected by double staining in proline-4-hydroxylase-positive synovial fibroblasts and in $CD163^+$ macrophages (Ohki et al. 2007).

Current disease intervention is based on commencement of early treatment with evidence-based, methotrexate-achored combinations with disease-modifying anti-rheumatic drugs (DMARDs), modified according to patient response (saw tooth principle) and guided by fixed goals using the treat-to-target (T2T) principles, in modern times strongly supported by biological anti-rheumatic drugs (biologicals). Biologicals are highly selective drugs with targeting entities such as TNF (infliximab, etanercept, adalimumab, certolizumab pegol, golimumab), IL-1 (anakinra), IL-6Rα (tocilizumab), CD80/CD86 costimulatory molecules (abatacept) or CD20/B-cells (rituximab). Janus kinase 3 tyrosine kinase inhibitor tofacitinib is the latest newcomer.

Synovitis is characterized by increased numbers of MC, which are also located at the site of cartilage destruction in the pannus tissue. However, early studies using adjunctive treatment with antihistamines (H_1R-antagonists) (Wilson 1953) and H_2R antagonists (Permin et al. 1981) failed, resulting in the relegation of histamine to an epiphenomenon rather than active molecular participant.

It was therefore notable that inhibition of c-kit tyrosine kinase, typically present in MCs, diminished disease activity including the number of tender and swollen joints, ESR and CRP (Eklund & Joensuu 2003). C-kit is a tyrosine kinase receptor (mast/stem cell growth factor receptor) for stem cell factor (SCF), previously annotated as Steel factor or kit-ligand.

In spite of early observations suggesting increased plasma and in particular synovial fluid histamine levels, implying intra-articular release (Frewin et al. 1986), a more recent finding suggests decreased plasma levels and significantly decreased synovial fluid levels of histamine (Adlesic et al. 2007). RA serum histamine was 0.93± 0.16 ng/mL compared to 1.89 ± 0.45 ng/mL in controls ($p<0.001$) and the RA synovial fluid histamine levels were only 0.37 ± 0.16 ng/mL ($p<0006$). Histamine levels increased in aTNF-treated RA patients ($p<0.01$). These authors also injected 20 µL fluid containing 10, 1, 0.1 or 0.01 ng histamine intra-articularly into healthy joints or joints challenged with a preceding intra-articular injection of high mobility group box-1 (HMGB1, an endogenous DAMP) or peptidoglycan (a pathogen/microbial-asssociated danger signal, PAMP/MAMP). Histamine produced no effects by itself or in induced arthritis. This led the authors to propose that histamine lacks harmful properties in RA (Adlesic et al. 2007).

What is clear from the above experiments is that histamine levels in RA are at nanomolar levels (0.11 ng/mL histamine corresponds to 1 nM) and $10^{-8.1}$ M histamine is sufficient to half-maximally activate H_4R. MC activation may locally and temporarily increase histamine levels to micromolar concentration range, but due to histamine degradation, uptake, binding and diffusion this increase is transient. In contrast, how the low but constitutitively present nanomolar histamine levels rather than micromolar levels affect processes such as DC_0 and T_h0 cell polarization and Ag presentation in RA remain open at present. In RA tissues, $CD304^+$ (BDCA4) pDCs producing IL-18 and IFN-α/β are more common than $CD1c^+$ (BDCA1) cDCs expressing IL12 and IL-23 and their numbers correlate with RF and ACPA (Lebre et al. 2008). Indeed, pDCs express H_4R. H_4R-agonists diminished pDC mediated TFN-α, IFN-γ and CXCL8 production. H_2R had similar effects, but it is not known whether the histamine levels in RA joints reach H_2R-stimulatory concentrations (Gschwandtner et al. 2011).

7.16. SLE

Systemic lupus erythematosus (SLE) is a systemic autoimmune and immune complex disease in which clearance of immune complexes and apoptotic cell rests seems to be impaired with nuclear auto-Ags being especially targeted. American College of Rheumatology classification criteria define 12 different symptoms and signs where if at least four of these symptoms are positive, the condition can be classified as SLE (Hochberg 1997).

HRH4 gene transcripts (H_4R coding mRNA) were high in SLE compared to healthy controls. Interestingly, amplification in HRH4 copy numbers (>2 HRH4 copies) was associated with increased incidence of anti-nuclear antibodies, arthritis and proteinuria, whereas copy number deletions (<2 HRH4 gene copies) were found to be protective against proteinuria (Yu et al. 2010). This seems to demonstrate the influence of gene copy-number variation (CNV) on the disease phenotype. In SLE, an increased HRH4 gene copy number seems to lead to higher H_4R mRNA copy numbers, which can cause disturbances in the nanomolar histamine-H_4R interactions which subsequently affect the disease severity.

7.17. Sjögren's Syndrome

Sjögren's syndrome (SS) is a female dominant (90%) autoimmune disease of the exocrine glands, which can occur in primary form but also as a secondary form in disorders such as RA and SLE. It is characterized by keratoconjunctivitis sicca and xerostomia in association with focal sialadenitis and SS-autoantibodies (Shiboski et al. 2012).

It is well known that local sialadenitis and local numbers of MC correlate to each other in SS salivary glands (Konttinen et al. 1990). At that time, histamine was expected to perhaps play a role via H_1R and/or H_2R, although it had earlier been shown that intra-arterial injection of 100 nM histamine causes an early, short-term (~60 s) increase in blood flow and salivation, followed by a later, longer lasting (15-20 minutes) response, which was partially inhibited by H_1R-antagonist but not affected by H_2R-antagonist and was found to be neutrally-mediated without any direct effect on acinar cells (Shimizu & Taira 1980).

More recently, H_4R were found on some infiltrating leukocytes as well as acinar and ductal epithelial cells in healthy human salivary glands. In contrast, H_4R expression on these glandular cells was highly decreased in SS (Stegaev et al. 2012), suggesting that this epithelial H_4R downregulation is due to repeated exposure to micromolar, burst release histamine in diseased SS salivary glands. This may also explain the relatively weak H_4R expression in the infiltrating lymphocytes. Epithelial H_4R was shown at the mRNA, receptor protein and functional level and it seems that these cells are non-professional histamine producing cells containing 74-kDa HDC, which may participate in the modulation of the off-state nanomolar histamine concentrations apparently necessary for the tissue homeostasis. Furthermore, acinar and ductal epithelial cells contained OCT3, which seems to allow histamine uptake for intracellular degradation at 100 nM concentrations but also allows intracellular histamine release during off-state. In SS, OCT3 levels were low, suggesting that the cells are incapable of buffering against burst-released histamine, which would partially explain the local downregulation of H_4R on resident and immigrant cells (Stegaev et al. 2013).

7.18. Ulcerative Colitis

Ulcerative colitis (UC) is an IBD considered as an autoimmune disease, in part because of several disease-activity-associated or independent extra-intestinal autoimmune manifestations and the immunosuppressive treatment used in the disease management. It affects the colon by causing superficial Th2 helper T-cell-dominated mucosal infiltrates, inflammatory cryptitis and ulcerative lesions. Th17 helper T-cells might also participate. Upon increasing severity, the number of blood and mucus mixed stools per day increases, malaise, fatige and fever appear as general symptoms, the inflammatory parameters CRP and ESR increase and blood hemoglobin decreases, possibly resulting in toxic megacolon and rupture. Although basically affecting the colon, some back-wash ileatitis often also occurs.

In a steroid-resistent UC, a combination of H_1R-antagonists, MC stabilizers and a hypoallergenic diet have been preliminarily found to be useful and suggest a role for burst release histamine in severe disease and disease exacerbation (Raithel et al. 2007). ECL cell and MC-derived histamine is supposed to play an important role in the initial increase of the vascular permeability characteristic for UC (Coron et al. 2012). Urinary secretion of the HNMT-produced histamine degradation product N-methylhistamine is increased in UC and Crohn's disease and correlates with disease activity (Winterkamp et al. 2002). Polymorphic, mutated DAO is associated with the severity of UC (García-Martin et al. 2006). Disturbances in nerve-derived and bacterially-produced histamine may further disturb the local histamine milieu and interfere with DC-T-cell co-operation but also with angiogenesis and fibrosis. H_4R antagonists diminished macroscopic damage, myeloperoxidase, TNF-α, wall thickness and neutrophil infiltratation when administered 24 hours before induction of an acute colitis model in rats using exposure to trinitrobenzene sulphonic acid (Varga et al. 2005).

7.19. Crohn's Disease

Crohn's disease was earlier called terminal ileitis due to the common involvement of this segment of the intestine. However, this disease can affect any part of the human gastro-intestinal tract, often as skip lesions with intact and apparently uninvolved bowel segments in between the focal and segmental leasions. Crohn's disease is characterized by abdominal pain, weight loss, fever, diarrhea, transmural involvement of the bowel wall (with peri-intestinal abscesses, fistuli and bowel stenosis. and extra-intestinal symptoms as well as anti-inflammatory and immunosuppressive medication. Its histology is characterized by intraepithelial neutrophils, cryptitis, Th1/Th17 response and non-caseating granulomas. Although H_4R is supposed to play a role in the regulation of the polarization of T_h0 cells, there is at present little knowledge on this aspect in Crohn's disease. Interactions between PAMP/MAMP and bacterial Ags with DCs and their subsequent interactions with T_h0 T-helper cells have been proposed to regulate the balance between T_{reg} and T_{eff} in a context-dependent manner in IBD (Himmel et al. 2008).

The H_1R-agonist HTMT-dimaleat, H_2R-agonist dimaprit, H_3R-agonist (R)-(-)-α-methylhistamine and H_4R-agonist 4-methylhistamine can excite enteric neurons in human submucosal plexus, where the receptor specificity of this excitation was proven by demonstrating blockage with a corresponding antagonist. Considering the role of the autonomic enteric nerves of the intestine as regulators of the bowel function, this implies a novel and interesting aspect for IBD (Breunig et al. 2007) and emphasizes the role of the neuro-immuno-endocrine aspects of autoimmune diseases (Lomax et al. 2005; Wood 2004).

In addition to nerves, H_1R, H_2R and some H_4R have been described in enterocytes, H_1R and H_2R in muscle layer and H_1R, H_2R and some H_4R in immune cells (Sander et al. 2006).

7.20.Glomerulonephritis

One rare form of rapidly progressive glomerulonephritis is the anti-glomerular basement membrane-induced Goodpasture's syndrome. An experimental model of Goodpasture's syndrome can be treated with histamine, which decreased proteinuria, macrophage infiltration and crescent formation. It also decreased concentrations of IL-12, a well-known Th1 polarizing cytokine. These effects were specifically inhibited by an H_4R-antagonist but not affected by H_1R- and H_2R-antagonists (Tanda et al. 2007), although an earlier study published when only H_1R and H_2R were known used only H_1R- and H_2R-antagonists and argued that histamine plays no role in Goodpasture's syndrome (Wilson et al. 1981). This once more reveals how earlier work needs to re-evaluated in light of new findings. There are currently no observations on H_4R status and function in the more common forms of glomerulonephritides.

7.21. Autoimmune Diseases of the CNS

Several lines of evidence suggest a key regulatory role of histamine in the widely used experimental autoimmune encephalomyelitis (EAE) murine model, with CNS myelin proteins as potential auto-Ags (Musio et al. 2006; Lu et al. 2010; Passani & Ballerini 2012). Therefore, the regulatory functions of histamine relevant to the onset and progression of neuroinflammatory diseases and EAE in particular are being studied in genetically modified mice lacking histamine receptors and with selective agonists and antagonists. It is becoming evident that histamine plays a complex role with variable and occasionally contradictory effects, depending on the receptor subtypes being activated and the specific targeted tissue (Table 7.2).

All histamine receptors are expressed on cells involved in autoimmune diseases, with the exception of the H_3R that is normally not expressed by hematopoietic cells and is mostly confined to the CNS (Passani & Blandina 2011). Susceptibility to EAE requires expression of Hrh1, the gene encoding H_1R in mice (Ma et al. 2002). H_1R is expressed on Th1 cells in EAE mice brain lesions (Pedotti et al. 2003) where its presence is necessary for full encephalitogenic expression (Noubade et al. 2007). Furthermore, expression of H_1R is upregulated on encephalitogenic PLP139-151-specific Th1 compared to Th2 cell lines. Unsurprisingly, specific pharmacological antagonists targeting H_1R result in amelioration of EAE (Pedotti et al. 2003; El Behi et al. 2007) and H_1R-knockout mice exhibit a significant delay in the onset of EAE and a reduction in the severity of clinical signs compared with WT mice. Indeed, $CD4^+$ T-cells from H_1R-knockout mice produce significantly less IFN-γ and more IL-4 (which induces differentiation of naïve $CD4^+$ T cells to Th2 cells) in *in vitro* assays compared to wild-type controls, indicating that H_1R signaling in $CD4^+$ T cells plays a central role in regulating pathogenic T-cell responses (Ma et al. 2002).

H_2R also seems to partially regulate encephalitogenic Th1-cell responses and EAE susceptibility, as H_2R-knockout mice develop a less severe disease than wild-type littermates during the acute, early phase (Teuscher et al. 2004). The failure of H_2R-knockout mice to generate encephalitogenic Th1 effector cell responses is attributed to H_2R-mediated regulation of cytokine production by DCs, which affects T-cell-polarizing activity. In conclusion, H_1R and H_2R seem to have pro-inflammatory and disease-promoting effects, but H_1R or H_2R activation may also play an important role in limiting autoimmune responses, meaning their biological action is probably more complex. This complexity is further suggested in an EAE mouse model, where H_1R and H_4R produced pro-inflammatory effects but H_2R and H_3R produced anti-inflammatory effects on the disease process (Jadidi-Niaragh & Mirshafiey 2010).

As mentioned above, H_3R are normally not expressed by hematopoietic cells and are mostly confined to the CNS where they limit histamine synthesis and release (Passani & Blandina 2011), and regulate release of other neurotransmitters (Blandina et al. 2010). Deletion of the H_3R gene leads to more severe EAE, an effect associated with altered BBB permeability and increased expression of chemokines/chemokine receptors that promote the entry of peripheral T cells

Table 7.2
Histamine receptors and EAE. APC, antigen presenting cells; BBB, blood brain barrier; mDC, myeloid dendritic cells; MOG_{35-55}, Myelin Oligodendrocyte Glycoprotein; $PLP_{139-151}$, Myelin Proteolipid Protein.

Histamine receptor	EAE	Investigated Cell Types	Pharmacological target	Disease outcome	Reference
H_1	SJL mice $PLP_{139-151}$	Increased H_1R expression on Th1 cells Humoral immune responses	H_1R antagonism	Less severe disease	Pedotti et al., 2003 El Behi et al., 2007
	MOG_{35-55} H_1R-KO mice	$CD4^+$ T cells		Reduced IFN-g, increased IL-4 Less severe disease	Ma et al., 2002 Noubade et al., 2007
	MOG_{35-55} H_1R-KO mice	Endothelial cells	H_1R overexpression	Restored BBB integrity Less severe disease	Lu et al., 2010
	$PLP_{139-151}$ SJL mice	$CD3^+$ T cells	H_1R activation	Reduced IFN-g Decreased endothelial adhesiveness	Lapilla et al., 2011
H_2	MOG_{35-55} H_2R-KO mice	APC Th1		Reduced cytokines Inhibition of cell polarization Less severe disease	Teuscher et al., 2004
	MOG_{35-55} C57/Bl6	Proinflammatory cells	H_2R activation	Less severe disease	Emerson et al., 2002
	$PLP_{139-151}$ SJL mice	$CD3^+$ T cells	H_2R activation	Reduced IFN-g Decreases endothelial adhesiveness	Lapilla et al., 2011
H_3	MOG_{35-55} H_3R-KO mice	Th1 Endothelial cells		Increased expression of chemokines/chemokine receptors BBB deregulation More severe disease	Teuscher et al., 2007
H_4	MOG_{35-55} H_4R-KO mice	Treg Th17		Lower frequency Higher frequency More severe disease	del Rio et al., 2012
	MOG_{35-55} C57/Bl6 mice	Th1 mDC	H_4R antagonism	Increased IFN-g, reduced IL-10 More severe disease	Passani et al., 2011

Modified from Passani and Ballerini (2012)

not expressing H_3R into the CNS (Teuscher et al. 2007). The authors suggest that neuronal H_3R may serve as a central control of cerebrovascular tone and decreases susceptibility to neuroinflammatory diseases. These results indicate that activation of H_3R may have beneficial effects.

Recent evidence has mapped the topological and functional localisation of H_4R in the CNS of humans and rodents, respectively (Strakhova et al. 2009; Connelly et al. 2009).

H_4R is detected on hematopoietic progenitor cells that enter the cell cycle upon stimulation (Petit-Bertron et al. 2009). Activation of H_4R by agonists before exposure to growth factors leads to a profound decrease in the percentage of cycling cells (Schneider et al. 2011). The H_4R expression is dynamic, as it is upregulated during the differentiation of human monocytes to dendritic cells (Gutzmer et al. 2005). In addition, receptor levels change with the progression of pathophysiological responses, such as the upregulation of H_4R expression in monocytes in response to inflammatory stimuli (Dijkstra et al. 2007), in kidney putative tubule cells in diabetic rats (Chazot, Rosa et al., manuscript submitted) and in putative immune cells at the early stage of inflammatory pain states (Chazot et al., unpublished). Furthermore, pilot data revealed that H_4Rs are expressed on the soma of both Aδ and C-fibre sensory neurons through intense staining of small and medium diameter neurons as well as lamina I-III of the rat lumbar spinal cord, where the immunoreactivity pattern suggests localisation with terminals of primary afferent neurons (Katebe M et al. 2012; Lethbridge & Chazot 2010).

Surprisingly, H_4R-knockout mice develop a more severe EAE together with increased neuroinflammatory signs compared to WT mice (del Rio et al., 2012). As genetically modified mice may carry alterations of systems other than the targeted ones and activation of vicarious mechanisms may hinder the effects related to the deleted gene, we recently studied myelin oligodendrocyte glycoprotein $(MOG)_{35-55}$-induced EAE in C57BL/6 mice treated with JNJ-7777120 (Ballerini et al., submitted to Br J Pharmacol). JNJ-7777120 was injected i.p. daily starting at day 10 post-immunisation. Disease severity was monitored by clinical and histopathological evaluation of inflammatory cells infiltrating into the spinal cord, anti-MOG_{35-55} antibody (Ab) production, T cell proliferation by $[^3H]$-thymidine incorporation, mononuclear cell phenotype by flow cytometry, cytokine production by ELISA assay and transcription factor quantification by mRNA expression. We found that treatment with JNJ-7777120 worsened the severity of EAE and increased inflammation and demyelination in the spinal cord of EAE mice. We also showed that during disease progression, T lymphocytes of JNJ-7777120-treated EAE mice produce more IFN-γ than vehicle-treated controls and fewer regulatory cytokines such as IL-4 and IL-10, despite unchanged frequency and proliferative capacity in response to MOG_{35-55}. Furthermore, JNJ-7777120 did not affect anti-MOG_{35-55} Ab production or mononuclear cell phenotypes. Hence, our results are in agreement with the recent report that mice which do not express H_4R exhibit an exacerbated disease and immunopathology (del Rio et al. 2012). This was ascribed to impaired formation of T_{reg} T-cells and their impaired chemotaxis and suppressor activity, leading to an increase in the Th17 T-cells.

7.22. Future Visions

Histamine is produced slowly by 73-kD HDC and released gradually via OCT2/3 and PMAT ion channels without storage phase by non-professional histamine-producing cells at low nanomolar concentrations, which only affect the high affinity H_3R/H_4R-equipped DCs, T-cells and other leukocytes.

This basal level H_3R/H_4R effect is during the resting "off-state" probably maintained within narrow limits by the high constitutive activation state of H_3R/H_4Rs, histamine produced by non-professional cells, histamine released by professional cells, histamine produced by intestinal microflora on one hand, and by the DAO-mediated extracellular and HNMT-mediated intracellular degradation as well as VMAT-mediated re-uptake and storage.

As this "off-state" appears to be prevailing in healthy individual during most of the time, it can be assumed that it participates in DC-T-cell interactions by favouring maintenance of tolerance, i.e. contributing to the active tolerogenic immune responses against self-Ag, which leads to production of iT_{reg} cells.

If a foreign Ag and an adjuvant (danger signal) are introduced to the system, in the context of danger a immunogenic response is produced, which leads to production of T_{eff} cells and provision of T-cell help to B-cells, which initiates T-cell-dependent, B-cell-mediated immune responses and production of antibodies. If directed against a transient foreign pathogen or vaccine, this provides immunity. If however this immune activation occurs against continuously present self-Ag, an autoimmune disease ensues.

Immediate type I immune responses cause a lot of morbidity in form of allergic responses and diseases. These diseases are mediated by professional histamine-producing cells, such as MCs and basophils. However, type I immune responses are only possible after the sensitization phase, which required cell-mediated, delayed type immune responses and production of MC/basophil-sensitizing IgE. When the IgE-sensitized MCs and basophils degranulate and release their stored histamine, short-term and locally high micromolar histamine concentrations are attained during an "on-state". The concentrations are high enough to activate the low affinity H_1R/H_2Rs.

Non-immune activation of MCs and basophils might in healthy individuals occur as a result of local irritation, in particular in sensitive mucosal surfaces or where the skin is penetrated, leading to a local reaction helping to get rid of or to dilute the local irritant. Autoinflammatory responses evoked by danger signals, *viz.* pathogen/microbe-associated molecular patterns and/or damage-associated molecular patterns (or alarmins), might enhance this non-immune responsiveness of tissues. If the exposure continues, these naïve responses are after sensitization enhanced by immediate type I hypersensitivity responses.

It is commonly accepted that in the treatment of these allergic diseases, if possible, elimination of the allergen by sanitary measures is an effective and logic form of intervention. In both autoimmune diseases and in allergic disease modulation of the DC-T-cell interactions seems to be a useful roadmap to new therapies. Induction of immune tolerance and a shift from IgE to IgG production seem to be tempting therapeutic targets. Due to the putative role of low nanomolar histamine and high-affinity H_3R/H_4Rs, modulation of these interactions by systemic and/or local synthetic small molecular weight H_3R- and H_4R-modulators to prevent sensitization or to induce tolerance seems to be the way to go. They good replace the currently widely used and expensive biologic drugs and biosimilars, which usually influence down-stream effector effects, for example by neutralizing pro-inflammatory cytokines, and extend the spectrum of non-steroidal anti-inflammatory drugs (NSAIDs) and corticosteroids to biologic anti-inflammatory drugs. Interventions targeting more proximal pathogenic events at the root of the disease, treating the cause rather than the symptoms, seem more logical. The clinical and commercial success of antihistamines and H_2R blockers encourage further research of H_3Rs and H_4Rs.

References

Abraham, S., & St. John, A. (2010). Mast cell-orchestrated immunity to pathogens. *Nature Rev Immunol, 10*, 440-452.

Adlesic, M., Verdrengh, M., Bokarewa, M., *et al.* (2007). Histamine in rheumatoid arthritis. *Scand J Immunol, 65*, 530-537.

Akdis, C., & Blaser, K. (2003). Histamine in immune regulation of allergic inflammation. *J Allergy Clin Immunol, 112*, 15-22.

Akdis, C., & Simons, E. (2002). Histamine receptors are hot in immunopharmacology. *Eur J Pharmacol, 533*, 69-76.

Amaral, M. M., Alvarez, C., Langellotti, C., *et al.* (2010). Histamine-treated dendritic cells improve recruitment of type 2 CD8 T cells in the lungs of allergic mice. *Immunology, 130*, 589-596.

Amaral, M. M., Davio, C., Ceballos, A., *et al.* (2007). Histamine improves antigen uptake and cross-presentation by dendritic cells. *J Immunol, 179*, 3425-3433.

Arstila, T. P., Casrouge, A., Baron, V., *et al.* (1999). A direct estimate of the human $\alpha\beta$ T cell receptor diversity. *Science, 286*, 958-961.

Arstila, T. P., Casrouge, A., Baron, V., *et al.* (2000). Diversity of human $\alpha\beta$ T cell receptors. *Science, 288*, 1135.

Ash, A., & Schild, H. (1966). Receptors mediating some actions of histamine. *Br J Pharmacol Chemother, 27*, 427-439.

Bakker, R. (2004). Histamine H_3-receptor isoforms. *Inflamm Res, 53*, 509-516.

Bakker, R., Wieland, K., Timmerman, H., *et al.* (2000). Constitutive activity of the histamine H receptor reveals inverse agonism of histamine H receptor antagonists. *Eur J Pharmacol, 387*, R5-R7.

Barnard, R., Barnard, A., Salmon, G., *et al.* (2008). Histamine-induced actin polymerization in human eosinophils: an imaging approach for histamine H_4 receptor. *Cytometry Part A, 73*, 299-304.

Bäumer, W., Stahl, J, Sander, K., *et al.* (2011). Lack of preventing effect of systemically and topically administered histamine H_1 or H_4 receptor antagonists in a dog model of acute atopic dermatitis. *Exp Dermatol, 20*, 577-581.

Bäumer, W., Wendorff, S., Gutzmer, R., et al. (2008). Histamine H_4 receptors modulate dendritic cell migration through skin-immunomodulatory role of histamine. *Allergy, 63*, 1387-1394.

Bielory, B. P., Mainardi, T., & Rottem, M. (2013). Evolutionary immune response to conserved domains in parasites and aeroallergens. *Allergy Asthma Proc, 34*, 93-102.

Biosse-Duplan, M., Baroukh, B., Dy, M., *et al.* (2009). Histamine promotes osteoclastogenesis through the differential expression of histamine receptors on osteoclasts and osteoblasts. *Am J Pathol, 174*, 1426-1434.

Bischoff, S. C. (2007). Role of mast cells in allergic and non-allergic immune responses: comparison of human and murine data. *Nat Rev Immunol, 7*, 93-104.

Blandina, P., Munari, L., Giannoni, P., *et al.* (2010). Histamine neuronal system as a therapeutic target for the treatment of cognitive disorders. *Future Neurol, 5*, 543-555.

Bonini, St., Sgrulletta, R., Coassin, M., & Bonini, Se. (2009). Allergic conjunctivitis: update on its pathophysiology and perspectives for future treatment. In: R. Pawankar, & T. Holgate (Eds.), Allergy Frontiers: Clinical Manifestations (pp. 25-48). Basel, Switzerland: Springer.

Bousquet, J, Jeffery, P., Busse, W., *et al.* (2000). Asthma. From bronchoconstriction to airways inflammation and remodelling. *Am J Resp. Crit Care Med, 161*, 1720-1745.

Boutin, Y., Jobin, M., Bernier, D., *et al.* (1993). Biological activity of monoclonal anti-idiotypic antibody representing the internal image of the major allergenic component of Lolium perenne pollen. *Int Arch Allergy Immunol, 102*, 10-14.

Breunig, E., Michel, K., Zeller, F., *et al.* (2007). Histamine excites neurones in the human submucous plexus through activation of H_1, H_2, H_3 and H_4 receptors. *J Physiol, 583,* 731-742.

Buckland, K., Williams, T., & Conroy, D. (2003). Histamine induces cytoskeletal changes in human eosinophils via the H_4 receptor. *Br J Pharmacol, 140,* 1117-1127.

Burrell, B. E., Nakayama, Y., Xu, J, *et al.* (2012). Regulatory T cell induction, migration, and function in transplantation. *J Immunol, 189,* 4705-4711.

Cao, J, Cetrulo, C. L., & Theoharides, T. C. (2006). Corticotropin-releasing hormone induces vascular endothelial growth factor release from human mast cells via the cAMP/protein kinase A/p38 mitogen-activated protein kinase pathway. *Mol Pharmacol, 69,* 998–1006.

Caron, G., Delneste, Y., Roelandts, E., *et al.* (2001). Histamine polarizes human dendritic cells into Th2 cell-promoting effector dendritic cells. *J Immunol, 167,* 3682-3686.

Čeponis, A., Konttinen, Y.T., Takagi, M., *et al.* (1998) Expression of stem cell factor (SCF) and SCR receptor (c-kit) in synovial membrane in arthritis: correlation with synovial mast cell hyperplasia and inflammation. *J Rheumatol, 25:* 2304-2314.

Chan, N., Robador, P., & Levi, R. (2012). Natriuretic peptide-induced catecholamine release from cardiac sympathetic neurons: Inhibition by histamine H_3 and H_4 receptor activation. *J Pharmacol Exp Ther, 343,* 568-577.

Chu, C., Di Meglio, P., & Nestle, F. (2011). Harnessing dendritic cells in inflammatory skin diseases. *Sem Immunol, 23,* 28-41.

Clark, R., Gallin, J, & Kaplan, A. (1975). The selective eosinophil chemotactic activity of histamine. *J Exp Med, 142,* 1462-1476.

Cogé, F., Guénin, S., Audinot, V., *et al.* (2001). Genomic organization and characterization of splice variants of the human histamine H_3 receptor. *J Biochem, 355,* 279-288.

Cogé, F., Guénin, S., Rique, H., *et al.* (2001). Structure and expression of the human histamine H_4-receptor gene. *Biochem Biophys Ress Commun, 284,* 301-309.

Cohen, I. R. (2007). Biomarkers, self-antigens and the immunological homunculus. *J Autoimmun, 29,* 246-249.

Connelly, W. M., Shenton, F. C., Lethbridge, N., *et al.* (2009). The histamine H_4 receptor is functionally expressed on neurons in the mammalian CNS. *Br J Pharmacol, 157,* 55-63.

Coron, E., Mosnier, J F., Ahluwalia, A., *et al.* (2012). Colonic mucosal biopsies obtained during confocal endomicroscopy are pre-stained with fluorescein in vivo and are suitable for histologic evaluation. *Endoscopy, 44,* 148-153.

Coruzzi, G., Adami, M., Guaita, E., *et al.* (2007). Antiinflammatory and antinociceptive effects of the selective histamine H_4-receptor antagonists JNJ7777120 and VUF6002 in a rat model of carrageenan-induced acute inflammation. *Eur J Pharmacol, 563,* 240-244.

Couillin, I., Maillet, I., Vargaftig, B.B., *et al.* (2004). Arthropod-derived histamine-binding protein prevents murine allergic asthma. *J Immunol,* 173, 3281-3286.

Cowden, J., Riley, J., Ma, J. *et al.* (2010a). Histamine H_4 receptor antagonism diminishes existing airway inflammation and dysfunction via modulation of Th2 cytokines. *Resp Res,* 11, 86.

Cowden, J., Zhang, M., Dunford, P. *et al.* (2010b). The histamine H4 receptor mediates Inflammation and pruritus in Th2-dependent dermal inflammation. *J Invest Dermatol,* 130, 1023-1033..

Cruikshank, W., Kornfeld, H., & Center, D. *et al.* (2000). Interleukin-16. *J Leukoc Biol, 67,* 757-766.

Daley-Yates, P., Ambery, C., Sweeney, L., *et al.* (2012). The efficacy and tolerability of two novel H_1/H_3 receptor antagonists in seasonal allergic rhinitis. *Int Arch Allergy Immunol, 158,* 84-98.

Damaj, B., Becerra, C., Esber, H., *et al.* (2007). Functional expression of H_4 histamine receptor in human natural killer cells, monocytes, and dendritic cells. *J Immunol, 179,* 7907-7915.

de Backer, M., Loonen, I., Verhasselt, P., *et al.* (1998). Structure of the human histamine H_1 receptor gene. *J Biochem, 335,* 663-70.

de Vries, V. C., Wasiuk, A., Bennett, K. A., *et al.* (2009). Mast cell degranulation breaks peripheral tolerance. *Am J Transpl, 9*, 2270-2280.

del Rio, R., Noubade, R., Saligrama, N., et al. (2012). Histamine H_4 receptor optimizes T regulatory cell frequency and facilitates anti-inflammatory responses within the central nervous system. *J Immunol, 188*, 541-547.

Desmadryl, G., Gaboyard-Niay, S., Brugeaud, A., *et al.* (2012). Histamine H_4 receptor antagonists as potent modulators of mammalian vestibular primary neuron excitability. *Br J Pharmacol, 167*, 905-916.

Dijkstra, D., Leurs, R., Chazot, P., *et al.* (2007). Histamine downregulates monocyte CCL2 production through the histamine H_4 receptor. *J Allergy Clin Immunol, 120*, 300-307.

Dijkstra, D., Stark, H., Chazot, P., *et al.* (2008). Human inflammatory dendritic epidermal cells express a functional histamine H_4 receptor. *J Invest Dermatol, 128*, 1696-1703.

Duan, H., & Wang, J (2010). Selective transport of monoamine neurotransmitters by human plasma membrane monoamine transporter and organic cation transporter 3. *J Pharmacol Exp Ther, 335*, 743-753.

Dunford, P., O'Donnell, N., Riley, J, *et al.* (2006). The histamine H_4 receptor mediates allergic airway inflammation by regulating the activation of $CD4^+$ T cells. *J Immunol, 176*, 7062-7070.

Dunford, P., Williams, K., Desai, P., *et al.* (2007). Histamine H_4 receptor antagonists are superior to traditional antihistamines in the attenuation of experimental pruritus. *J Allergy Clin Immunol, 119*, 176-183.

Dy, M., & Schneider, E. (2004). Histamine–cytokine connection in immunity and hematopoiesis. *Cytokine Growth Factor Rev, 15*, 393-410.

Eklund, K. K., & Joensuu, H. (2003). Treatment of rheumatoid arthritis with imatinib mesylate: clinical improvement in three refractory cases. *Ann Med, 35*, 362-367.

El Behi, M., Zéphir, H., Lefranc, D., *et al.* (2007). Changes in self-reactive IgG antibody repertoire after treatment of experimental autoimmune encephalomyelitis with anti-allergic drugs. *J Neuroimmunol, 182*, 80-88.

Emerson, M.R., Orentas, D.M., Lynch, S.G., *et al.* (2002). Activation of histamine H_2 receptors ameliorates experimental allergic encephalomyelitis. *Neuroreport, 13,* 1407-1410.

Feng, Z., Hou, T., & Li, Y. (2013). Docking and MD study of histamine H_4R based on the crystal structure of H_1R. *J Mol Graph, 39,* 1-12.

Firestein, G. (2003). Evolving concepts of rheumatoid arthritis. *Nature, 423,* 356-361.

Förster, R., Braun, A., & Worbs, T. *et al.* (2012). Lymph node homing of T cells and dendritic cells via afferent lymphatics. *Trends Immunol, 33,* 271-280.

Fox, C. C., Jewell, S. D., & Whitacre, C. C. (1994). Rat peritoneal mast cells present antigen to a PPD-specific T cell line. *Cellular Immunol, 158,* 253-264.

Frewin, D., Cleland, L., Jonsson, J, *et al.* (1986). Histamine levels in human synovial-fluid. *The J Rheum, 13,* 13-14.

Gantner, F., Sakai, K., Tusche, M., *et al.* (2002). Histamine H_4 and H_2 receptors control histamine-induced interleukin-16 release from human CD8+ T cells. *J Pharmacol Exp Ther, 303,* 300-307.

García-Martin, E., Mendoza, J L., Martínez, C., *et al.* (2006). Severity of ulcerative colitis is associated with a polymorphism at diamine oxidase gene but not at histamine N-methyltransferase gene. *World J Gastroenterol, 12,* 615-620.

Geng, S., Gao, Y.D., Yang, J, et al. (2012). Potential role of store-operated Ca^{2+} entry in Th2 response induced by histamine in human monocyte-derived dendritic cells. *Int Immunopharmacol, 12,* 358-367.

Goodnow, C., Sprent, J, Fazekas de St Groth, B., *et al.* (2005). Cellular and genetic mechanisms of self tolerance and autoimmunity. *Nature, 435,* 590-597.

Gordon, S. (2003). Alternative activation of macrophages. *Nature Rev Immunol, 3,* 23-35.

Gschwandtner, M., Mommert, S., Köther, B., *et al.* (2011). The histamine H_4 receptor is highly expressed on plasmacytoid dendritic cells in psoriasis and histamine regulates their cytokine production and migration. *J Invest Dermatol, 131,* 1668-1676.

Gschwandtner, M., Rossbach, K., Dijkstra, D., *et al.* (2010). Murine and human Langerhans cells express a functional histamine H_4 receptor: modulation of cell migration and function. *Allergy, 65*, 840-849.

Gutzmer, R., Diestel, C., Mommert, S., *et al.* (2005). Histamine H_4 receptor stimulation suppresses IL-12p70 production and mediates chemotaxis in human monocyte-derived dendritic cells. *J Immunol, 174*, 5224-5232.

Gutzmer, R., Gschwandtner, M., Rossbach, K., *et al.* (2011). Pathogenetic and therapeutic implications of the histamine H_4 receptor in inflammatory skin diseases and pruritus. *Front Biosci, 3*, 985-994.

Gutzmer, R., Mommert, S., Gschwandtner, M., *et al.* (2009). The histamine H_4 receptor is functionally expressed on Th2 cells. *J Allergy Clin Immunol, 123*, 619-625.

Haas, H., Sergeeva, O., & Selbach, O. (2008). Histamine in the nervous system. *Physiol Rev, 88*, 1183-1241.

Hancock, A., Esbenshade, T., Krueger, K., *et al.* (2003). Genetic and pharmacological aspects of histamine H_3 receptor heterogeneity. *Life Sci, 73*, 3043-3072.

Hashizume, H., Horibe, T., Yagi, H., *et al.* (2005). Compartmental imbalance and aberrant immune function of blood CD123^{+} (plasmacytoid) and CD11c^{+} (myeloid) dendritic cells in atopic dermatitis. *J Immunol, 174*, 2396-2403.

Himmel, M. E., Hardenberg, G., Piccirillo, C. A., *et al.* (2008). The role of T-regulatory cells and Toll-like receptors in the pathogenesis of human inflammatory bowel disease. *Immunology, 125*, 145-153.

Hishinuma, S., Komazaki, H., Fukui, H., *et al.* (2010). Ubiquitin/proteasome-dependent down-regulation following clathrin-mediated internalization of histamine H_1-receptors in Chinese hamster ovary cells. *J Neurochem, 113*, 990-1001.

Hochberg, M. C. (1997). Updating the American College of Rheumatology revised criteria for the classification of systemic lupus erythematosus. *Arthritis Rheum, 40*, 1725-1734.

Hofstra, C., Desai, P., Thurmond, R., *et al.* (2003). Histamine H_4 receptor mediates chemotaxis and calcium mobilization of mast cells. *J Pharmacol Exp Ther, 305*, 1212-1221.

Huang, J, & Thurmond, R. (2008). The new biology of histamine receptors. *Curr Allergy Asthma Rep, 8,* 21-27.

Hukkanen, M., Grönblad, M., Rees, R., *et al.* (1991). Regional distribution of mast cells and peptide-containing nerves in normal and adjuvant arthritic rat synovium. *J Rheumatol, 18,* 177-183.

Ichikawa, A., & Sugimoto, Y. (2010). Molecular biology of histidine decarboxylase and prostaglandin receptors. *Proc Jap Acad B, 86,* 848-866.

Idoyaga, J, Fiorese, C., Zbytnuik, L., *et al.* (2013). Specialized role of migratory dendritic cells in peripheral tolerance induction. *J Clin Invest, 123,* 844–854

Ihler, F., Bertlich, M., Sharaf, K., *et al.* (2012). Betahistine exerts a dose-dependent effect on cochlear stria vascularis blood flow in guinea pigs In vivo. *PLoS One, 7,* e39086.

Ikawa, Y., Shiba, K., Ohki, E., *et al.* (2008). Comparative study of histamine H_4 receptor expression in human dermal fibroblasts. *J Toxicol Sci, 33,* 503-508.

Ikawa, Y., Suzuki, M., Shiono, S., *et al.* (2005). Histamine H_4 receptor expression in human synovial cells obtained from patients suffering from rheumatoid arthritis. *Biol Pharm Bull, 28,* 2016-2018.

Imamura, T., Di Virgilio, S., Bagarozzi, D.A. Jr., *et al.* (1996) Induction of histamine release from non-immune guinea pigs: a possible involvement of lectin-like factor(s) in pollinosis. *International Arch Allergy Immunol, 111,* 161-165.

Jablonowski, J, Grice, C., Chai, W., *et al.* (2003). The first potent and selective non-imidazole human histamine H_4 receptor antagonists. *J Med Chem, 46,* 3957-3960.

Jadidi-Niaragh, F., & Mirshafiey, A. (2010). Histamine and histamine receptors in pathogenesis and treatment of multiple sclerosis. *Neuropharmacology, 59,* 180-189.

Jongbloed, S. L., Kassianos, A. J, McDonald, K. J, *et al. (2010).* Human $CD141^+$ $(BDCA\text{-}3)^+$ dendritic cells (DCs) represent a unique myeloid DC subset that cross-presents necrotic cell antigens. *J Exp Med, 207,* 1247-1260.

Jutel, M., Akdis, M., & Akdis, C. (2009). Histamine, histamine receptors and their role in immune pathology. *Clin Exp Allergy, 39,* 1786-1800.

Jutel, M., Watanabe, T., Klunker, S., *et al.* (2001). Histamine regulates T-cell and antibody responses by differential expression of H_1 and H_2 receptors. *Nature, 413*, 420-425.

Kajihara, Y., Murakami, M., Imagawa, T., *et al.* (2010). Histamine potentiates acid-induced responses mediating transient receptor potential v1 in mouse primary sensory neurons. *J NeuroSci, 166*, 292-304.

Kassianos, A. J, Hardy, M. Y., Ju, X., *et al.* (2012). Human $CD1c^+$ (BDCA-1) myeloid dendritic cells secrete IL-10 and display an immuno-regulatory phenotype and function in response to Escherichia coli. *Eur J Immunol, 42*, 1512-1522.

Katebe M, Lethbridge N, Chazot PL: Differential changes in H_4R expression in acute and chronic inflammatory pain model. In: 41st EHRS Meeting, Belfast, UK, 2012.

Kempuraj, D., Papadopoulou, N. G., Lytinas, M., *et al.* (2004). Corticotropin-releasing hormone and its structurally related urocortin are synthesized and secreted by human mast cells. *Endocrinology, 145*, 43–48.

Kitawaki, T., Kadowaki, N., Sugimoto, N., *et al.* (2006). IgE-activated mast cells in combination with pro-inflammatory factors induce Th2-promoting dendritic cells. *Int Immunol, 18*, 1789-1799.

Konttinen, Y. T., Kemppinen, P., Segerberg, M., *et al.* (1994). Peripheral and spinal neural mechanisms in arthritis, with particular reference to treatment of inflammation and pain. *Arthritis Rheum., 37*, 965-982.

Konttinen, Y., Tuominen, S., Segerberg-Konttinen, M., *et al.* (1990). Mast cells in the labial salivary glands of patients with Sjögren's syndrom: a histochemical, immunohistochemical, and electron microscopical study. *Ann Rheum Dis, 49*, 685-689.

Kouri, V-P., Olkkonen, J, Ainola, M., *et al.* (2013). Circadian timekeeping is disturbed in rheumatoid arthritis at molecular level. *PLoS One, 8*, e54049.

Kozma, G. T., Losonczy, G., Keszei, M., *et al.* (2003). Histamine deficiency in gene-targeted mice strongly reduces antigen-induced airway hyper-responsiveness, eosinophilia and allergen-specific IgE. *Int Immunol*, 15, 963-973.

Kubo, Y., & Nakano, K. (1999). Regulation of histamine synthesis in mouse $CD4^+$ and $CD8^+$ T lymphocytes. *Inflamm Res, 48*, 149-153.

Kurts, C., Kosaka, H., Carbone, F. R., *et al.* (1997). Class I-restricted cross-presentation of exogenous self-antigens leads to deletion of autoreactive CD8+ T cells. *J Exp Med, 186,* 239-245.

Laitinen, L. A., Laitinen, A., & Haahtela, (1993). Airway mucosal inflammation even in patients with newly diagnosed asthma. *Am Rev Resp Dis, 147,* 697-704.

Lapilla, M., Gallo, B., Martinello, M., *et al.* (2011). Histamine regulates autoreactive T cell activation and adhesiveness in inflamed brain microcirculation. *J Leukoc. Biol, 89,* 259-267.

Lebre, M. C., Jongbloed, S. L., Tas, S. W., *et al.* (2008). Rheumatoid arthritis synovium contains two subsets of CD83-DC-LAMP- dendritic cells with distinct cytokine profiles. *Am J Pathol, 172,* 940-950.

Leff, P. (1995). The two-state model of receptor activation. *Trends Pharmacol Sci,* 16, 89-97.

Leite-de-Moraes, M. C., Diem, S., Michel, M. L., *et al.* (2009). Cutting edge: histamine receptor H_4 activation positively regulates in vivo IL-4 and IFN-gamma production by invariant NKT cells. *J Immunol, 182,* 1233-1236.

Leite-de-Moraes, M., Diem, S., Michel, M., *et al.* (2009). Cutting Edge: Histamine receptor H_4 activation positively regulates in vivo IL-4 and IFN-gamma production by invariant NKT cells. *J Immunol, 182,* 1233-1236.

Lenert, P., Brummel, R., Field, E. H., *et al.* (2005). TLR-9 activation of marginal zone B cells in lupus mice regulates immunity through increased IL-10 production. *J Clin Immunol,* 25, 29-40.

Leonardi, A., Di Stefano, A., Vicari, C., *et al.* (2011). Histamine H_4 receptors in normal conjunctiva and in vernal keratoconjunctivitis. *Allergy, 66,* 1360-1366.

Lethbridge, N. L., & Chazot, P. L. (2010). Immunological identification of the mouse H4 histamine receptor on spinal cord motor neurons using a novel anti-mouse H_4R antibody. *Inflamm Res, 59,* S197-198.

Leurs, R., Bakker, R., Timmerman, H., *et al.* (2001). The histamine H_3 receptor: from gene cloning to H_3 receptor drugs. *Nat Rev Drug Disc, 4,* 107-120.

Leurs, R., Church, M., & Taglialatela, M. (2002). H_1-antihistamines: inverse agonism, anti-inflammatory actions and cardiac effects. *Clin Exp Allergy, 32,* 489-498.

Li, T. F., Mandelin, J, Hukkanen, M., *et al.* (2002). Dendritic cells in rheumatoid synovial membrane after total removal of the hyaline articular cartilage. *Rheumatology, 41,* 319-323.

Li, T. F., Santavirta, S., Waris, V., *et al.* (2001). No lymphokines in T-cells around loosened hip prostheses. *Acta Orthop Scand,* 72, 241-247.

Lim, H., van Rijn, R., Ling, P., *et al.* (2005). Evaluation of histamine H_1-, H_2-, and H_3-receptor ligands at the human histamine H_4 receptor: Identification of 4- methylhistamine as the first potent and selective H_4 receptor agonist. *J Pharmacol Exp Ther, 314,* 1310-1321.

Ling, P., Ngo, K., Nguyen, S., *et al.* (2004). Histamine H_4 receptor mediates eosinophil chemotaxis with cell shape change and adhesion molecule upregulation. *Br J Pharmacol, 142,* 161-171.

Lippert, U., Artuc, M., Grützkau, A., *et al.* (2004). Human skin mast cells express H_2 and H_4, but not H_3 receptors. *J Invest Dermatol, 123,* 116-123.

Little, F., de Bie, J, van Oosterhout, A., *et al.* (2003). Immunomodulatory effect of interleukin-16 on allergic airway inflammation. *Chest, 123,* 431S-432S.

Little, F., Lynch, E., Fine, G., *et al.* (2003). Tumor necrosis factor-alpha - Induced synthesis of interleukin-16 in airway epithelial cells - Priming for serotonin stimulation. *Am J Resp Cell Mol Biol, 28,* 354-362.

Liu, C., Ma, X., Jiang, X., *et al.* (2001). Cloning and pharmacological characterization of a fourth histamine receptor (H_4) expressed in bone marrow. *Mol Pharmacol, 59,* 420-426.

Liu, F., Goodarzi, H., & Chen, H. (2011). IgE, Mast Cells, and Eosinophils in Atopic Dermatitis. *Clin Rev Allergy Immun, 41,* 298-310.

Lomax, A. E., Fernandez, E., & Sharkey, K. A. (2005). Plasticity of the enteric nervous system during intestinal inflammation. *Neurogastroenterol Motility, 17,* 4–15.

Lovenberg, T., Roland, B., Wilson, S., *et al.* (1999). Cloning and functional expression of the human histamine H3 receptor. *Mol Pharmacol, 55,* 1101-1107.

Lu, C., Diehl, S. A., Noubade, R., *et al.* (2010). Endothelial histamine H_1 receptor signaling reduces blood-brain barrier permeability and susceptibility to autoimmune encephalomyelitis. *Proc Nat Acad Sci USA, 107,* 18967-18972.

Lundberg, K., Broos, S., Greiff, L., *et al.* (2011). Histamine H_4 receptor antagonism inhibits allergen-specific T-cell responses mediated by human dendritic cells. *Eur J Pharmacol, 651*, 197-204.

Ma, R. Z., Gao, J, Meeker, N. D., *et al.* (2002). Identification of Bphs, an autoimmune disease locus, as histamine receptor H_1. *Science, 297*, 620-623.

Maintz, L., & Novak, N. (2007). Histamine and histamine intolerance. *Am J Clin Nutr, 85*, 1185-1196.

Mann, M. K., Maresz, K., Shriver, L. P., *et al.* (2007). B cell regulation of $CD4^{+}CD25^{+}$ T regulatory cells and IL-10 via B7 is essential for recovery from experimental autoimmune encephalomyelitis. *J Immunol, 178*, 3447-3456.

Mantovani, A., Sica, A., Sozzani, S., *et al.* (2004). The chemokine system in diverse forms of macrophage activation and polarization. *Trends Immunol, 25*, 677-686.

Martinez, F. O., Helming, L., & Gordon, S. (2009). Alternative activation of macrophages: an immunologic functional perspective. *Ann Rev Immunol, 27*, 451-483.

Matsushita, A., Seike, M., Okawa, H., *et al.* (2012). Advantages of histamine H_4 receptor antagonist usage with H_1 receptor antagonist for the treatment of murine allergic contact dermatitis. *Exp Dermatol, 21*, 714-715.

Mauri, C., Gray, D., Mushtaq, N., *et al.* (2003). Prevention of arthritis by interleukin 10-producing B cells. *J Exp Med, 197*, 489-501.

McFadden, C., Morgan, R., Rahangdale, S., *et al.* (2007). Preferential migration of T regulatory cells induced by IL-16. *J Immunol, 179*, 6439-6445.

Mekori, Y. A., & Hershko, A. Y. (2012). T cell-mediated modulation of mast cell function: heterotypic adhesion-induced stimulatory or inhibitory effects. *Front Immunol, 3*, 6.

Mellanby, K. (1946) Man's reaction to mosquito bites. *Nature, 158*, 554.

Milligan, G., Bond, R., & Lee, M. (1995). Inverse agonism: pharmacological curiosity or potential therapeutic strategy? *Trends Pharmacol Sci, 16*, 10-13.

Mizoguchi, A., Mizoguchi, E., Takedatsu, H., *et al.* (2002). Chronic intestinal inflammatory condition generates IL-10 producing regulatory B cell subset characterized by CD1d upregulation. *Immunity, 16*, 219-230.

Mommert, S., Gschwandtner, M., Gutzmer, R., *et al.* (2011). The role of the histamine H_4 receptor in atopic dermatitis. *Curr Allergy Asthma Rep., 11*, 21-28.

Mommert, S., Gschwandtner, M., Koether, B., *et al.* (2012). Human memory Th17 cells express a functional histamine H_4 receptor. *Am J Pathol, 180*, 177-185.

Morgan, R., McAllister, B., Cross, L., *et al.* (2007). Histamine 4 receptor activation induces recruitment of FoxP3 + T cells and inhibits allergic asthma in a murine model. *J Immunol, 178*, 8081-8089.

Morisset, S., Rouleau, A., Ligneau, X., *et al.* (2000). High constitutive activity of native H_3 receptors regulates histamine neurons in brain. *Nature, 408*, 860-864.

Morse, K., Behan, J, Laz, T., *et al.* (2001). Cloning and characterization of a novel human histamine receptor. *J Pharmacol Exp Ther, 296*, 1058-1066.

Musio, S.,Gallo, B.,Scabeni, S., *et al.* (2006). A key regulatory role for histamine in experimental autoimmune encephalomyelitis: disease exacerbation in histidine decarboxylase-deficient mice. *J Immunol, 176*, 17-26.

Nakamura, T., Itadani, H., Hidaka, Y., *et al.* (2000). Molecular cloning and characterizationvof a new human histamine receptor, HH_4R. *Biochem Biophys Res Commun, 279*, 615-620.

Nakano, Y., Takahashi, Y., Ono, R., *et al.* (2009). Role of histamine H_4 receptor in allergic conjunctivitis in mice. *Eur J Pharmacol, 608*, 71-75.

Nakaya, M., Takeuchi, N., & Kondo, K. (2004). Immunohistochemical localization of histamine receptor subtypes in human inferior turbinates. *Ann Otol. Rhinol. Laryngol., 113*, 552-557.

Nguyen, T., Shapiro, D., George, S., *et al.* (2001). Discovery of a novel member of the histamine receptor family. *Mol Pharmacol, 59*, 427-433.

Niijima-Yaoita, F., Tsuchiya, M., Ohtsu, H., *et al.* (2012). Roles of histamine in exercise-induced fatigue: favouring endurance and protecting against exhaustion. *Biol Pharm. Bull., 35*, 91-97.

Noubade, R., Milligan, G., Zachary, J F., *et al.* (2007). Histamine receptor H_1 is required for TCR-mediated p38 MAPK activation and optimal IFN-gamma production in mice. *J Clin Invest, 117*, 3507-3518.

Nowak, E. C., de Vries, V. C., Wasiuk, A., *et al.* (2012). Tryptophan hydroxylase-1 regulates immune tolerance and inflammation. *J Exp Med,* 209, 2127-2135.

Oda, T., Morikawa, N., Saito, Y., *et al.* (2000). Molecular cloning and characterization of a novel type of histamine receptor preferentially expressed in leukocytes. *J Biol Chem, 275,* 36781-36786.

Ohki, E., Suzuki, M., Aoe, T., *et al.* (2007). Expression of histamine H_4 receptor in synovial cells from rheumatoid arthritic patients. *Biol Ppharm Bull, 30,* 2217-2220.

Ohsawa, Y., & Hirasawa, N. *et al.* (2012). The antagonism of histamine H_1 and H_4 receptors ameliorates chronic allergic dermatitis via anti-pruritic and anti-inflammatory effects in NC/Nga mice. *Allergy, 67,* 1014-1022.

Paesen, G., Adams, P., Harlos, K., *et al.* (1999). Tick histamine-binding proteins: isolation, cloning, and three-dimensional structure. *Mol Cell, 3,* 661-671.

Passani, M. B., & Ballerini, C. (2012). Histamine and neuroinflammation: insights from murine experimental autoimmune encephalomyelitis. *Front Systems Neurosci, 6,* 1-6.

Passani, M., & Blandina, P. (2011). Histamine receptors in the CNS as targets for therapeutic intervention. *Trends Pharmacol Sci, 32,* 242-249.

Pedotti, R., DeVoss, J J, Youssef, S., *et al.* (2003). Multiple elements of the allergic arm of the immune response modulate autoimmune demyelination. *PNAS, 100,* 1867-1872.

Permin, H., Skov, P. S., Norn, S., *et al.* (1981). Possible role of histamine in rheumatoid arthritis. Treatment with cimetidine and mepyramine. *Allergy, 36,* 435-436.

Petit-Bertron, A. F., Machavoine, F., Defresne, M. P., *et al.* (2009). H_4 histamine receptors mediate cell cycle arrest in growth factor induced murine and human hematopoietic progenitor cells. *PLoS One, 4,* 1-13.

Pramod, S.N., Vnekatesh, Y.P., Mahesh, P.A., (2007) Potato lectin activates basophils and mast cells of atopic subjects by its interaction with core chitobiose of cell-bound non-specific immunoglobulin E. *Clin Exp Immunol, 148,* 391-401.

Pudendran, B., Artis, D., (2012). New paradigms in type 2 immunity. *Science, 337,* 431-435.

Raithel, M., Winterkamp, S., Weidenhiller, M., *et al.* (2007). Combination therapy using fexofenadine, disodium cromoglycate, and a hypoallergenic amino acid-based formula induced remission in a patient with steroid-dependent, chronically active ulcerative colitis. *Int J Colorectal Dis, 22*, 833-839.

Rossbach, K., Nassenstein, C., Gschwandtner, M., *et al.* (2011). Histamine H_1, H_3 and H_4 receptors are involved in pruritus. *J Neurosci, 190*, 89-102.

Sander, L. E., Lorentz, A., Sellge, G., *et al.* (2006). Selective expression of histamine receptors H_1R, H_2R, and H_4R, but not H_3R, in the human intestinal tract. *Gut, 55*, 498-504.

Schneider, E., Bertron, A. F., & Dy, M. (2011). Modulation of hematopoiesis through histamine receptor signaling. *Front Biosci, 3*, 467-473.

Seike, M., Furuya, K., Omura, M., *et al.* (2010). Histamine H_4 receptor antagonist ameliorates chronic allergic contact dermatitis induced by repeated challenge. *Allergy, 65*, 319-326.

Shan, L., Bossers, K., Luchetti, S., *et al.* (2012). Alterations in the histaminergic system in the substantia nigra and striatum of Parkinson's patients: a postmortem study. *Neurobiol Aging, 33*, 2585-2598.

Shi, Y., Sheng, R., Zhong, T., *et al.* (2012). Identification and characterization of ZEL-H16 as a novel agonist of the histamine H3 receptor. *PLoS One, 7*, e42185.

Shiboski, S. C., Shiboski, C. H., Criswell, L., *et al.* (2012). American College of Rheumatology classification criteria for Sjögren's syndrome: a data-driven, expert consensus approach in the Sjögren's International Collaborative Clinical Alliance cohort. *Arthritis Care Res (Hoboken), 64*, 475-487.

Shimuzu, T. & Taira, N. (1980). Pharmacological analysis of salivary and blood flow responses to histamine of the submandibular gland of the dog. *Br J Pharmacol, 68*, 651-661.

Shiraishi, Y., Jia, Y., Domenico, J, *et al.* (2013). Sequential engagement of FcεRI on mast cells and basophil histamine H_4 receptor and FcεRI in allergic rhinitis. *J Immunol, 190*, 539-548.

Simon, D., Braathen, L., & Simon, H. (2004). Eosinophils and atopic dermatitis. *Allergy, 59*, 561-570.

Simon, T., Laszlo, V., Lang, O., *et al.* (2011). Histamine regulates relevant murine dendritic cell functions via H_4 receptor. *Front Biosci, 1,* 1414-1424.

Simoni, Y., Diana, J, Ghazarian, L., *et al.* (2013). Therapeutic manipulation of natural killer (NK) T cells in autoimmunity: are we close to reality? *Clin Exp Immunol, 171,* 8-19.

Simons, F. (1999). Is antihistamine (H_1-receptor antagonist) therapy useful in clinical asthma? *Clin Exp Allergy, 29,* 98-104.

Simons, F. E. (2004). Drug therapy-advances in H_1-antihistamines. *New Eng J Med, 351,* 2203-2217.

Simons, F. E. (2008). Anaphylaxis. *J Allergy Clin Immunol, 121,* S402-S407.

Skoner, D. (2001). Allergic rhinitis: Definition, epidemiology, pathophysiology, detection, and diagnosis. *J Allergy Clin Immunol, 108,* S2-S8.

Smit, M., Leurs, R., Alewijnse, A., *et al.* (1996). Inverse agonism of histamine H_2 antagonists accounts for upregulation of spontaneously active histamine H_2 receptors. *Pharmacology, 93,* 6802-6807.

Spies, C. M., Straub, R. H., & Buttgereit, F. *et al.* (2012). Energy metabolism and rheumatic diseases: from cell to organism. *Arthritis Res Ther, 14,* 216.

Ständer, S., & Steinhoff, M. (2002). Pathophysiology of pruritus in atopic dermatitis: an overview. *Exp Dermatol, 11,* 12-24.

Stegaev, V., Sillat, T., Porola, P., *et al.* (2012). First identification of H_4 histamine receptor in healthy salivary glands and in focal sialadenitis in Sjögren's syndrome. *Arthritis Rheum, 64,* 2663-2668.

Stegaev, V., Nies, A.T., Porola, P. et al. (2013). Histamine transport and metabolism are deranged in salivary glands in Sjogren's syndrome. *Rheumatology* (Oxford), 52, 1599-1608.

Steinman, R. M., & Banchereau, J. (2007). Taking dendritic cells into medicine. *Nature, 449,* 419-426.

Steinman, R. M., Hawiger, D., & Nussenzweig, M. C. (2003). Tolerogenic dendritic cells. *Ann Rev Immunol, 21,* 685-711.

Stelekati, E., Bahri, R., D'Orlando, O., *et al.* (2009). Mast cell-mediated antigen presentation regulates CD8⁺ T cell effector functions. *Immunity, 31*, 665-676.

Sternberg, E. M., Young, W. S., Bernardini, R., *et al.* (1989). A central nervous system defect in biosynthesis of corticotropin-releasing hormone is associated with susceptibility to streptococcal cell wall-induced arthritis in Lewis rats. *PNAS, 6*, 4771-4775.

Strakhova, M. I.,Nikkel, A. L.,Manelli, A. M., *et al.* (2009). Localization of histamine H_4 receptors in the central nervous system of human and rat. *Brain Res, 1250*, 41-48.

Sugata, Y., Okano, M., Fujiwara, T., *et al.* (2007). Histamine H_4 receptor agonists have more activities than H_4 agonism in antigen-specific human T-cell responses. *J Immunol, 121*, 266-275.

Suwa, E., Yamaura, K., Oda, M., *et al.* (2011). Histamine H_4 receptor antagonist reduces dermal inflammation and pruritus in a hapten-induced experimental model. *Eur J Pharmacol, 667*, 383-388.

Suzuki, S., Takeuchi, K., & Majima, Y. (2008). Localization and function of histamine H_3 receptor in the nasal mucosa. *Clin Exp Allergy*, 38, 1476-1482.

Szeberényi, J, Pállinger, É., Zsinkó, M., et *al.* (2001). Inhibition of effects of endogenously synthesized histamine disturbs in vitro human dendritic cell differentiation. *Immunol Lett, 76*, 175-182.

Szegedi, G., Kávai, M., Horváth, S., *et al.* (1979). Histamine-binding receptors and Fc receptors of T lymphocytes in SLE. *Ann Immunol Hung, 19*, 137-141.

Takahashi, Y., Kagawa, Y., Izawa, K. et al. (2009). Effect of histamine H(4) receptor antagonist on allergic rhinitis in mice. *Int Immunopharmacol, 9*, 734-738.

Takeshita, K., Bacon, K., & Gantner, F. (2004). Critical role of L-selectin and histamine H_4 receptor in zymosan-induced neutrophil recruitment from the bone marrow: Comparison with carrageenan. *J Pharmacol Exp Ther, 310*, 272-280.

Takeshita, K., Sakai, K., Bacon, K., *et al.* (2003). Critical role of histamine H_4 receptor in leukotriene B4 production and mast cell-dependent neutrophil recruitment induced by zymosan in vivo. *J Pharmacol Exp Ther, 307*, 1072-1078.

Tanaka, S., & Ichikawa, A. (2011). Regulation of mammalian histamine synthesis: histidine decarboxylase. In: M. Shahid, N. Khardori, R.A. Khan, & T. Tripathi. (Eds.), Biomedical Aspects of Histamine. Current Perspectives (pp. 15-30). New York, NY: Springer.

Tanda, S., Mori, Y., Kimura, T., *et al.* (2007). Histamine ameliorates anti-glomerular basement membrane antibody-induced glomerulonephritis in rats. *Kidney Int, 72*, 608-613.

Tardivel-Lacombe, J, Morriset, S., Gbhaou, F., *et al.* (2001). Chromosomal mapping and organization of the human histamine H_3 receptor gene. *Mol Cellular Neurosci, 12*, 321-324.

Teuscher, C., Poynter, M. E., Offner, H., *et al.* (2004). Attenuation of Th1 effector cell responses and susceptibility to experimental allergic encephalomyelitis in histamine H_2 receptor knockout mice is due to dysregulation of cytokine production by antigen-presenting cells. *Am J Pathol, 164*, 883-892.

Teuscher, C., Subramanian, M., Noubade, R., *et al.* (2007). Central histamine H_3 receptor signaling negatively regulates susceptibility to autoimmune inflammatory disease of the CNS. *Proc Nat Acad Sci USA, 104*, 10146-10151.

Thurmond, R. L., Desai, P. J, Dunford, P. J, *et al.* (2004). A potent and selective histamine H_4 receptor antagonist with anti-inflammatory properties. *J Pharmacol Exp Ther, 309*, 404-413.

Thurmond, R., Gelfand, E., & Dunford, P. (2008). The role of histamine H_1 and H_4 receptors in allergic inflammation: the search for new antihistamines. *Nat. Rev Drug Disc., 7*, 41-53.

Toebak, M. J, Gibbs, S., Bruynzeel, D. P., *et al.* (2009). Dendritic cells: biology of the skin. *Contact Dermatitis, 60*, 2-20.

Traiffort, E., Vizuete, M., Tardivel-Lacombe, J, *et al.* (1995). The guinea pig histamine H_2 receptor: Gene cloning, tissue expression and chromosomal localization of its human counterpart. *Biochem Biophys Res Commun, 211*, 570-577.

Tsumiyama, K., Miyazaki, Y., & Shiozawa, S. (2009). Self-organized criticality theory of autoimmunity. *PLoS One, 4*, e8382.

van Rijn, R., van Marle, A., Chazot, P., *et al.* (2008). Cloning and characterization of dominant negative splice variants of the human histamine H_4 receptor. *J Biochem, 414*, 121-131.

Varga, C., Horvath, K., Berko, A., *et al.* (2005). Inhibitory effects of histamine H_4 receptor antagonists on experimental colitis in the rat. *Eur J Pharmacol, 522*, 130-138.

Walter, M., Kottke, T., & Stark, H. (2011). The histamine H_4 receptor: Targeting inflammatory disorders. *Eur J Pharmacol, 668*, 1-5.

Wellendorph, P., Goodman, M., Burstein, E., *et al.* (2002). Molecular cloning and pharmacology of functionally distinct isoforms of the human histamine H_3 receptor. *Neuropharmacology, 42*, 929-940.

Wiedemann, P., Bönisch, H., Oerters, F., *et al.* (2002). Structure of the human histamine H_3 receptor gene (HRH3) and identification of naturally occurring variations. *J Neural Trans, 109*, 443-453.

Wilson, C. B., Gushwa, L. C., Peterson, O. W., *et al.* (1981). Glomerular immune injury in the rat: effect of antagonists of histamine activity. *Kidney Int*, 20, 628-635.

Wilson, D. C. (1953). Effect of an anti-histamine in rheumatoid arthritis. *Ann Rheum Dis*, 12, 38-39.

Wilson, M.S., Mentink-Kane, M.M., Pesce, JT., *et al.* (2007). Immunopathology of schistosomiasis. *Immun Cell Biol*, 85, 148-154.

Winterkamp, S., Weidenhiller, M., Otte, P., *et al.* (2002). Urinary excretion of N-methylhistamine as a marker of disease activity in inflammatory bowel disease. *Am J Gastroenterol, 97*, 3071-3077.

Wollenberg, A., Kraft, S., Hanau, D., *et al.* (1996). Immunomorphological and ultrastructural characterization of Langerhans cells and a novel, inflammatory dendritic epidermal cell (IDEC) population in lesional skin of atopic eczema. *J Invest Dermatol*, 106, 446-453.

Wollenberg, A., Sharma, S., von Bubnoff, D., *et al.* (2001). Topical tacrolimus (FK506) leads to profound phenotypic and functional alterations of epidermal antigenpresenting dendritic cells in atopic dermatitis. *J Allergy Clin Immunol, 107*, 519-525.

Wood, J D. (2004). Enteric neuroimmunophysiology andpathophysiology. *Gastroenterology, 127,* 635–657.

Xu, A., Kuramasu, A., Maeda, K., *et al.* (2008). Agonist-induced internalization of histamine H_2 receptor and activation of extracellular signal-regulated kinases are dynamin-dependent. *J Neurochem, 107,* 208-217.

Yamaura, K., Oda, M., Suwa, E., *et al.* (2009). Expression of histamine H_4 receptor in human epidermal tissues and attenuation of experimental pruritus using H_4 receptor antagonist. *J Toxicol Sci, 34,* 427-431.

Yamaura, K., Oda, M., Suzuki, M., *et al.* (2012). Lower expression of histamine H_4 receptor in synovial tissues from patients with rheumatoid arthritis compared to those with osteoarthritis. *Rheumatol Int, 32,* 3309-3313.

Yamaura, K., Shigemori, A., Suwa, E., *et al.* (2013). Expression of the histamine H_4 receptor in dermal and articular tissues. *Life Sci, 92,* 108-113.

Yu, B., Shao, Y., Li, P., *et al.* (2010). Copy number variations of the human histamine H_4 receptor gene are associated with systemic lupus erythematosus. *Br J Dermatol, 163,* 935-940.

Yu, B., Shao, Y., Zhang, J, *et al.* (2010). Polymorphisms in human histamine receptor H_4 gene are associated with atopic dermatitis. *Br J Dermatol, 162,* 1038-1043.

Yuan, X., & Malek, T. R. (2012). Cellular and molecular determinants for the development of natural and induced regulatory T cells. *Human Immunol, 73,* 773-782.

Zhu, Y., Michalovich, D., Wu, H., *et al.* (2001). Cloning, expression and pharmacological characterization of a novel human histamine receptor. *Mol Pharmacol, 59,* 434-441.

Chapter 8

Histamine in Cancer

Vanina A. Medina[1,2], Gabriella Coruzzi[3], Diego J. Martinel Lamas[1], Noelia Massari[1], Maristella Adami[3], Francesca Levi-Schaffer[4], Micha Ben-Zimra[4], Hubert Schwelberger[5] and Elena S. Rivera[1]

[1]Laboratory of Radioisotopes, School of Pharmacy and Biochemistry, University of Buenos Aires, e-mail: elesurivera@gmail.com

[2]National Scientific and Technical Research Council (CONICET), Argentina

[3]Department of Neuroscience, University of Parma, Italy

[4]Pharmacology Unit, School of Pharmacy, Institute For Drug Research, Faculty of Medicine, the Hebrew University of Jerusalem, Jerusalem, Israel

[5]Molecular Biology Laboratory, Department of Visceral, Transplant and Thoracic Surgery, Medical University Innsbruck, Innsbruck, Austria

Abstract

This chapter addresses the latest evidence regarding the involvement of histamine and histamine receptors in cancer development and progression with particular attention to H_4R. The expression of H_4R in different cell lines, the signal transduction pathways and biological responses associated with H_4R as well as the *in vivo* treatment of experimental tumours with H_4R ligands will be reviewed. The controversial aspects of histaminergic involvement in gastrointestinal physiopathology will be discussed, with particular reference to gastric mucosal damage and intestinal inflammation. In addition, the existing evidence of the presence of mast cells in human and murine lung cancer and the effect of histamine and other mast cell derived mediators in this cancer will be presented. Recent findings support the therapeutic potential of H_4R ligands in melanoma, breast, colon, pancreatic and lung cancer and the exploitation of the H_4R as a molecular target for novel approaches in the treatment of cancer.

8.1. Introduction: Histamine and Cell Proliferation

The role of histamine in cell growth and proliferation has been extensively investigated. The association between histamine and malignant tumours was first proposed in the 1960s (Khalson & Rosegreen, 1968) and still remains controversial. Since the first report (Bartholeyns & Bouclier, 1984) showing that the inhibition of the histamine-synthesizing enzyme L-histidine decarboxylase (HDC, EC 4.1.1.22) with monofluormethylhistidine lead to anti-tumoural effects on experimental tumours in rodents, a large body of experimental evidence has been collected indicating that histamine can modulate the proliferation of different normal and malignant cells (for review see Medina & Rivera, 2010b). Moreover, most malignant cell lines and experimental tumours express this histamine-synthesizing enzyme and contain high concentrations of endogenous histamine. In this regard, histamine concentration is higher in diverse human tumours compared to surrounding normal tissue, including melanomas, colon and breast cancer (Garcia-Caballero et al., 1994; Reynolds et al., 1997; Hegyesi et al., 2001; Sieja et al., 2005; von Mach-Szczypiński et al., 2009). Additionally, numerous human cancer cell lines, as well as tumoural tissues, express histamine receptor subtypes with the ability to regulate cell proliferation (Tilly et al., 1990; Davio et al., 1993; Rivera et al., 1993; Cricco et al., 1994; Lemos et al., 1995; Davio et al., 1996; Wang et al., 1997; Falus et al., 2001; Molnar et al., 2001; Molnar et al., 2002; Cianchi et al., 2005; Hegyesi et al., 2005; Medina et al., 2006; Cricco et al., 2008; Davenas et al., 2008; Medina et al., 2008; Medina & Rivera, 2010b).

Histamine may regulate diverse biological responses related to tumour growth with opposite effects depending on the concentration and/or the type of receptor (Rivera et al., 2000), including cell proliferation, migration, differentiation and apoptosis. In addition, it can act as an angiogenic factor and induce vascular endothelial growth factor (VEGF) production, thus influencing the process of tumour invasion and metastasis (Sorbo et al., 1994; Ghosh et al., 2001). Furthermore, it has been reported that histamine is involved in the modulation of the immune response (Falus et al., 2001; Medina & Rivera, 2010a).

The identification of the human histamine H_4 receptor (H_4R) more than 10 years ago by several groups has helped refine our understanding of the roles of histamine, including the modulation of immune function. Increasing evidence supports the major role of H_4R in inflammatory diseases such as pruritus, asthma, and allergic rhinitis and as a result, growing attention is directed toward the therapeutic development of H_4R ligands for inflammation and immune disorders. However, the functional expression of H_4R in different types of tumours suggests novel functions for histamine in the carcinogenesis and further indications of H_4R modulators in cancer, leading to reconsideration of new perspectives in histamine pharmacology research (Huang & Thurmond, 2008; Leurs et al., 2009; Tiligada et al., 2009; Zampeli & Tiligada, 2009, Medina & Rivera, 2010b; Kiss & Keserű, 2012).

H_4R isoforms with different ligand binding and signalling characteristics, hetero-oligomerization and post-translational changes of this receptor coupled with the recently described biased GPCR signalling contribute to additional pharmacological complexity for H_4R ligand development, while the design of specific biased ligands could potentially increase the therapeutic index of these drugs (Rosethorne & Charlton, 2011; Seifert et al., 2011; Nijmeijer et al., 2012).

The recent observation that H_4R expression is altered in neoplastic tissue and that H_4R activation by histamine decreases malignant cell proliferation may shed new light on the controversial role of histamine in cancer development and progression.

8.2. Histamine in Breast Cancer

8.2.1. Introduction

Approximately 1 million cases of breast cancer are diagnosed annually worldwide. Breast cancer is the second most common cancer worldwide after lung cancer and the leading cause of cancer death in women. Even with advances in early detection, about 30% of patients with early-stage breast cancer have recurrent disease, which is metastatic in most cases, showing a 5-year survival rate of 20% (Gonzalez-Angulo et al., 2007; Ferlay et al., 2010; Jemal et al., 2010).

Breast cancer is a heterogeneous disease in terms of presentation, morphology, molecular profile, and clinical response to therapy. The identification of genes and biochemical pathways involved in breast carcinogenesis are of utmost importance for the development of therapeutic approaches that offer increased efficacy and low toxicity (Camidge & Jodrell, 2005; Fentiman, 2005; Medina & Rivera, 2010b).

8.2.2. Histamine in Normal and Malignant Mammary Gland

Histamine is involved in growth regulation, differentiation and the functioning of the mammary gland during development, pregnancy and lactation. Moreover, there is increasing evidence to demonstrate that histamine plays a significant role in breast cancer progression, since functional histamine receptors and HDC activity are present in breast tissue (Malinski et al., 1993; Davio et al., 1994; Wagner et al., 2003; Pós, et al., 2004; Medina & Rivera, 2010b). Histamine is increased in plasma and cancerous tissue derived from breast cancer patients compared to a healthy group and its plasma concentration is dependent on the number of involved lymph nodes and the grade of histologic malignancy. The significant elevation of histamine concentrations in the cancerous tissues of women with ductal breast cancers is associated with an increased HDC activity and decreased activity of the catabolic enzyme diamine oxidase (DAO, EC 1.4.3.6), suggesting that changes in the activity of the enzymes involved in the biotransformation of histamine are involved in ductal breast cancer development (Reynolds et al., 1998; Sieja et al., 2005; von Mach-Szczypiński et al., 2009). Endogenous histamine levels were also evaluated in cell lines derived from human mammary glands. The triple negative MDA-MB-231 breast cancer cell line exhibited higher histamine content than the estrogen-responsive human breast cancer cell line MCF-7, itself higher than the non-tumourigenic breast epithelial HBL-100 cells, suggesting a direct correlation of endogenous histamine levels with malignancy of breast cells (Medina et al, 2006).

To further explore the importance of histamine catabolism in breast cancer cells, in this study the histamine catabolising enzymes were investigated in MDA-MB-231 and MCF-7 cell lines. Histamine *N*-methyltransferase (HMT, DC 2.1.1.8) was expressed in MDA-MB-231 and MCF-7 cells, with HMT activity detected in both cell lines and exogenous histamine treatments decreasing the activity of HMT. DAO was not detected in both cell lines either at the protein or mRNA levels and non-significant modification of intracellular histamine levels was observed upon histamine treatment in both cell lines. Immunohistochemical analysis indicated that HMT or DAO enzymes were not detected in the xenograft tumours of the highly invasive human breast cancer cells MDA-MB-231 in immune deficient *nude* mice, which is in agreement with a high observed histamine intracellular content. These results demonstrate that HMT is the only histamine inactivating enzyme active in these cells, suggesting a crucial role of this enzyme in histamine catabolism in breast cancer cells (Figure 8.1).

In addition, histamine is an important mediator of immunologic reactions of breast glands. Experiments in syngeneic breast cancer induced in histamine-free (HDC knock-out) and wild-type mice demonstrated that endogenous histamine stimulates the growth of breast adenocarcinoma tumour implants by suppressing anti-tumour immunity (Hegyesi et al., 2007).

8.2.3. Histamine Receptors and Regulation of Breast Cancer Cell Growth

Previous data demonstrate that histamine exerts a regulatory function on cell growth by acting directly on specific H_1R and H_2R expressed on the cell membrane of N-nitro-N-methylurea (NMU)-induced experimental mammary carcinomas. In this experimental model, astemizole (an H_1R antagonist, 2 mg/kg.day) not only increases the number of tumours per rat but also significantly decreases their latency period. In addition, *in vivo* treatment with H_2R antagonists (ranitidine or cimetidine) produces a significant decrease in tumoural incidence and in the number of tumours developed per rat, as well as an increase in the latency period (Rivera et al., 1993, Cricco et al., 1994; Davio et al., 1995; Rivera et al., 2000).

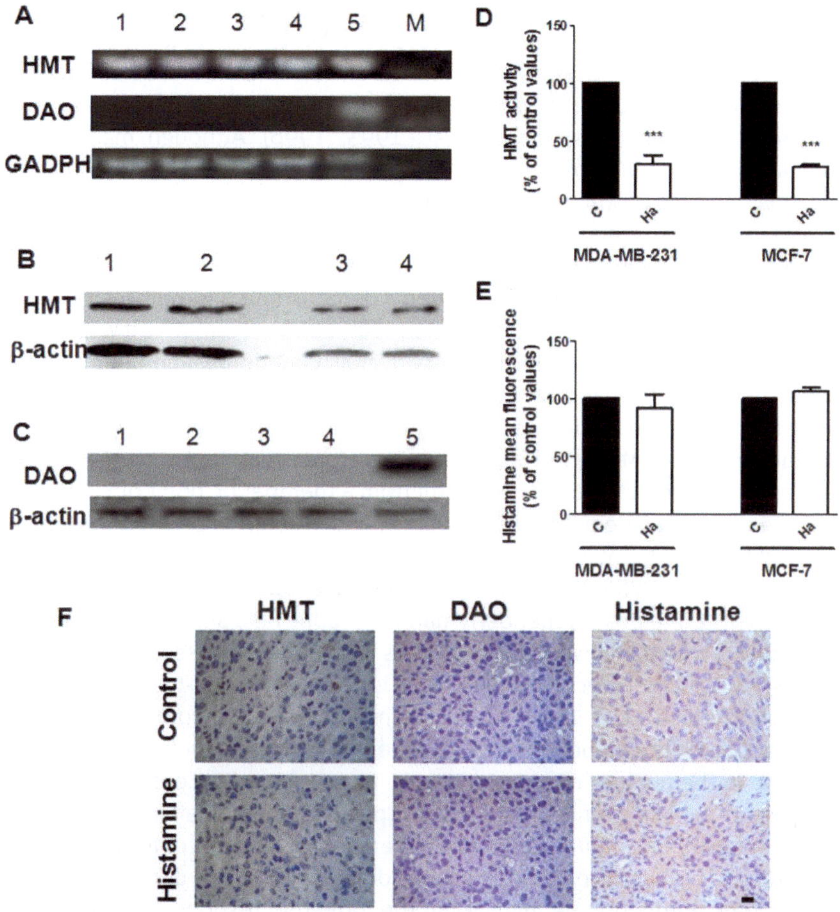

Figure 8.1 Histamine catabolism in MDA-MB-231 and MCF-7 breast cancer cells. (A) Cells were left untreated or were treated with histamine (Ha, 10 μM) for 24 h. **(A) HMT (555 bp) and DAO (558 bp) expression was determined by RT-PCR.** Lanes: 1, untreated MCF-7 cells; 2, histamine-treated MCF-7 cells; 3, untreated MDA-MB-231 cells; 4, histamine-treated MDA-MB-231 cells; 5, positive control (kidney lysate); M, molecular size marker. Glyceraldehyde 3-phosphate dehydrogenase (GAPDH, 576 bp) was used as load control **(B) Western blot analysis of HMT (33 kDa).** Lanes: 1, untreated MCF-7 control; 2, histamine-treated MCF-7 cells; 3, untreated MDA-MB-231 cells; 4, histamine-treated MDA-MB-231 cells. **(C) Western blot analysis of DAO (100 kDa)**, lanes: 1, untreated MCF-7 control; 2, histamine-treated MCF-7 cells; 3, untreated MDA-MB-231 cells; 4, histamine-treated MDA-MB-231 cells; 5, positive control (kidney lysate); β-actin (42 kDa) was used as load control **(D) HMT activity determination by a radiometric assay based on the transmethylation of histamine with S-adenosyl-L-[methyl-^{14}C] methionine.** Error bars represent the means ± SEM. (***$P<0.0001$, vs. Control (C); t-Test). **(E) Histamine intracellular content was determined by immunofluorescence analysis and flow cytometry.** Error bars represent the means ± SEM. **(F) Immunohistochemistry of HMT, DAO and histamine on the triple-negative MDA-MB-231 xenografted tumour induced in *nude* mice.** Representative immunohistochemical images in paraffin-embedded tumour specimens of untreated and histamine (1mg/kg, *s.c.*) treated animals. Original magnification, x630. Scale bar = 20 μm.

The presence of H_1R and H_2R in normal and malignant tissues was previously demonstrated. H_2R produces an increase in cAMP levels while H_1R is coupled to phospholipase C (PLC) activation in benign lesions. On the other hand, H_1R is solely linked to the PLC pathway but H_2R stimulates both transductional pathways in carcinomas (Davio et al., 1993; 1996). Although many reports indicate the expression of these two histamine receptor subtypes in normal and malignant tissues as well as in different cell lines derived from human mammary gland (Davio et al., 1993; Lemos et al., 1995; Medina & Rivera, 2010b), preclinical studies with H_1R antagonists (Rivera et al., 1993) and the clinical trials carried out with H_2R antagonists in cancer patients demonstrate unpromising results for breast cancer (Bolton et al., 2000; Parshad et al., 2005). In addition, it was recently shown that the use of the H_2R blockers overall, cimetidine, and famotidine are not associated with an increased risk of either invasive ductal or invasive lobular breast cancer, while current users of ranitidine have a 2.2-fold increased risk of ductal carcinoma (Mathes et al., 2008). Therefore, treatment with H_1R and H_2R antagonists does not seem to be useful from a therapeutic point of view.

Recently, it was demonstrated that H_3R and H_4R are expressed in cell lines derived from the human mammary gland (Medina et al., 2006; 2008). Histamine modulates the proliferation of the triple negative human breast cancer cell line MDA-MB-231 in a dose-dependent manner, producing a significant decrease at 10 $\mu mol \cdot L^{-1}$, where at lower concentrations it moderately increases proliferation through H_3R and no effect on proliferation is observed in non-tumourigenic HBL-100 cells (Medina et al., 2006; 2008). The negative effect on proliferation is also associated with the induction of cell cycle arrest, differentiation and a significant increase in the number of apoptotic cells. Furthermore, it is associated with an increase in the levels of reactive oxygen species (ROS) and an imbalance in the activity of enzymes implicated on their metabolism (Medina et al., 2006).

Through the use of pharmacological compounds, it was demonstrated that the main receptor subtype involved in the inhibitory response on proliferation is H_4R (Medina et al., 2008; 2011a). In addition, the effect of clobenpropit (H_4R agonist and H_3R antagonist), the putative H_4R agonist VUF8430 and/or the combined treatment with the specific antagonist JNJ7777120 was evaluated on the proliferation and apoptosis of MDA-MB-231 cells. Both agonists reduce the incorporation of BrdU, induced cell cycle arrest and increase cell apoptosis and senescence while treatment with JNJ7777120 reverses the histamine inhibitory effect on cell proliferation and senescence (Medina et al., 2008; 2011a). To further confirm the involvement of H_4R in the reduction of proliferation, H_4R siRNA was employed to knockdown H_4R expression resulting in the blockade of the inhibitory effect of histamine on proliferation (Martinel Lamas et al., 2013).

Furthermore, histamine differentially regulates the expression and activity of matrix metalloproteinases, cell migration and invasiveness through H_2R and H_4R in MDA-MB-231 cells modulating H_2O_2 intracellular levels (Cricco et al., 2011).

In addition, histamine at all doses tested decreases the proliferation of a more differentiated breast cancer cell line, MCF-7, through the stimulation of the four histamine receptor subtypes, exhibiting a higher effect through the H_4R. Treatment of MCF-7 cells with H_4R agonists augments the number of cells in the G0/G1 phase of the cell cycle and also exerts an anti-proliferative effect, augmenting the number of apoptotic and senescent cells (Medina et al., 2011a; Martinel Lamas et al., 2013).

These results represent the first reports on the expression of H_3R and H_4R in human breast cells and interestingly show that H_4R is involved in the regulation of breast cancer cell proliferation, apoptosis, senescence, migration and invasion.

8.2.4. Histamine in Human Breast Cancer and Therapeutic Approaches

In agreement with the above results, recent reports indicate that H_3R and H_4R are expressed in human biopsies of benign lesions and breast carcinomas, with the level of their expression being significantly higher in carcinomas, confirming that H_3R and H_4R are present not only in cell lines but also in human breast tissue. Furthermore, H_3R was positively correlated with a marker of proliferation in breast carcinomas while 50% of malignant lesions expressed H_4R (Medina et al., 2008).

The triple-negative tumours are undeniably one of the most relevant subgroups of breast cancer given the lack of targeted therapies for this group and their aggressive clinical behaviour (Sotiriou et al., 2003; Badve et al., 2011; Pal et al., 2011). In this regard, in order to better understand the role of H_4R in breast cancer and to explore novel therapeutic approaches that could offer increased efficacy and low toxicity, the anti-tumour effect of other H_4R agonists (Clozapine and JNJ28610244) was evaluated *in vivo* and *in vitro* in a triple negative human breast cancer model. Clozapine is an atypical antipsychotic that has been used for a long time, is clinically available, approved for use in humans and shows high affinity for H_4R, with several works supporting its use as an H_4R agonist *in vitro* and *in vivo* (Oda et al., 2000; Lim *et al.*, 2005; van Rijn et al., 2008; Leurs et al., 2009; Vera et al., 2012). The compound JNJ28610244 has excellent potency and selectivity for H_4R, serving as a useful pharmacological tool for exploring and better understanding the function of H_4R (Yu et al., 2010).

Treatments with clozapine or JNJ28610244 also produce a concentration-dependent inhibitory effect on the proliferation of MDA-MB-231 cells and the anti-proliferative action of these H_4R agonists is fully blocked with the combined treatment with the H_4R antagonist JNJ7777120.

Xenograft tumours of the highly invasive human triple-negative breast cancer line MDA-MB-231 in immune deficient nude mice exhibited moderate expression of H_4R, high expression of a proliferation marker (PCNA) and reduced apoptotic cells. In agreement with the results described above, *in vivo* treatments with histamine (5mg/kg, *s.c.*, daily administration) or H_4R agonists (clozapine, 1mg/kg; JNJ28610244, 10mg/kg, *s.c.*, daily administration) significantly diminish the tumour growth rate, evidenced by an increase in the exponential doubling time. This effect was associated with a decrease in PCNA expression levels and reduced intra-tumoural vessels in histamine- and clozapine-treated mice. Histamine also significantly increases median survival and tumoural apoptosis (Martinel Lamas et al., 2013). Results demonstrate the functional expression of H_4R in a breast cancer experimental model and show the anti-tumour properties of H_4R agonists, opening new perspectives in histamine pharmacology research aimed to develop a new generation of antihistamines targeting H_4R that may contribute to advances in the treatment of cancer.

Lastly, histamine through different histamine receptor subtypes produces a radiosensitising action involving enhanced radiation-induced oxidative DNA damage, apoptosis and senescence in both human estrogen-dependent MCF-7 cells and estrogen-independent MDA-MB-231 cells. Furthermore, these effects in MDA-MB-231 cells correlate with increased intracellular ROS and inhibition of the activity of antioxidant enzymes, decreasing antioxidant defence (Martinel Lamas et al., unpublished data). Histamine also was safely used in different experimental models as a radioprotective agent of normal radiosensitive tissues (Medina et al., 2007; 2010; 2011b; Medina & Rivera, 2010a). Therefore, the combined use of histamine with radiation could be an attractive strategy to enhance the efficacy of radiotherapy for both estrogen-dependent and estrogen-independent breast cancers and clinical trials are warranted.

8.2.5. Conclusion

The identification of histamine receptor subtypes and the elucidation of their role in the development and growth of human mammary carcinomas may represent an essential clue for advances in breast cancer treatment. The presented evidence contributes to the identification of molecules involved in breast carcinogenesis, confirming the role of H_4R in the regulation of breast cancer growth and progression, representing a novel molecular target for new therapeutic approach.

8.3. Histamine in Melanoma

8.3.1. Introduction

Melanoma arises from epidermal melanocytes and is a major concern for health-care providers. The incidence continues to increase globally, especially among fair skinned people, by 3% to 7% annually (Lens & Dawes, 2004).

The increasing incidence of melanoma has also been associated with an increased mortality related to the disease. In the United States about 76,250 new cases of melanoma were diagnosed and about 9,180 deaths were estimated by The American Cancer Society in 2012. The mortality rate for melanoma is especially remarkable for those presenting the more severe forms (i.e., stage IV) of metastatic melanoma. Early detection is vital for long-term survival, given that there is a direct correlation between tumour thickness and mortality (Cummins et al., 2006).

Histamine has been implicated as one of the mediators involved in regulation of proliferation in both normal and neoplastic tissues. Melanoma cells but not normal melanocytes contain large amounts of histamine, which has been found to accelerate malignant growth (Pós et al., 2004). In Mel-5 positive melanocytes isolated from skin samples of healthy persons, the absence of HDC expression suggests that the level of this enzyme is strongly associated with malignancy in the skin (Haak-Frendscho et al., 2000). In some experiments, the use of specific anti-sense oligonucleotides designed to inhibit HDC protein synthesis strongly (>50%) decreased the proliferation rate of human melanoma cells (WM938/B and HT168/91). In accordance with these results, similar effects were found with other two melanoma cell lines, WM35 and M1/15, suggesting that endogenous histamine may act as an autocrine growth factor (Hegyesi et al., 2001).

On the other hand, over-expression of HDC markedly accelerates tumour growth and increases metastatic colony-forming potential along with rising levels of local histamine production that was correlated to tumour H_2R and rho-C expression in mouse melanoma (Pós et al., 2005).

8.3.2. Histamine Receptors and Melanoma Cell Proliferation

The expression of H_1R, H_2R, H_3R and H_4R in human melanoma cell lines has been previously reported (Hegyesi et al., 2005; Massari et al., 2011). Histamine acting through H_1R decreases cell proliferation but enhances growth when acting through H_2R (Lázar-Molnar et al., 2002). Second messenger measurement indicates that H_1R function is exerted via PLC activation and its subsequent intracellular calcium mobilization while H_2R is linked to cAMP production, suggesting an involvement of protein kinase A (PKA) in the mitogenic pathway triggered in this system. This is corroborated by the fact that forskolin and permeable cAMP analogues also produce a dose-dependent increase in cell proliferation (Molnar et al., 2002). Conversely, histamine has been shown to induce melanogenesis via H_2R and growth-differentiation factor-15 in melanoma cell lines (SK-MEL-2, B16F10 and Melan-a), suggesting that growth-differentiation factor-15 is a new player in histamine-induced melanogenesis and that histamine could act as a re-pigmenting agent for the treatment of vitiligo (Lee et al., 2012).

No evidence of mitogenic signalling through H_3R in human melanoma has been detected (Hegyesi et al., 2005). Interestingly, recent results demonstrate that melanoma cells express H_4R at the mRNA and protein levels and by using agonists, antagonists and siRNA, it was shown that the inhibitory effect of histamine on proliferation are partially mediated through the stimulation of H_4R (Massari et al., 2011). This histamine-induced reduction of proliferation is associated to a two-fold induction of premature or accelerated cell senescence, a biological program of terminal growth arrest, and an increase in melanogenesis, which is a differentiation marker on these cells (Medina et al., 2009; Massari et al., 2011). In accordance with these results, treatment with histamine and H_4R agonists produces an accumulation of melanoma cells in the G0/G1 phase of the cell cycle, which was clearly observed at 24 h and continued up to 72 h of treatment (Massari et al., unpublished data).

In WM35 and M1/15 cells, histamine produces a significant increase in cAMP levels at high doses and the activation of H_4R reduces the forskolin-induced increase in cAMP levels but only in highly metastatic M1/15 cell line (Massari et al., unpublished data).

8.3.3. Histamine in Human Melanoma and Therapeutic Approaches

It is important to point out that H_4R is expressed in human melanoma biopsies, confirming that the H_4R is present not only in experimental cell lines but also in human melanoma tissue. Immunohistochemical analysis showed that H_4R is detected in 42% (8/19) of melanoma samples and in 83% (15/18) of nevi, showing cytoplasmic localization. H_4R is present in different histopathological types, including superficial spreading, nodular and acral-lentiginous types (Massari et al., 2011). The immunostaining of the proliferation marker PCNA showed that only melanomas express PCNA, which is inverse correlated with H_4R expression (***$p < 0.0001$; Spearman $r = -0.61$) (Figure 8.2)

Numerous *in vivo* studies demonstrate that both endogenous and exogenous histamine has the ability to stimulate tumour growth while H_2R antagonists (such as cimetidine, famotidine, roxatidine) inhibit this effect (Uçar, 1991; Szincsák et al., 2002; Pós et al., 2005; Tomita & Okabe, 2005). It was also found that melanoma tumour growth is not modulated by *in vivo* histamine treatment while the H_1R antagonist terfenadine in vitroinduces melanoma cell death by apoptosis and significantly inhibits tumour growth in murine models *in vivo* (Blaya et al., 2010). Furthermore, the *in vivo* daily treatment with histamine (1mg/kg, *s.c*) or clozapine (1mg/kg, *s.c.*) of *nude* mice bearing M1/15 melanoma cell xenografted tumours showed an increased survival compared to control group. The enhanced survival is associated with an anti-tumour effect of histamine and clozapine, including suppression

A

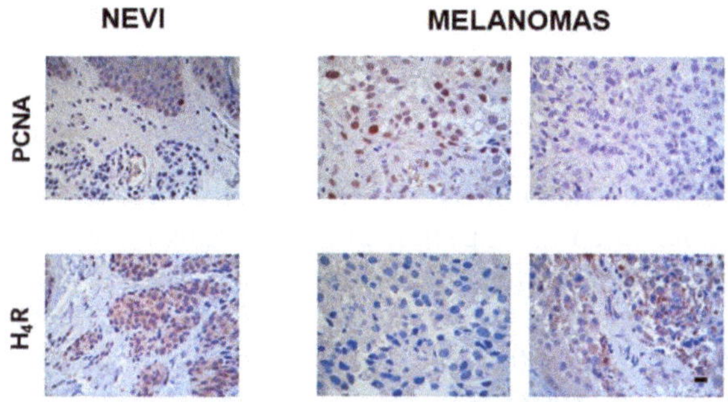

B

	PCNA +	PCNA -	TOTAL
H_4R+	4	20	24
H_4R -	11	3	14
TOTAL	15	23	38

Figure 8.2 Correlations between H_4R and PCNA protein expression determined by immunohistochemistry in human nevi and melanoma biopsies. (A) Representative immunohistochemical images in paraffin-embedded specimens. Original magnification, x630. Scale bar = 20 µm. **(B)** Contingency table: includes total immunohistochemical analysis of nevi and melanomas. 16/19 nevus samples were positive for the expression of H_4R and 18/19 were negative for PCNA. 8/19 melanoma samples were positive for the expression of H_4R and 5/19 were negative for PCNA. Chi square Test ***p= 0.0002 and Fisher exact Test ***p= 0.0003.

of tumour growth (Massari et al., unpublished data). Differences between melanoma cells in their capacity to produce and catabolise histamine could explain the different sensitivities of melanoma cell types to exogenous histamine treatment.

Moreover, several phase II and III clinical trials in metastatic melanoma demonstrated the clinical benefit of histamine (Ceplene) as an adjuvant to immunotherapy with IL-2 (Agarwala et al., 2002). The addition of histamine dihydrochloride to an outpatient regimen of IL-2 is safe, well tolerated and demonstrates a survival advantage over IL-2 alone (9.4 vs. 5.1 months) in melanoma patients with liver metastases (Agarwala et al., 2002). However, a second confirmatory phase III study failed to show any survival benefit for those patients (Naredi, 2002).

Further studies are needed to corroborate the H_4R importance as potential target for new drug development for the treatment of this disease.

8.4. Histamine in Leukaemia and Lymphoma

8.4.1. Introduction

Approximately every 10 minutes, one person in the US dies from a blood cancer while an estimated 1,012,533 people in the US are living with, or are in remission from, leukaemia, lymphoma or myeloma (Cancer Facts & Figures, 2012). This point is clearly illustrated by acute myeloid leukaemia (AML), a disease with incidence ranging from 2 to 4 per 100,000 persons in Europe and the United States (Redaelli et al., 2003).

It is well documented that histamine participates in the regulation of haematopoiesis by stimulating both primitive multipotent and more lineage-restricted myeloid progenitors. After an initial work in the late 1970s showing that histamine is able to induce haematopoietic stem cell proliferation via H_2R (Byron, 1977), a rush broke out in searching for further effects of histamine in haematopoiesis and haematological neoplasias. The histamine levels have been determined in lymph nodes of patients with malignant lymphomas, Hodking´s disease (HD) or non-Hodking lymphomas (NHL), and in all cases the values were higher than in controls. In patients with NHL, these levels show dependence on the grade of malignancy as they are found to be significantly higher in those classified as high-grade malignant (Belcheva & Mishkova, 1995). Immunostaining and ELISA methods also confirmed the presence of histamine in the cytoplasm of acute lymphocytic leukaemia (ALL) cells, and H_1R antihistamines inhibit their clonogenic growth. There is no correlation between the clonogenic growth of ALL cells and their histamine content, suggesting that while histamine may be important for the clonogenic growth of ALL cells, other factors also affect their clonogenity (Malaviya et al., 1996). Furthermore, the promonocytic U-937 cell line, derived from a histiocytic lymphoma, expresses histamine receptors and the switching histamine receptor expression from H_2R to H_1R during differentiation of monocytes into macrophages is observed (Wang et al., 2000). Forskolin treatment induced U-937 cell differentiation through a sustained rise in cAMP levels whereas histamine or H_2R agonists, which increase cAMP levels, failed to promote differentiation due to rapid homologous and GRK2 dependent desensitization of H_2R (Brodsky et al., 1998; Fernández et al., 2002). In addition, in U937 and other AML cell lines, amthamine (H_2R agonist) augments intracellular cAMP level with a concomitant increase in the efflux regulated by multidrug resistance-associated proteins (MRPs), particularly MRP4 (Copsel et al., 2011).

Recent reports demonstrated that HDC is expressed primarily in D11b+Ly6G+ immature myeloid cells (IMCs) that are recruited early on in chemical carcinogenesis and that HDC knockout mice exhibit a markedly increased rate of colon and skin carcinogenesis, indicating a key role of histamine and HDC in myeloid cell differentiation and CD11b+Ly6G+ IMCs in early cancer development (Yang et al., 2011).

8.4.2. Histamine in Human AML and Therapeutic Approaches

Most patients with AML achieve complete remission after induction chemotherapy. Despite ensuing courses of consolidation chemotherapy, a large fraction of patients experience relapses with poor prospects of long-term survival. Interleukin-2 (IL-2) and interferon-alpha (IFNa) are effective activators of lymphocytes with anti-neoplastic properties, such as T-cells or natural killer (NK) cells, constituting the basis for their widespread use as immunotherapeutic agents in human neoplastic disease. The functions of intratumoural lymphocytes in many human malignant tumours are inhibited by reactive oxygen species (ROS), which are generated by adjacent monocytes/macrophages. *In vitro* data suggest that those immunotherapeutic cytokines only weakly activate T cells or NK cells in a reconstituted environment of oxidative stress and inhibitors of ROS formation or ROS scavengers synergize with IL-2 and IFNa to activate T cells and NK cells. Recently, IL-2 therapy for solid neoplastic diseases and haematopoietic cancers has been supplemented with histamine dihydrochloride (Ceplene), a synthetic derivative of histamine, with the aim of counteracting immunosuppressive signals from monocytes and macrophages. Histamine dihydrochloride inhibits the formation of ROS, which suppresses the activation of T cells and NK cells by suppressing the activity of NADPH oxidase via H_2R. When administered in addition to IL-2, histamine dihydrochloride enables the activation of these lymphocytes by the cytokine, resulting in tumour cell apoptosis. This combination was recently approved within the EU as a remission maintenance immunotherapy in AML, as histamine dihydrochloride reduces myeloid cell-derived suppression of anti-leukemic lymphocytes, improving NK and T-cell activation. Data from a randomized multinational phase III trial of 320 adults with AML demonstrated that maintenance therapy with 3-week cycles of histamine dihydrochloride plus low-dose IL-2 for up to 18 months significantly improved leukaemia-free survival (LFS) but lacked power to detect an overall survival (OS) difference. Analyses of consistency and robustness may aid interpretation of data from multicenter trials, as the size of randomized clinical trials is limited and show that LFS may be an acceptable surrogate for OS in future AML trials (Hellstrand et al., 2000; Martner et al., 2010; Buyse et al., 2011; Yang & Perry, 2011). In recent study, 45 trial patients with morphological forms of AML defined by the French-American-British classification showed strongly improved LFS in M4/M5 (myelomonocytic/monocytic) leukaemia but not in M2 (myeloblastic) leukaemia. H_2R is typically absent from M2 cells but frequently expressed by M4/M5 cells and in the latter produced ROS triggered apoptosis in adjacent NK cells; events that are significantly inhibited by histamine dihydrochloride (Aurelius et al., 2012).

A different study indicates that both murine and human progenitor cell populations express H_4R and respond to its agonists by reducing growth factor-induced cell cycle progression, leading to decreased myeloid, erythroid and lymphoid colony formation (Petit-Bertron et al., 2009). Further research in this area will shed light on the role of histamine with the aim to improve cancer immunotherapy efficacy.

8.5. Histamine and Digestive Diseases: Ulcer, Inflammation and Cancer

8.5.1 Introduction

Histamine is a pleiotropic biogenic amine with a broad range of activities in both physiological and pathological conditions. Both histamine producing cells and receptors are extensively distributed within the digestive system (Bertaccini & Coruzzi, 1992). Moreover, the enzyme HDC is elevated during inflammatory response in several gastrointestinal (GI) disorders, such as food allergy, inflammatory bowel disease (IBD) and irritable bowel syndrome (IBS), as well as in GI cancer (Raithel et al., 1995; Backhaus et al., 2004; Medina & Rivera, 2010b). In the digestive system, the effects of histamine are mediated by the four types of G-protein linked receptors: H_1R is mainly involved in the regulation of vascular contraction, permeability, visceral pain and hypersensitivity, H_2R has a central role in the regulation of parietal cell acid secretion and H_3R is functionally related to gastric mucosal defence. In the digestive system, H_4R expression has been reported in different areas and in a variety of cell types, including immune, inflammatory, neural, endocrine and epithelial cells (Table 8.1). Furthermore, H_4R expression can be altered under several pathological conditions (Table 8.2). As reported for other systems, histamine has long been known to participate in the pathogenesis and progression of cancer in the digestive tract (Chandra & Ganguly, 1987; Garcia-Caballero et al., 1988; Reynolds et al., 1997; Masini et al., 2005; Kennedy et al., 2012). Protective effects mediated by H_4R ligands have been reported in some experimental models of GI and liver damage, thus suggesting that this novel histamine receptor subtype might represent a potential drug target in the treatment of digestive diseases (for review, see Coruzzi et al., 2012a).

In the present review we report the available data from the literature together with original data from our laboratory concerning the location and functional role of histamine H_4R in the digestive system and the potential clinical implications for human GI inflammation and cancer.

8.5.2. Salivary Glands

H_4R expression has been recently identified in salivary glands of healthy subjects, localized to the acinar and ductal epithelial cells (Table 8.1).

Table 8.1

Expression of histamine H_4 receptors in the digestive system.

Species	Tissue	Reference
Mouse	Peritoneal exudates	Strakhova et al., 2009a
	Small intestine	Takagaki et al., 2009
	Distal colon	Sutton et al., 2008
	Colon	Kumawat et al., 2010
Rat	Stomach (ghrelin producing cells)	Morini et al., 2008
	Whole intestine (myenteric neurons)	Chazot et al., 2007
	Cholangiocytes	Francis et al., 2012
Pig	Colon	Oda et al., 2002
Dog	Small intestine	Jiang et al., 2008
	Liver and colon	Eisenschenk et al., 2011
Monkey	Colon	Oda et al., 2005; Kim et al., 2011
Human	Salivary glands	Stegaev et al., 2012
	Stomach	Liu et al., 2001
	Colon	Cianchi et al., 2005, Boer et al., 2008
	Small intestine	Nakamura et al., 2000; Oda et al., 2000; Cogè et al., 2001; Liu et al., 2001; Morse et al., 2001
	Whole intestine	Sander et al., 2006

Unlike H_1R and H_2R, H_4R was stimulated by the low levels of histamine present in the saliva (0.3-12.4 ng/ml) and by the H_4R agonist ST-1006 (Sander et al., 2009) to produce interleukin-8 and VEGF (Stegaev et al., 2012). The strong expression of H_4R in tubuloacinar cells of salivary glands was not detected in patients with primary Sjögren's syndrome, suggesting a receptor downregulation induced by the increased histamine levels (Table 8.2). These findings agree with the suggested role of histamine in autoimmune diseases and suggest future therapeutic alternatives in the treatment of Sjögren's syndrome.

Table 8.2
Histamine H_4 receptor expression alteration in pathological tissues.

Species	Pathological tissue	H_4R density	Reference
Mouse	TNBS colitis	increase	Sutton et al., 2008
Mouse	Cholangiocarcinoma	increase	Meng et al., 2011
Mouse*	Spontaneous colitis	increase	Kumawat et al., 2010
Guinea pig	Eosinophilic esophagitis	increase	Yu et al., 2008
Human	Colorectal carcinoma	decrease	Cianchi et al., 2005
			Boer et al., 2008
			Fang et al., 2011
Human	Gastric carcinoma	decrease	Zhang et al., 2012
Human	Sjörgen´s syndrom	decrease	Stegaev et al., 2012

TNBS = trinitrobenzensulphonic acid; * Gαi2-deficient mouse

8.5.3. Oesophagus

The presence of H_4R in the oesophagus was not identified in early studies describing H_4R localization within the human body (Table 8.1), however, histamine H_4R-like immunoreactivity was recently detected on oesophageal mast cells and eosinophils of guinea pigs actively sensitised with inhaled ovalbumin (Table 8.2) (Yu et al., 2008). The sensitised guinea pig is a good model for human eosinophilic oesophagitis to study the role of histamine and H_4R in inflammatory response, because unlike rat and mouse models, mast cells from guinea pigs and humans contain primarily histamine (Metcalfe et al., 1997). Moreover, histamine released by mast cell activation regulates mast cell redistribution and eosinophilic infiltration into the epithelial mucosa through an H_4R-mediated pathway (Metcalfe et al., 1997).

8.5.4. Stomach

The major role of histamine in the stomach is regulation of acid secretion by parietal cells, as demonstrated in humans by the clinical efficacy of H_2R antagonists in various clinical settings (Feldman & Burton, 1990a; Feldman & Burton, 1990b). Whereas gastric H_1R is mainly involved in the vasodilatation and reactive hyperaemia in response to acid challenge (Bertaccini & Coruzzi, 1992; Rydning et al., 2001), the role of H_3R seems to be predominantly related to the enhancement of gastric mucosal defence, inhibition of enteric neurotransmission and feedback regulation of histamine release (for a review, see Coruzzi & Adami, 2008). Indeed, it has been demonstrated by various groups that H_3R activation by (R)-alpha-methylhistamine protects the rat gastric mucosa against the acute mucosal damage induced by absolute ethanol, non-steroidal anti-inflammatory drugs (NSAIDs), ammonia, concentrated HCl or stress (Morini et al., 1995; Morini et al., 1997; Warzecha et al., 2000), by increasing mucus production, gastric mucosal blood flow, epithelial cell proliferation and the activity of sensory nerves (Dembinski et al., 2005; Grandi et al., 2006).

Whereas early studies reported low H_4R expression in the human and rat stomach, more recent information about the cell distribution of H_4R in the rat GI tract reveal a selective location of H_4R in myenteric neurons and in endocrine cells (A-like cells) of the fundic mucosa producing the orexigenic peptide ghrelin (Table 8.1), leading to speculation about a possible role of histamine in the secretion of this peptide (Chazot et al., 2007; Morini et al., 2008). However, functional experiments carried out by our group using the reference H_4R antagonist JNJ7777120 (Thurmond et al., 2004) did not demonstrate a major role of H_4R in the regulation of gastric acid secretion or motility, at least in rats (Coruzzi et al., 2012a). By contrast, this receptor would seem to participate in the ulcerogenic effect of histamine, since JNJ7777120 was able to protect rat and mouse gastric mucosa from the damage induced by NSAIDs (Table 8.3).

Table 8.3

Functional effects of H_4R ligands on experimental gastric ulcer models.

Species	Ulcerogen	Ligand	Effect	Reference
Rat	Indomethacin	JNJ7777120	Protection	Coruzzi et al., 2010
		VUF6002	Protection	" "
		VUF8430	Protection	" "
	0.6 HCl	JNJ7777120	None	Coruzzi et al.,2010
		VUF6002	" "	" "
	Compound 48/80	JNJ7777120	Protection	Adami, unpublished
Mouse	IND/BET	JNJ7777120*	Protection	Adami et al., 2012
		VUF8430	None	" "
	Cold/restraint stress	JNJ7777120	None	Adami, unpublished

IND/BET = Indomethacin+bethanechol; *CD-1, NMRI and Balb/C mice; no effect on C57BL/6J mice.

The gastric safety of H_4R antagonists could be of major interest, given that these compounds could be proposed as new anti-inflammatory drugs. Nevertheless, a functional role of H_4R in the stomach remains to be proven, since the H_4R agonist VUF8430 (Lim et al., 2006) is paradoxically effective in reducing indomethacin-induced lesions in rats (Table 8.3) (Coruzzi et al., 2010). To further complicate matters, the gastroprotective effects of JNJ7777120 against indomethacin in a mouse ulcer model are dependent on the specific strain used, no gastroprotection was observed when the H_4R antagonist was administered to C57BL/6J mice (Adami et al., 2012), in line with results obtained in other inflammation models (Coruzzi et al., 2012b). Since mRNA expression and H_4R protein levels did not differ among various murine strains (Baumer et al., 2008), the contrasting data obtained with H_4R ligands could be explained by the recently hypothesized concept of *functional selectivity* or *ligand-specific receptor conformation* (Kenakin & Miller, 2010; Schnell et al., 2011). In this context, the "reference" H_4R antagonist JNJ7777120 can behave as a partial or full agonist in some assays (Seifert et al., 2011) while both H_4R agonists and antagonists can display the same effects (Morgan et al., 2007; Clarke et al., 2008; Coruzzi et al., 2010; Neumann et al., 2010), rendering the interpretation of animal experiments very difficult.

Several studies have shown that histamine induces proliferative effects in various gastric tumour cell lines (Hahm et al., 1996) and clinical trials show increased survival of gastric cancer patients after treatment with certain H_2R antagonists (Tonnesen et al., 1988; Watson et al., 1993). So far, only one study has examined the expression of histamine H_4R in gastric carcinoma (GC) and showed that expression levels of H_4R is reduced mainly in advanced GC samples compared to the adjacent normal tissue (Table 8.2). In addition, both histamine and the H_4R agonist/H_3R antagonist clobenpropit are able to induce cell cycle arrest in AGS cell lines by blocking the cell-cycle progression through G1 to S phase (Zhang et al., 2012). Based on the study of Sander et al. (2006), which did not detect H_3R expression in the human intestine, the effect of clobenpropit in the regulation of cell-cycle was attributed to specific involvement of H_4R and a potential role of H_4R in mediating cell growth arrest was hypothesized. However, some experimental data does not entirely support this conclusion: H_4R expression is not observed in the human stomach (Sander et al., 2006), the absence of H_3R in the human intestine is challenged by the stimulatory effect of a H_3R agonist on human sub-mucosal neurons (Breunig et al., 2007) and clobenpropit is not the most suitable ligand to unravel H_4R-mediated effects (Leurs et al., 2009).

8.5.5. Intestine

In recent years, it has become apparent that intestinal mast cell mediators and the enteric nervous system are key players in the intricate neuroimmune network that regulates intestinal homeostasis and the inflammatory response to noxious stimuli (Wood, 2006; Buhner & Schemann, 2012). Histamine can influence neurotransmission at both submucous and myenteric plexus, thereby modifying intestinal secretion and motility through the activation of H_1R, H_2R and H_3R (Bertaccini & Coruzzi, 1992; Stack et al., 1995; Wood, 2006; Coruzzi & Adami, 2008). Several groups have underlined the increase in histamine content in mucosal biopsies from Crohn's disease, ulcerative colitis (UC) and food allergy, but the efficacy of antihistamines or mast cell stabilizers in the therapy of IBD is unproven (Knutson et al., 1990; Raithel et al., 1995; Barbara et al., 2004; Bischoff & Crowe, 2005; Maintz & Nowak, 2007). Although some recent data have suggested a role for H_3R in the bowel inflammation because of the protective effects exerted by the H_3R agonist (R)-alpha-methylhistamine (Nosalova et al., 2001; Coruzzi et al., 2002; Fogel et al., 2007), conclusive evidence for the involvement of this receptor subtype in the regulation of intestinal functions is lacking. The existence of a H_3-like receptor in the porcine small intestine was initially hypothesized by Schwörer et al. (1994) and subsequently confirmed by several independent groups in different animal species, including humans (Table 8.1). According to other studies, however, no detectable levels of H_4R mRNA were found in the GI tract (Nguyen et al., 2001; Kim et al., 2011).

In this context, it is important to note that the specificity of commercially available H_4R antibodies has been recently criticized and the parallel use of highly selective H_4R ligands has been considered mandatory in order to unravel H_4R localization and function (Beerman et al., 2012; Neumann et al., 2012). However, despite the fact that H_4R has been recently detected on neurons of the mammalian and rodent central and peripheral nervous system (Breunig et al., 2007; Chazot et al., 2007; Strakhova et al., 2009b), we were unable to detect any effect of either H_4R agonists or antagonists on cholinergic neurotransmission in the isolated rat duodenum or in surgical specimens from human colon (Pozzoli et al., 2009). Kim et al. (2011) recently confirmed the lack of H_4R-mediated effects in murine colon strips while observing an increase in colon contractility in response to H_4R agonist in monkey colonic muscle. To the best of our knowledge, no study has yet examined whether H_4R is located on afferent fibers of the enteric nervous system. In view of the inhibitory effects of H_4R antagonists in different pain models (Coruzzi et al., 2007; Cowart et al., 2008; Liu et al., 2008; Hsieh et al., 2010), it might be of interest to explore the effects of H_4R ligands in animal models of visceral pain and hypersensitivity.

As for intestinal transport, histamine increases intestinal ion and water secretion in antigen-sensitised guinea pigs via activation of H_2R located on epithelial cells and on colonic submucous plexus (Frieling et al., 1993). In addition, pre-junctional H_3R negatively modulates cholinergically-mediated intestinal secretion by removing the inhibitory control exerted by the adrenergic system (Frieling et al., 1993; Wood, 2006). As opposed to animal findings, early studies in the human intestine revealed that histamine-induced increase in chloride secretion by colonic epithelium is exclusively related to the activation of H_1R (Keely et al., 1995). A recent study in human submucous plexus from surgical specimens suggests, however, that histamine may induce excitation of enteric neurons through activation of all four histamine receptors (H_1R-H_4R) (Breunig et al., 2007). The H_3R-mediated excitatory effects on secretory neurons reported in this study are unexpected, in view of previous literature showing lack of H_3R expression in the human bowel or of H_3R-mediated effects on intestinal contractility (Sander et al., 2006; Pozzoli et al., 2009). The pathophysiological significance of the excitatory action of histamine on secretory neurons is uncertain, as hyperactivity of these neurons leads to neurogenic secretory diarrhoea, as observed in various pathological conditions including UC, Crohn's disease, allergic enteropathy and parasitic infection (Wood, 2006; Maintz & Novak, 2007). Mast cells in colonic mucosal biopsies from IBS patients with diarrhoea release more histamine than in normal subjects (Barbara et al., 2004). Therefore, H_4R antagonists may be of therapeutic value in these GI disorders in the same way as mast cell stabilizers or H_2R antagonists (Santos et al., 2006).

The discovery of H_4R has opened new horizons on the role of histamine in the pathophysiology of the bowel. Indeed, both *in vitro* and *in vivo* studies provided evidence for beneficial effects of H_4R antagonists as anti-inflammatory agents (Table 8.4).

The selective H_4R antagonist JNJ7777120 is effective in ameliorating the colitis induced by TNBS, a hapten which provokes in rodents many of the macroscopic, histological and immunological hallmarks of the human IBD (Elson et al., 1995). JNJ7777120 is also able to reduce macroscopic damage, neutrophil infiltration and the production of both TNFa and IL-6, two cytokines that play a critical role in the pathogenesis of the human disease (Fiocchi, 1998; Varga et al., 2005; Dunford et al., 2006).

Interestingly, H_4R expression pattern is altered in the colonic mucosa of mice deficient in the $G\alpha i2$ protein, a mouse knockout line which develops a spontaneous colitis similar to human UC, including development of colon cancer with a T helper 1 (Th1)-dominated immune response (Table 8.2). Whereas H_1R and H_2R expression remained constant at various stages of colitis, H_4R expression levels were slightly increased in early and late colitis for $G\alpha i2$-deficient mice compared to wild type mice (Kumawat et al., 2010).

In the mouse model of zymosan-induced peritonitis, the blockade of H_4R induced by the H_3R/H_4R antagonist thioperamide reduces neutrophil recruitment and the production of leukotriene B4 (LTB4), one of the most potent chemoattractant for neutrophils (Takeshita et al., 2003). The efficacy of H_4R antagonists in this model was confirmed by the use of JNJ7777120 and more recent ligands (Table 8.4), which are all able to reduce neutrophil recruitment as well as the production of inflammatory mediators, such as PGD2, PGE2 (Thurmond et al., 2004; Cowart et al., 2008; Liu et al., 2008; Strakhova et al., 2009a). The analysis of peritoneal cell exudate in zymosan-treated mice revealed seven-fold

Table 8.4

Functional effects of H_4R ligands on experimental models of intestinal inflammation.

Species	Assay	Ligand	Effect	Reference
Mouse	IR-damage	Thioperamide	↓ neutrophil infiltration	Ghizzardi et al., 2009
Balb/C mouse	Zymosan-peritonitis	Thioperamide	↓ neutrophil infiltration ↓ LTB4 production	Takeshita et al., 2003
CD-1 mouse	Zymosan-peritonitis	JNJ7777120	↓ neutrophil infiltration	Thurmond et al., 2004
Balb/C mouse	Zymosan-peritonitis	JNJ7777120	↓ neutrophil infiltration	Hsieh et al., 2010
C57BL/6J mouse	Thioglycollate-peritonitis	JNJ7777120	None	Thurmond et al., 2004
Balb/C mouse	Zymosan-peritonitis	A-943931	↓ neutrophil infiltration	Cowart et al., 2008
Balb/C mouse	Zymosan-peritonitis	A-987306	↓ neutrophil infiltration	Liu et al., 2008
Balb/C mouse	Zymosan-peritonitis	A-940894	↓ neutrophil infiltration ↓ PGD2, PGE2 production	Strakhova et al., 2009b
Rat*	TNBS	JNJ7777120, JNJ10191584	↓ macroscopic damage ↓ TNFα production	Varga et al., 2005
	TNBS	JNJ7777120	↓ neutrophil infiltration ↓ IL-6 production	Dunford et al., 2006
	TNBS	Thioperamide	↓ macroscopic damage	Fogel et al., 2007
	IR-liver damage	Clobenpropit	↓ macroscopic damage	Adachi et al., 2006

IR = ischaemia-reperfusion; LTB4 = Leukotriene B4; TNBS = trinitrobenzensulphonic acid; TNF = tumour necrosis factor; * Wistar

higher expression of H_4R mRNA in naive animals as compared to genetically modified mice devoid of mast cells, suggesting that resident mast cells may be the predominant H_4R-expressing cell in the peritoneum (Hsieh et al., 2010). The observation that JNJ7777120 is effective in peritonitis induced by zymosan (mast cell-dependent) but not by thioglycollate (mast cell-independent) has been considered consistent with JNJ7777120 acting on mast cells (Takeshita et al., 2003; Thurmond et al., 2004). However, JNJ7777120 is not effective on the thioglycollate model (Thurmond et al., 2004) in C57BL6/J mice, which unlike Balb/C, CD-1 or NMRI mouse strains, is unresponsive to the anti-inflammatory effect of JNJ7777120 on croton-oil induced ear oedema (Coruzzi et al., 2012b) and the gastro-protective effects against indomethacin (Adami et al., 2012). The choice of assay, species and ligand seems to be highly relevant in defining H_4R-mediated effects. Keeping with this, the specific involvement of H_4R in ischemia/

reperfusion-induced damage recently reported in mice (Ghizzardi et al., 2009) is only based on the effect of the mixed H_3R/H_4R blocker thioperamide. Previous data obtained in rats show that the effect of histamine on intestinal ischemia is related to activation of H_1R (Tsunada et al., 1994).

A summary of the functional effects induced by H_4R ligands in the different models of GI inflammation is shown in Table 8.4.

Finally, it has recently been shown that histamine can be protective against cellular damage induced by ionising radiation in the mouse small intestine (Medina et al., 2005). The mechanism underlying this effect was also investigated by the same group and it appears to be related to an increase in cell proliferation of damaged intestinal mucosa and suppression of apoptosis. Although no selective H_4R ligand was tested, these data would suggest a potential clinical value of histaminergic drugs in patients undergoing radiotherapy (Medina et al., 2007).

Colorectal cancer (CRC) is one of the most commonly diagnosed cancers in the world. Surgery is the primary intervention in early stages of cancer, but anticancer drugs are used as adjuvant therapy to control the growth and spread of undetected cancer cells. Histamine and its receptors are involved in the regulation of CRC growth (Suonio et al., 1994; Rajendra et al., 2004; Masini et al., 2005) and several clinical trials with H_2R antagonists show a trend towards improved survival of CRC patients (Bolton et al., 2000; Nielsen et al., 2002; Deva & Jameson, 2012). A recent study investigated the distribution of histamine receptor subtypes in colorectal tumour cells compared to the normal mucosa by different techniques, such as RT-PCR, Western blot analysis and immunostaining (Breunig et al., 2007). This study demonstrated the presence of H_1R, H_2R and H_4R, but not H_3R expression in adenoma and human colon carcinoma at protein level. As reported in GC cells (Table 8.2), histamine receptor expression pattern in CRC cells is altered as compared to normal adjacent tissue, where downregulation of H_4R has been described in several studies with H_4R reduction paralleling the progression of cancer (Cianchi et al., 2005; Boer et al., 2008; Fang et al., 2011).

Contradicting findings have been reported on the effects of histamine on cell proliferation (Rivera et al., 2000), with both stimulation and/or inhibition being observed depending on the receptor. In colon cancer cell lines, the stimulatory effects of histamine on cell proliferation are antagonised by both zolantidine and JNJ7777120, suggesting the involvement of both H_2R and H_4R (Cianchi et al., 2005). Moreover, both receptors mediate the increased production of PGE2 and VEGF in cyclooxygenase-2 (COX-2)-expressing colon cancer cells (Cianchi et al., 2005), confirming the pro-angiogenic effects of histamine in tumour cells (Table 8.5). However, H_4R activation by clozapine and clobenpropit has recently been reported to induce growth arrest and activation of apoptosis in CRC carcinoma cell lines (Fang et al., 2011).

Table 8.5

Functional effects of H_4R ligands on GI cancer experimental models.

Species	Assay	Ligand	Effect	Reference
Mouse	CCA	Clobenpropit	↓ tumour growth	Meng et al., 2011
Human	AGS cell line	Clobenpropit	cell cycle arrest	Zhang et al., 2012
Human	CRC cell line	Histamine*	↑ cell growth ↑ VEGF, PGE2	Cianchi et al., 2005 " "
Human	CRC cell line	Clozapine	↑ apoptosis ↓ cell growth	Fang et al., 2011 " "
Human	PANC-1 cell line	Clobenpropit	↓ cell proliferation	Cricco et al., 2008

CCA = cholangiocarcinoma; AGS = gastric cancer; CRC = colorectal cancer; PANC-1 = pancreatic carcinoma; *effects are blocked by either H_2R or H_4R antagonists

Again, it has to be considered that these compounds are not highly selective H_4R ligands and are able to activate or block other histamine receptor subtypes (Lim et al., 2009). Thus, despite the alteration of H_4R expression in CRC cells and the effects of histaminergic ligands on proliferation and angiogenesis, a clear understanding of H_4R role in CRC cancer is still lacking.

8.5.6. Pancreas

Pancreatic cancer and adenocarcinoma in particular accounts for approximately 80% of all pancreatic malignancies and is a highly aggressive malignant tumour characterised by high incidence of mortality worldwide (Zakharova et al., 2012). The extremely poor prognosis of pancreatic cancer is linked to a very rapid local spread, the early distant metastases and the lack of a suitable therapy. Based on the well-recognised involvement of histamine in tumour growth, some groups have focused on the histaminergic system in search of new therapeutic strategies. Indeed, some studies have reported an over-expression of H_1R and H_2R and more recently H_3R and H_4R in human pancreatic carcinoma cell line (PANC-1) (Cricco et al., 2004; Cricco et al., 2008) (Table 8.2). Histamine displays opposite effects depending on the concentrations used and the receptor subtype involved, where low concentrations of histamine (0.01 μM) increase cell proliferation and PANC-1 cell growth via H_1R and H_3R, and higher amounts (10 μM) inhibit cell proliferation via H_2R and H_4R (Cricco et al., 2004; Cricco et al., 2008) (Table 8.5). The inhibitory effect mediated by H_4R on pancreatic cell proliferation agrees with data obtained in human haematopoietic progenitor cells (Petit-Bertron et al., 2009) and in other human cancer cells (Massari et al., 2011; Medina et al., 2008).

8.5.7. Liver

Data on H_4R in the liver is relatively small. Expression of H_4R in normal hepatic tissue was detected in rats and dogs (Table 8.1), where in normal rats treated with clobenpropit for one week, no difference was found in bile duct mass or cholangiocyte proliferation, which was increased by H_1R or H_2R activation (Francis et al., 2012). A recent study in rats reported that ischemia/reperfusion liver injury was reduced by H_4R stimulation and not blockade, as would be expected from the supposed inflammatory activity mediated by H_4R (Adachi et al., 2006). As discussed earlier, this raises many questions in the understanding of H_4R pharmacology (Seifert et al., 2011).

Cholangiocarcinoma (CCA) is a devastating type of cancer which arises from both intra- and extra-hepatic bile ducts and is able to metastasise to other organs (Sirica, 2005). CCA is characterized by a difficult diagnosis and a poor prognosis due to very limited therapeutic options. As observed for other tumours, H_4R expression is present in CCA cells (Meng et al., 2011) and the H_3R antagonist/H_4R agonist clobenpropit is able to reduce CCA growth and metastases *in vivo* without altering cell proliferation and apoptosis (Table 8.5). The anti-tumour effect of clobenpropit was confirmed in H_3R-knockout mice, leading the authors to suggest a specific involvement of H_4R (Meng et al., 2011). An inhibitory effect on cell proliferation induced by clobenpropit and the H_4R agonist VUF8430 (Lim et al., 2006) has been demonstrated in M1/15 cells (derived from liver metastasis) (Massari et al., 2011) and in MDA-MB-231 and MCF-7 breast cancer cells (Medina et al., 2008; Medina et al., 2011a). Although these findings suggest an H_4R-mediated anti-cancer effect, it must be considered that neither clobenpropit nor VUF8430 can be considered as "reference" H_4R agonists (Leurs et al., 2009; Lim et al., 2009; Coruzzi et al., 2011).

8.5.8. Conclusion

The current data available on the new H_4R subtype suggests that H_4R is expressed in the digestive system and is functionally relevant under pathological conditions, including gastric ulcer, GI inflammation and malignancies. However, functional data with selective H_4R ligands is limited and difficult to interpret, where the selectivity profile of most H_4R compounds varies greatly according to the experimental assay and/or species and some ligands behave in intact animals as "protean" ligands, displaying functional tissue selectivity (antagonism as well as partial, total or inverse agonism) (Kenakin, 1995; Rosethorne & Charlton, 2011; Seifert et al., 2011). Finally, the specificity of commercially available H_4R antibodies has been recently criticised (Beerman et al., 2012; Neumann et al., 2012). A careful validation of experimental assays, the selectivity of H_4R ligands and the specificity of H_4R antibodies seem to be of key importance in order to unravel H_4R localization/function and to assess a potential role of the H_4R as a target for new therapeutic GI drugs.

8.6. Mast Cell and Histamine in Lung Cancer

8.6.1. Introduction

Mast cells are FceRI-expressing, highly granulated, long-living immune cells. Mast cells are generated from haematopoietic stem cells circulating as immature progenitors that enter tissues and differentiate into fully mature functional mast cells. The distribution of mast cells throughout vascularised tissues exposed to the external environment, including the lungs, allows for interaction with environmental antigens, toxins, or invading pathogens (Galli & Tsai, 2010). Mast cells not only have a prominent role in allergy, but also in many other inflammatory states. They are often present in tumour inflammatory infiltrates, although their role in tumour biology remains controversial. They have been attributed with both pro- and anti-tumourigenic functions depending on the exact type and stage of the tumour (Theoharides & Conti, 2004). Mast cells mediators such as histamine, tryptase, bFGF, VEGF and IL-8 (Boldrini et al., 2006) can enhance tumour outgrowth and/or vascularisation while other mast cells mediators such as TNFa and heparin can act as tumour suppressors (Noble, 2010).

Lung cancer is the most common cause of cancer-related death in men and women, and in the United States an estimated 160,340 people are expected to die from lung cancer in 2012, accounting for approximately 28 percent of all cancer deaths.

General subtypes of lung cancer are small cell lung cancer that spreads through vascularisation, forms metastases at early stages of the disease and responds better to chemotherapy and radiotherapy, as well as non-small cell lung carcinoma (NSCLC). NSCLC accounts for about 80% of lung cancer cases and usually appears as longer term solid lung tumours. NSCLC includes 3 subtypes – adenocarcinoma, squamous-cell lung carcinoma, and large-cell lung carcinoma.

8.6.2. Histamine and Mast Cells in Growth of Lung Cancer

Prominent lung-residing and inflammatory influx participant mast cells respond to and influence outgrowing tumour cells and therefore could be manipulated as part of the medical effort to defeat the disease.

Most of the studies directed to discover the involvement of mast cells in lung cancer are "retrospective" ones analyzing large amounts of lung tumour samples by histochemical and immunohistochemical methods, followed by statistical analyses. A partial summary of these studies are presented in Table 8.6. Almost all studies conclude with positive correlation between mast cells

Table 8.6

Retrospective studies correlating mast cells and lung cancer angiogenesis and prognosis.

Correlation between MC and angiogenesis	Correlation between MC and favourable prognosis	Correlation between angiogenesis and prognosis	Reference
Positive	Negative	NT	Takanami et al., 2000
Positive	Negative	NT	Imada et al., 2000
Positive	NT	NT	Tomita et al., 2000
Positive	Negative with MCTC-type	NT	Nagata et al., 2003
No	Positive between MC and early stage	No Between angiogenesis and stage	Tataroğlu et al., 2004
Positive	NT	NT	Ibaraki et al., 2005
	Weak positive	NT	Welsh et al., 2005
Positive Between MC and microvessel density	No	No	Dundar et al., 2008
Weak Adenocarcinoma and stage II disease	No A minor negative trend	No A minor negative trend	Niczyporuk et al., 2012

MC = mast cells; NT=not tested; MCTC = mast cells tryptase and chymase positive

and angiogenesis, but concerning the nature of mast cells and prognosis, different and sometimes contradictory results are observed. The discrepancy between the different studies can be attributed to different cancer cells/stages and normal control samples, different detection methods for both mast cells and angiogenesis (which actually probe different mast cells subtypes and aspects of angiogenesis) and different statistical analysis methods.

It should be noted that in some of these studies, there were attempts, using the same methods, to suggest a mechanism underlying the observed correlations. For instance, it was suggested that expression of TNFa in mast cells and macrophages localized within tumour islets but not tumour stroma is a favourable prognostic marker (Ohri et al., 2010). In the same way, the apparent co-expression of tryptase and VEGF suggests a mechanism for mast cell-angiogenesis positive correlation in lung cancer (Imada et al., 2000). In order to use mast cell activity in situ to treat lung cancer, there is an urgent need for functional studies. These are needed both *in vitro*, to find pathways in which mast cells and lung cancer cells can crosstalk and *in vivo*, where mast cells function can be analyzed in knockout mice.

However, out of the limited data on the *in vitro* function of mast cells and lung cancer cells, some notable observations have been made. Following the observation that mast cells are located within fibrotic intra- and peri-tumoural regions in many tumours including lung and breast carcinomas, *in vitro* cross talk was investigated. This revealed that in breast cancer-fibroblast co-culture, heparin, which is produced exclusively by mast cells, significantly reduces the size and number of cancer cell colonies (Samoszuk et al., 2005), where a similar mechanism is also probably active in lung carcinoma.

PGD_2, a well-known mast cell mediator, inhibits proliferation and induces apoptosis of A549 and H2199 NSCLC cells *in vitro*. The active molecule in this effect is apparently 15d-PGJ_2, a spontaneously occurring metabolite of PGD_2 that activates an intrinsic apoptotic pathway in the cancer cells (Wang & Mak, 2011). This mechanism was further confirmed by an *in vivo* study performed with mast cell-deficient $Kit^{W\text{-}sh}$ mice injected with LLC NSCLC cells that formed solid tumours. Mast cell-deficient $Kit^{W\text{-}sh}$ mice reconstituted with PGD_2-deficient mast cells developed much larger tumours than mice reconstituted with wild type mast cells (Murata et al., 2011).

Histamine is a typical mast cell mediator and modulates cell proliferation through four G-protein-coupled receptors (H_1R, H_2R, H_3R and H_4R), which differ in tissue expression profiles and functions. Histamine receptors are also expressed in numerous cancers, where they can bind mast cell-released histamine and induce cancerous cell proliferation (Lampiasi et al., 2007). In order to analyze interactions between mast cells and lung cancer cells with an emphasis on histamine, cord blood derived mast cells (CBMC) were co-cultured with A549 NSCLC cells (Stoyanov et al., 2012). These co-cultures resulted in degranulation of CBMC and enhanced proliferation of the cancerous cells, implying on a pro-tumourigenic activity for mast cells. Enhancement of A549 cell proliferation was also achieved by the supernatant of CBMC, activated by an IgE-dependent mechanism. In particular, addition of histamine to A549 culture medium resulted in a significant increase in DNA synthesis and cell proliferation. Involvement of H_1R, H_2R and H_4R was confirmed by the use of specific antagonists. Similar results were obtained *in vitro* with LLC (murine NSCLC cell line) cells. To confirm these results *in vivo*, wild type and mast cell-deficient $Kit^{W\text{-}sh}$ mice were injected with LLC cells and resulted in much more cancer cell foci and larger lung area (28.0±5 versus 2.0±0.9 and 275.2±10.8 mm^2 versus 158.7±7.3 mm^2, respectively) in the mast cell-deficient Sash mice (Figure 8.3). These results suggest that an accumulation of mast cells in the periphery of the early tumours in the wild type mice have a protective barrier function against cancer cells. In agreement with these results, injection of wild type mice with LLC cells together with the mast cell stabilizer nedochromil sodium result in much more aggressive tumourogenesis, similar to mast cell-deficient Sash mice. The discrepancy between *in vitro* and *in vivo* experiments can arise from different reasons. Firstly, CBMC differ from fully mature tissue dwelling mast cells, leading to contrary data. Also, interaction of tissue-embedded mast cells with other cell types can lead to the protective function observed *in vivo*. Furthermore, an "early stage tumour" trying to implant

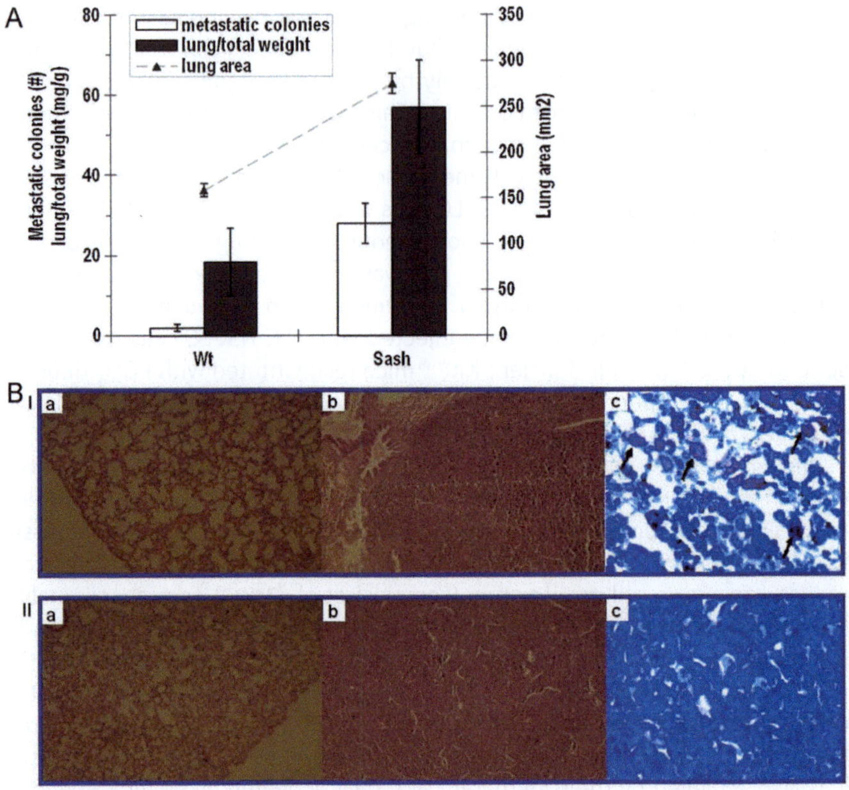

Figure 8.3 LLC model in mast cell deficient Kit^{W-sh} mice. LLC development in mast cell-deficient and wild type mice was checked 14 days after the induction of the disease. **(A)** The number of metastatic colonies (empty bars), lung area and lung/total weight ratio (black bars and dotted line) were compared. Data are the mean ± SD from mast cell-deficient and wild type mice. **(B)** Lung sections of LLC-injected mast cell-deficient (I) and wild type (II) mice. H&E staining of normal looking areas (a) and cancerous areas (b) of the lung tissue sections (10×) and (c) toluidine blue staining of cancerous lung tissue sections (40×). Arrows indicate mast cells. Representative pictures obtained from mast cell-deficient and wild type mice are shown. **Figure legend, reproduced from:** Stoyanov E, Uddin M, Mankuta D, Dubinett SM & Levi-Schaffer F. (2012). "Mast cells and histamine enhance the proliferation of non-small cell lung cancer cells". Lung Cancer, 75(1):38-44. Fig. 5.

itself in the lung can be confronted by the mast cells, whereas *in vitro* mature cancer cells, growing in their "typical niche" in suspension are much more resistant.

Regarding the use of Kit^{W-sh} mice to discover mast cells function in cancer, most studies resulted in an apparent pro-tumourigenic mast cell function. In two of these studies, skin squamous carcinoma or b-cell carcinoma, which are both related to lung squamous carcinoma, were induced by different genetic manipulations. The execution of the same genetic manipulations in $Kit^{W\text{-}sh}$ mice resulted in an attenuated pre-malignant angiogenesis, confirming pro-tumourigenic mast cells involvement in the first stages of tumour growth (Coussens et al., 1999; Soucek et al., 2007). It might be that in the NSCLC model, in wild type mice, the mast cells face an "exogenous" invasion of tumour cells from the blood in a way they would face invading metastatic cells, recognizing them as "non self" and attack. In the skin/b-cell carcinoma models, the tissue-intrinsic genetic switch is accepted as normal and therefore receives full "cooperation" from the mast cells. In another two studies, exogenous introduction of cancer cells into wild type/$Kit^{W\text{-}sh}$ also resulted in an apparent pro-tumourigenic effect for mast cells (Starkey et al., 1988; Oldford et al., 2010). Nevertheless, in the more recent of these (Oldford et al., 2010) it was shown that although originally tissue-dwelling mast cells are pro-tumourigenic, as indicated by decreased tumour growth in mast cells deficient mice, activation of these mast cells with the synthetic TLR-2 agonist $Pam3CSK_4$ turns them into aggressive anti-tumourigenic immune cells.

8.6.3. Conclusion

In vivo studies performed with mast cell-deficient mice retain the dual possibilities with which mast cells function influences cancer cell. It is also evident that additional functional and mechanistic research can be applied to develop new tools to manipulate these cells. In continuation of this line of research, information about the expression of both activating and inhibitory cell surface receptors in mast cells is currently growing (Migalovich-Sheikhet et al., 2012). This data also raises new options to regulate mast cell function specifically, for example, by the use of functional monoclonal Abs (mAbs) to inhibit tumour growth.

With additional research on the exact spatiotemporal involvement of mast cells in the outgrowth of cancer, the malignant course of tumour growth could be intervened by either activating or down-regulating mast cell effector functions via anti-receptor mAbs or specific receptor agonists or antagonists.

8.7. Concluding Remarks

Although research over the last decade has led to new and improved therapies for a variety of different diseases, anticancer drug therapy continues to have undesirable outcomes, including both poor response and severe toxicity. In the context of the complexity of cancer disease processes, future anti-cancer treatments will have to take into account the tumour microenvironment and aim to target the different cellular and molecular participants encompassed in a tumour as well as their specific interactions. Cell proliferation is crucial for tumour development and progression and in this regard, the involvement of histamine in cancer has been extensively investigated. In the present chapter, we have presented major findings of the most recent research on histamine and the involvement the preposition of histamine receptors in cancer, focusing on the remarkable role of H_4R. These data clearly indicate that histamine plays a key role as a mediator in most human tumours and that histamine is not only involved in cancer cell proliferation, migration and invasion, but also that the tumour microenvironment and immune system responses are tightly affected. In human neoplasias, H_4R seemed to be one of the main histamine receptor subtypes involved in the control of the metabolic pathways responsible for tumour growth and progression, suggesting that H_4R represent a potential molecular target and avenue for cancer drug development.

References

Adachi, N., Liu, K., Motoki, A., et al. (2006). Suppression of ischemia/reperfusion liver injury by histamine H4 receptor stimulation in rats. *Eur J Pharmacol, 544*, 181-187.

Adami, M., Pozzoli, C., Menozzi, A., et al. (2012). Effects of histamine H4 receptor ligands in a mouse model of gastric ulceration. *Pharmacology, 89*, 287-294.

Agarwala, S.S., Glaspy, J, O'Day, S.J, et al. (2002). Results from a randomized phase III study comparing combined treatment with histamine dihydrochloride plus interleukin-2 versus interleukin-2 alone in patients with metastatic melanoma. *J Clin Oncol, 20*(1), 125-133.

Aurelius, J., Martner, A., Brune, M., et al. Remission maintenance in acute myeloid leukemia: impact of functional histamine H2 receptors expressed by leukemic cells. *Haematologica 97*(12), 1904-1908.

Backhaus, B., Weidenhiller, M., Bijlsma, P., et al. (2004). Evaluation of spontaneous histamine release from colorectal mucosa in patients with gastrointestinal mediated allergy and in a healthy control group. *Inflamm Res, 53*, Suppl. 1, S87-S88.

Badve, S., Dabbs, D.J, Schnitt, S.J, et al. (2011). Basal-like and triple-negative breast cancers: a critical review with an emphasis on the implications for pathologists and oncologists. *Modern Pathol, 24*(2), 157-167.

Barbara, G., Stanghellini, V., De Giorgio, R., et al. (2004). Activated mast cells in proximity to colonic nerves correlate with abdominal pain in irritable bowel syndrome. *Gastroenterology, 126*, 693-702.

Bartholeyns, J & Bouclier, M. (1984). Involvement of histamine in growth of mouse and rat tumors: antitumoral properties of monofluoromethylhistidine, an enzyme-activated irreversible inhibitor of histidine decarboxylase. *Cancer Res, 44*(2), 639-645.

Baumer, W., Wendorff, S., Gutzmer, R., et al. (2008). Histamine H4 receptors modulate dendritic cell migration through skin--immunomodulatory role of histamine. *Allergy, 63*, 1387-1394.

Beerman, S., Seifert, R., & Neumann, D. (2012). Commercially available antibodies against human and murine histamine H4-receptor lack specificity. *Naunyn-Schniedeberg's Arch Pharmacol, 385*, 125-135.

Belcheva, A., & Mishkova, R. (1995). Histamine content in lymph nodes from patients with malignant lymphomas. *Inflamm Res, 44*(1), S86-S87.

Bertaccini, G., & Coruzzi, G. (1992). Histamine receptors in the digestive system. In JC. Schwartz & H.L. Haas (Eds), *The Histamine receptor* (pp. 193-230). New York, NY: Wiley-Liss.

Bischoff, S., & Crowe, S.E. (2005). Gastrointestinal food allergy: new insights into pathophysiology and clinical perspectives. *Gastroenterology, 128*, 1089-1113.

Blaya, B., Nicolau, G.F., Jangi, S.M., et al. (2010). Histamine and histamine receptor antagonists in cancer biology. *Inflamm Allergy Drug Tar, 9*(3), 146-157.

Boer, K., Helinger, E., Helinger, A., et al. (2008). Decreased expression of histamine H1 and H4 receptors suggests disturbance of local regulation in human colorectal tumours by histamine. *Eur J Cell Biol, 87*, 227-236.

Boldrini, L., Gisfredi, S., Ursino, S., et al. (2006). Tumor necrosis factor-alpha: prognostic role and relationship with interleukin-8 and endothelin-1 in non-small cell lung cancer. *Int J Mol Med, 17*, 887–892.

Bolton, E., King, J, & Morris, D.L. (2000). H_2-antagonists in the treatment of colon and breast cancer. *Sem Cancer Biol, 10*, 3-10.

Breunig, E., Michel, K., Zeller, F. et al. (2007). Histamine excites neurones in the human submucous plexus through activation of H1, H2, H3 and H4 receptors. *J Physiol, 583*, 731-742.

Brodsky, A., Davio, C., Shayo, C., et al. (1998). Forskolin-induced U-937 cell line differentiation: a result of a sustained cAMP elevation. *Eur J Pharmacol, 350,* 121–127.

Buhner, S., & Schemann, M. (2012). Mast cell-nerve axis with a focus on the human gut. *Biochim Biophys Acta, 1822,* 85-92.

Buyse, M., Squifflet, P., Lucchesi, K.J, (2011). Assessment of the consistency and robustness of results from a multicenter trial of remission maintenance therapy for acute myeloid leukemia. *Trials, 12,* 86.

Byron, JW. (1977). Mechanism for histamine H2-receptor induced cell-cycle changes in the bone marrow stem cell. *Agents Actions, 7*(2), 209-213.

Camidge, D.R. & Jodrell, D.I. (2005). Chemotherapy. In M. Knowles & P. Selby (Eds), *Introduction to the Cellular and Molecular Biology of Cancer* (pp. 399–413). New York, NY: Oxford University Press.

Cancer Facts & Figures 2012 (2012). NY: *Leukemia and Lymphoma Society.*

Chandra, R., & Ganguly, A.K. (1987). Diamineoxidase activity and tissue histamine content of human skin, breast and rectal carcinoma. *Cancer Lett, 34,* 207-212.

Chazot, P.L., Shenton, F.C., Waldvogel, H., et al. (2007). The H4 histamine receptor is expressed in both the human CNS and rodent PNS. *European Histamine Research Society, 36th Meeting, p.27.*

Cianchi, F., Cortesini, C., Schiavone, N.F., et al. (2005). The role of cyclooxygenase-2 in mediating the effects of histamine on cell proliferation and vascular endothelial growth factor production in colorectal cancer. *Clin Cancer Res, 11,* 6807-6815.

Clarke, N.P., Brown, C.D., Lane, C., et al. (2008). PF-2988403- an 'H4 antagonist' demostrating the full range of in vitro pharmacologies which translate in vivo in the rat. In: *European Histamine Research Society, 37th Annual Meeting,* Abstract No.07, p.34.

Cogè, F., Guenin, S.P., Rique, H.J, et al. (2001). Structure and expression of the human histamine H4-receptor gene. *Biochem Biophys Res Commun., 284,* 301-309.

Copsel, S., Garcia, C., Diez, F., et al. (2011). Multidrug Resistance Protein 4 (MRP4/ABCC4) Regulates cAMP Cellular Levels and Controls Human Leukemia Cell Proliferation and Differentiation. *J Biol Chem, 286*(9), 6979–6988.

Coruzzi, G., Poli, E., Pozzoli, C., et al., (2002). Effect of BP 2.94, a histamine H3-receptor agonist prodrug, on different models of colitis. *Inflamm Res, 51, Suppl. 1*, S31-S32.

Coruzzi, G., Adami, M., Guaita, E., et al. (2007). Antiinflammatory and antinociceptive effects of the selective histamine H4-receptor antagonists JNJ7777120 and VUF6002 in a rat model of carrageenan-induced acute inflammation. *Eur J Pharmacol, 563*, 240-244.

Coruzzi, G., & Adami, M. (2008). Peripheral actions and therapeutic potential in periphery. In D. Vohora (Ed), *The third Histamine Receptor: Selective Ligands as Potential Therapeutic Agents in CNS Disorders* (pp.167-209). New York, NY: CRC Press, Taylor and Francis Group.

Coruzzi, G., Adami, M., Pozzoli, C., et al. (2010). Gastroprotective effects of histamine H4 receptor ligands in rodent ulcer models. *Br J Pharmacol, 7*, 150P.

Coruzzi, G., Adami, M., Pozzoli, C. et al. (2011). Selective H_3 and H_4 receptor agonists exert opposite effects against the gastric lesions induced by HCl in the rat stomach. *Eur J Pharmacol, 669*, 121-127.

Coruzzi, G., Adami, M., & Pozzoli, C. (2012a). Role of histamine H4 receptors in the gastrointestinal tract. *Front Biosci, 4*, 226-239.

Coruzzi, G., Pozzoli, C., Adami, M., et al. (2012b). Strain-dependent effects of the histamine H4 receptor antagonist JNJ7777120 in a murine model of acute skin inflammation. *Exp Dermatol, 21*, 32-37.

Coussens, L.M., Raymond, W.W., Bergers, G., et al. (1999). Inflammatory mast cells up-regulate angiogenesis during squamous epithelial carcinogenesis. *Genes Develop, 13*(11), 1382-1397.

Cowart, MD., Altenbach, R.J, Liu, H., et al. (2008). Rotationally constrained 2,4-diamino-5,6-disubstituted pyrimidines: a new class of histamine H4 receptor antagonists with improved druglikeness and in vivo efficacy in pain and inflammation models. *J Med Chem, 51*, 6547-6557.

Cricco, G.P, Davio, C.A., Fitzsimons, C.P., et al. (1994). Histamine as an autocrine growth factor in experimental carcinomas. *Agents Actions, 43*(1-2), 11-94.

Cricco, G.P., Martin, G., Medina, V., et al. (2004). Histamine regulates the MAPK pathway via the H2 receptor in PANC-1 human cells. *Inflamm Res, 53*, Suppl. 1, S65-S66.

Cricco, G.P., Mohamad, N.A., Sambuco, L.A., et al. (2008). Histamine regulates pancreatic carcinoma cell growth through H_3 and H_4 receptors. Inflamm Res, 57, Suppl. 1, S23-S24.

Cricco, G., Mohamad, N., Sáez, M.S., et al. (2011). Histamine and Breast Cancer: a New Role for a Well Known Amine. In M. Gunduz (Ed.), Breast Cancer - Carcinogenesis, Cell Growth and Signalling Pathways (pp. 611-634). Rijeka: InTech.

Cummins, D.L., Cummins, JM., Pantle, H., et al. (2006). Cutaneous malignant melanoma. *Mayo Clinic proceedings, 81*(4), 500-507.

Davenas, E., Rouleau, A., Morisset, S., et al. (2008). Autoregulation of McA-RH7777 hepatoma cell proliferation by histamine H3 receptor. *J Pharmacol Exp Ther, 326,* 406–413.

Davio, C.A., Cricco, G.P., Andrade, N., et al. (1993). H1 and H2 histamine receptors in human mammary carcinomas. *Agents Actions, 38,* C172-C174.

Davio, C.A., Cricco, G.P., Martin, G., et al. (1994). Effect of histamine on growth and differentiation of the rat mammary gland. *Agents Actions, 41,* C115-C117.

Davio, C.A., Cricco, G.P., Bergoc, R.M., et al. (1995). H1 and H2 histamine receptors in experimental carcinomas with an atypical coupling to signal transducers. *Biochem. Pharmacol, 50,* 91-96.

Davio, C., Madlovan, A., Shayo, C., et al. (1996). Histamine receptors in neoplastic transformation. Studies in human cell lines. *Inflamm Res, 45*(1), S62-S63.

Dembinski, A., Warzecha, Z., Ceranowicz, P., et al. (2005). Role of capsaicin-sensitive nerves and histamine H1, H2, and H3 receptors in the gastroprotective effect of histamine against stress ulcers in rats. *Eur J Pharmacol, 508,* 211-221.

Deva, S., & Jameson, M. (2012). Histamine type 2 receptor antagonists as adjuvant treatment for resected colorectal cancer. *Cochrane Database of Systematic Reviews, 8,* CD007814.

Dundar, E., Oner, U., Peker B.C., et al. (2008). The significance and relationship between mast cells and tumour angiogenesis in non-small cell lung carcinoma. *J Int Med Res 36(1),* 88-95.

Dunford, P.J, Varga, C., Thurmond R.L., et al. (2006). Histamine H_4 receptor antagonism attenuates toll-like receptor signaling and inhibits experimental colitis in the rat. *Gastroenterology, 130,* W1597.

Eisenschenk, M.N.C., Torres, S.M.F., Oliveira, S., et al. (2011). The expression of histamine H4 receptor mRNA in the skin and other tissues of normal dogs. *Vet Dermatol, 22,* 396-400.

Elson, C.O., Sartor, R.B., Tennyson, G.S., & Been, C.S. (1995). Experimental models of inflammatory bowel disease. *Gastroenterology, 109,* 1344-1367.

Falus, A., Hegyesi, H., Lazar-Molnar, E., et al. (2001). Paracrine and autocrine interactions in melanoma: histamine is a relevant player in local regulation. *Trends Immunol, 22,* 648–652.

Fang, Z., Yao, W., Xiong, Y., et al. (2011). Attenuated expression of HRH4 in colorectal carcinomas: a potential influence on tumor growth and progression. *BMC Cancer, 11,* 195-205.

Feldman, M., & Burton, M.E. (1990a). Histamine2-receptor antagonists. Standard therapy for acid-peptic diseases. 1. part. *New Eng J Med, 323,* 1672-1680.

Feldman, M., & Burton, M.E. (1990b). Histamine2-receptor antagonists. Standard therapy for acid-peptic diseases. 2. part. *New Eng J Med, 323,* 1749-1755.

Fentiman, I.S. (2005). Local treatment of cancer. In M. Knowles & P. Selby (Eds), *Introduction to the Cellular and Molecular Biology of Cancer* (pp. 390–398). New York, NY: Oxford University Press.

Ferlay, J, Shin, H.R., Bray, F., et al. (2010). Estimates of worldwide burden of cancer in 2008: GLOBOCAN 2008. *Int J Cancer, 127*(12), 2893-2917.

Fernández, N., Monczor, F., Lemos, B., et al. (2002). Reduction of G protein-coupled receptor kinase 2 expression in U-937 cells attenuates H2 histamine receptor desensitization and induces cell maturation. *Mol Pharmacol, 62*(6),1506-1514.

Fiocchi, C. (1998). Inflammatory bowel disease: etiology and pathogenesis. *Gastroenterology, 115,* 182–205.

Fogel, W.A., Jochem J, & Lewinski, A. (2007). Influence of the H_3/H_4 receptor antagonist, thioperamide on regional haemodinamics in rats with trinitrobenzene sulfonic acid-induced colitis. *Inflamm Res, 56, Suppl. 1,* S21-S22.

Francis, H.L., Demorrow, S., Franchitto, A., et al. (2012). Histamine stimulates the proliferation of small and large cholangiocytes by activation of both IP3/$Ca2^+$ and cAMP-dependent signaling mechanisms. *Lab Invest, 92*, 282-294.

Frieling, T., Cooke, H.J, & Wood, JD. (1993). Histamine receptors on submucous neurons in the guinea pig colon. *Am J Physiol, 264*, G74-G80.

Galli, S.J, Tsai, M. (2010). Mast cells in allergy and infection: versatile effector and regulatory cells in innate and adaptive immunity. *Eur J Immunol, 40*(7), 1843–1851.

Garcia-Caballero, M., Neugebauer, E., Campos, R., et al. (1988). Increased histidine decarboxylase (HDC) activity in human colorectal cancer: results of a study on ten patients. *Agents Actions, 23*, 357-360.

Garcia-Caballero, M., Neugebauer, E., Campos, R., et al. (1994). Histamine synthesis and content in benign and malignant breast tumors. *Surgical Oncol, 3*, 167–173.

Gonzalez-Angulo, A.M., Morales-Vasquez, F., & Hortobagyi, G.N. (2007). Overview of resistance to systemic therapy in patients with breast cancer. *Adv. Exp Med Biol, 608*, 1-22.

Ghizzardi, P., Gobbetti, T., Bertoni, S., et al. (2009). Histamine H_4 receptor antagonism and mesenteric ischemia/reperfusion injury in mice. *Gastroenterology, 136*, Suppl. 1, A-397.

Ghosh, A.K., Hirasawa N., & Ohuchi K. (2001). Enhancement by histamine of vascular endothelial growth factor production in granulation tissue via H_2 receptor. *Br J Pharmacol, 134*, 1419-1428.

Grandi, D., Schunack, W., & Morini, G. (2006). Epithelial cell proliferation is promoted by the histamine H(3) receptor agonist (R)-alpha-methylhistamine throughout the rat gastrointestinal tract. *Eur J Pharmacol, 538*, 141-147.

Haak-Frendscho, M., Darvas, Z., Hegyesi, H., et al. (2000). Histidine decarboxylase expression in human melanoma. *J Invest Dermatol, 115*(3), 345-352.

Hahm, K.B., Park, I.S., Kim, H.C., et al. (1996). Comparison of antiproliferative effects of 1-histamine-2receptor antagonists, cimetidine, ranitidine, and famotidine, in gastric cancer cells. *Int J Immunopharmacol, 18*, 393-399.

Hegyesi, H., Somlai, B., Varga, V.L., et al. (2001). Suppression of melanoma cell proliferation by histidine decarboxylase specific antisense oligonucleotides. *J Invest Dermatol, 117*(1), 151-153.

Hegyesi, H., Horváth, B., Pállinger, E., et al. (2005). Histamine elevates the expression of Ets-1, a protooncogen in human melanoma cell lines through H2 receptor. *FEBS Lett, 579*(11), 2475-2479.

Hegyesi, H., Colombo, L., Pállinger, E., et al. (2007). Impact of systemic histamine deficiency on the crosstalk between mammary adenocarcinoma and T cells. *J Pharmacol Sciences, 105*(1), 66-73.

Hellstrand, K., Brune, M., Naredi, P., et al. (2000). Histamine: a novel approach to cancer immunotherapy. *Cancer Invest, 18*(4), 347-355.

Hsieh, G.C., Chandran, P., Salyers, A.K., et al. (2010). H4 receptor antagonism exhibits anti-nociceptive effects in inflammatory and neuropathic pain models in rats. *Pharmacol Biochem Behavior, 95*, 41-50.

Huang, JF., & Thurmond, R.L. (2008). The new biology of histamine receptors. *Curr. Allergy Asthma Rep., 8*, 21–27.

Ibaraki, T., Muramatsu, M., Takai, S., et al. (2005). The relationship of tryptase- and chymase-positive mast cells to angiogenesis in stage I non-small cell lung cancer. *Eur J Cardiothorac Surg 28(4)*, 617-621.

Imada, A., Shijubo, N., Kojima, H., et al. (2000). Mast cells correlate with angiogenesis and poor outcome in stage I lung. *Eur Resp. J, 15*(6), 1087-1093.

Jemal, A., Siegel, R., Xu, J, & Ward, E. (2010). Cancer statistics. *CA: Cancer J Clin, 60*, 227-300.

Jiang, W., Lim, H.D., Zhang, M., et al. (2008). Cloning and pharmacological characterization of the dog histamine H4 receptor. *Eur J Pharmacol, 592*, 26-32.

Kahlson, G., Rosengren, E. (1968). New approaches to the physiology of histamine. *Physiol Rev, 48*, 155-196.

Keely, S.J. Stack, W.A., O'Donoghue, D.P., et al. (1995). Regulation of ion transport by histamine in the human colon. *Eur J Pharmacol, 279*, 203-209.

Kenakin, T. (1995). Pharmacological proteus? *Trends Pharmacol Sci, 16*, 256-258.

Kenakin, T., & Miller, L.J (2010). Seven transmembrane receptors as shapeshifting proteins: the impact of allosteric modulation and functional selectivity on new drug discovery. *Pharmacol Rev, 62*, 265-304.

Kennedy, L., Hodges, K., Meng, F., et al. (2012). Histamine and histamine receptor regulation in gastrointestinal cancers. *Transl GastroInt Cancer, 1*, 138-150.

Kim, H., Dwyer, L., Song, JH., et al. (2011). Identification of histamine receptors and effects of histamine on murine and simian colonic excitability. *Neurogastroenterol Motility, 23*, 949-e409.

Kiss, R., Keserű, G.M. (2012). Histamine H4 receptor ligands and their potential therapeutic applications: an update. *Exp Opinion Ther Patents, 22*(3), 205-221.

Knutson, L., Ahrenstedt, O., Odlind, B., et al. (1990). The jejunal secretion of histamine is increased in active Crohn's disease. *Gastroenterology, 98*, 849-854.

Kumawat, A.K., Gotlind, Y.Y., Fredin, M.F., et al. (2010). Expression patterns of histamine receptors in the Galphai2-deficient mouse model of colitis. *Inflamm Res, 59*, Suppl. 4, S358-S359.

Lampiasi, N., Azzolina, A., Montalto, G., et al. (2007). Histamine and spontaneously released mast cells granules affect the cell growth of human hepatocellular carcinoma cells. *Exp Mol Med, 39*, 284–294.

Lemos, B., Davio, C., Gass, H., et al. (1995). Histamine receptors in human mammary gland, different benign lesions and mammary carcinomas. *Inflamm Res, 44*(S1), S68-S69.

Lee, H.J, Park, M.K., Lee, E.J, et al. (2012). Histamine receptor 2-mediated growth-differentiation factor-15 expression is involved in histamine-induced melanogenesis. *Int J Biochem Cell Biol, 2*, S1357-2725

Lens, M.B., Dawes, M. (2004). Global perspectives of contemporary epidemiological trends of cutaneous malignant melanoma. *Br J Dermatol, 150*, 179–185

Leurs, R., Chazot, P.L., Shenton, F.C., et al. (2009). Molecular and biochemical pharmacology of the histamine H4 receptor. *Br J Pharmacol, 157*, 14-23.

Lim, H.D., van Rijn, R.M., Ling, P., et al. (2005). Evaluation of histamine H1-, H2-, and H3-receptor ligands at the human histamine H4 receptor: identification of 4-methylhistamine as the first potent and selective H4 receptor agonist. *J Pharmacol Exp Ther, 314*(3), 1310-1321.

Lim, H.D., Smits, R.A., Bakker, R.A., et al. (2006). Discovery of S-(2-guanidylethyl)-isothiourea (VUF 8430) as a potent nonimidazole H4 receptor agonist. *J Med Chem., 49*, 6650-6651.

Lim, H.D., Adami, M., Guaita, E., et al. (2009). Pharmacological characterization of the new histamine H4 receptor agonist VUF 8430. *Br J Pharmacol, 157*, 34-43.

Liu, C., Ma, X., Jiang, X., et al. (2001). Cloning and pharmacological characterization of a fourth histamine receptor (H4) expressed in bone marrow. *Mol Pharmacol, 59*, 420-426.

Liu, R.J, Altenbach, T.L., Carr, P., et al. (2008). cis-4-(Piperazin-1-yl)-5,6,7a,8,9,10,11,11a-octahydrobenzofuro [2,3-h]quinazolin-2-amine (A-987306), a new histamine H_4R antagonist that blocks pain responses against carrageenan-induced hyperalgesia. *J Med Chem., 51*, 7094-7098.

Maintz, L., & Novak, N. (2007). Histamine and histamine intolerance. *Am J Clin Nutrition, 85*, 1185-1196.

Malaviya, R., Morrison, A.R., Pentland, A.P. (1996). Histamine in human epidermal cells is induced by ultraviolet light injury. *J Invest Dermatol, 106*(4), 785-789.

Malinski, C., Kierska, D., Fogel, W., et al. (1993). Histamine: its metabolism and localization in mammary gland. *Comp Biochem Physiol C, 105*(2), 269-273.

Martinel Lamas, D.J, Croci M., Carabajal E., et al. (2013). Therapeutic potential of histamine H4 receptor agonists in triple-negative human breast cancer experimental model. *Br J Pharmacol*, 170, 188-199.

Martner, A., Thorén, F.B., et al. (2010). Immunotherapy with histamine dihydrochloride for the prevention of relapse in acute myeloid leukemia. *Exp Rev Hematol, 3*(4), 381-391.

Masini, E., Fabbroni, V., Giannini, L., et al. (2005). Histamine and histidine decarboxylase up-regulation in colorectal cancer: correlation with tumor stage. *Inflamm Res, 54*, Suppl. 1, S80-S81.

Massari, N.A., Medina, V.A., Martinel Lamas, D.J, et al. (2011). Role of H4 receptor in histamine-mediated responses in human melanoma. *Melanoma Res, 21*, 395-404.

Mathes, R.W., Malone, K.E., Daling, JR., et al. (2008). Relationship between histamine2-receptor antagonist medications and risk of invasive breast cancer. *Cancer Epidem, Biomarkers and Prevention, 17(1),* 67-72.

Medina, V., Cricco, G., Mohamad, N., et al. (2005). Histamine is a selective protector against cellular damage produced by ionizing radiation. *Inflamm Res, 54,* Suppl. 1, S17-S18.

Medina, V., Cricco, G., Nuñez, M., et al. (2006). Histamine-mediated signaling processes in human malignant mammary cells. *Cancer Biol Ther, 5*(11), 1462-1471.

Medina, V., Croci, M., Mohamad, N., et al. (2007). Mechanisms underlying the radioprotective effect of histamine on small intestine. *Int J Radiat Biol, 83,* 653-663.

Medina, V., Croci, M., Crescenti, E., et al. (2008). The role of histamine in human mammary carcinogenesis: H3 and H4 receptors are potential therapeutic targets for breast cancer treatment. *Cancer Biol Ther, 7,* 28-35.

Medina, V.A., Massari, N.A., Cricco, G.P., et al. (2009). Involvement of hydrogen peroxide in histamine-induced modulation of WM35 human malignant melanoma cell proliferation. *Free Radic Biol Med, 46*(11), 1510-1515.

Medina, V.A., Croci, C., Carabajal, E., et al. (2010). Histamine protects bone marrow against cellular damage induced by ionizing radiation. *Int J Rad Biol, 86*(4), 283-290.

Medina, V.A., & Rivera, E.S. (2010a). Histamine as a potential adjuvant to immuno and radiotherapy for cancer treatment. Discovering new functions for the oldest biogenic amine. *Curr Immunol Rev, Special Issue; Advances in Immunopathology, 6*(4), 357-370.

Medina, V.A., & Rivera, E.S. (2010b). Histamine receptors and cancer pharmacology. *Br J Pharmacol, 161,* 755-767.

Medina, V.A., Brenzoni, P.G., Martinel Lamas, D.J,et al. (2011a). Role of histamine H4 receptor in breast cancer cell proliferation. *Front Biosci, 3,* 1042-1060.

Medina, V.A., Prestifilippo, JP., Croci, M., et al. (2011b). Histamine prevents functional and morphological alterations of submandibular glands induced by ionising radiation *Int J Rad. Biol, 87*(3), 284-292.

Meng, F., Han, Y, Staloch, D., et al. (2011). The H4 histamine receptor agonist, clobenpropit, suppresses human cholangiocarcinoma progression by disruption of epithelial mesenchymal transition and tumor metastasis. *Hepatology, 54*, 1718-1728.

Metcalfe, D.D., Baram, D., & Mekori, Y.A. (1997). Mast cells. *Physiol Rev, 77*, 1033-1079.

Migalovich-Sheikhet, H., Friedman, S., Mankuta, D., et al. (2012). Novel identified receptors on mast cells. *Front Immunol, 3*, 238.

Molnar, E.L., Cricco, G., Martin, G., et al. (2001). Histamine as a potential autocrine regulator of melanoma. *Inflamm Res, 50*(2), S102-S103.

Molnar, E.L., Hegyesi. H., Pállinger, E., et al. (2002). Inhibition of human primary melanoma cell proliferation by histamine is enhanced by interleukin-6. *Eur J Clin Invest, 32*, 743–749.

Morgan, R., McAllister, B., Cross, L., et al. (2007). Histamine 4 receptor activation induces recruitment of FoxP3+ T cells and inhibits allergic asthma in a murine model. *J Immunol, 178*, 8081–8089.

Morini, G., Grandi, D., & Bertaccini G. (1995). (R)-alpha-methylhistamine inhibits ethanol-induced gastric lesions in the rat: involvement of histamine H3 receptors? *Digestion, 56*, 145-152.

Morini, G., Grandi, D., Krause, M., et al. (1997). Gastric mucosal injury by nonsteroidal anti-inflammatory drugs is reduced by (R)-aplha-methylhistamine and its pro-drugs in the rat. *Inflamm Res, 46*, Suppl 1, S101-S102.

Morini, G., Becchi, G., Shenton, F.C., et al. (2008). Histamine H3 and H4 receptors are expressed on distinct endocrine cell type in the rat fundic mucosa. *Inflamm Res, 57*, Suppl. 1, S57-S58.

Morse, K.L., Behan, J, Laz, T.M., et al. (2001). Cloning and characterization of a novel human histamine receptor. *J Pharmacol Exp Ther, 296*, 1058-1066.

Murata T, Aritake K, Matsumoto S, Kamauchi S, Nakagawa T, Hori M, Momotani E, Urade Y, Ozaki H. (2011). Prostagladin D2 is a mast cell-derived antiangiogenic factor in lung carcinoma. *Proc Natl Acad Sci USA, 108*(49), 19802-7.

Nagata, M., Shijubo, N., Walls, AF., et al. (2003). Chymase-positive mast cells in small sized adenocarcinoma of the lung. *Virchows Arch 443(4)*, 565-573.

Nakamura, T., Itadani, H., Hidaka, Y., et al. (2000). Molecular cloning and characterization of a new human histamine receptor, HH4R. *Biochem Biophys Res Commun, 279*, 615-620.

Naredi, P. (2002). Histamine as an adjunct to immunotherapy. *Sem Oncol, 3*(7), 31-34.

Neumann, D., Beermann, S., & Seifert, R. (2010). Does the histamine H4 receptor have a pro- or anti-inflammatory role in murine bronchial asthma? *Pharmacology, 85*, 217-223.

Neumann, D., Beerman, S., Magel, L., et al. (2012). Problems associated with the use of commercial and non-commercial antibodies against the histamine H4 receptor. *Naunyn-Scmiedeberg's Arch Pharmacol, 385*, 855-860.

Nguyen, T., Shapiro, D.A., George, S.R., et al. (2001). Discovery of a novel member of the histamine receptor family. *Mol Pharmacol, 59*, 427-433.

Niczyporuk, M., Hermanowicz, A., Matuszczak, E., et al. (2012). A lack of correlation between mast cells, angiogenesis, and outcome in non-small cell lung cancer. *Exp Lung Res 38(6)*, 281-285.

Nielsen, H.J, Christensen, U., Moesgaard, F., & Kehlet, H. (2002). Ranitidine as adjuvant treatment in colorectal cancer. *Br J Surg, 89*, 1416-1422.

Nijmeijer, S., Vischer, H.F., Rosethorne, E.M., et al. (2012). Analysis of Multiple Histamine H4 Receptor Compound Classes Uncovers Gαi Protein- and β-Arrestin2-Biased Ligands. *Mol Pharmacol, 82*(6), 1174-1182.

Noble, S. (2010). Low-molecular-weight heparin and survival in lung cancer. *Thromb Res, 129*, Suppl 1, S114-S118.

Nosalova, V., Cerna, S., Schunack, W., et al. (2001). Effects of (R)alpha-methylhistamine on experimental colitis. *Inflamm Res, 50*, Suppl. 2, S108-S109.

Oda, T., Morikawa, N., & Saito, Y. (2000). Molecular cloning and characterization of a novel type of histamine receptor preferentially expressed in leukocytes. *J Biol Chem, 275*, 36781-36786.

Oda, T., Matsumoto, S., Masuho, Y., et al. (2002). cDNA cloning and characterization of porcine histamine H4 receptor. *Biochim Biophys Acta, 1575*, 135-138.

Oda, T., Matsumoto, S., Matsumoto, M., et al. (2005). Molecular cloning of monkey histamine H4 receptor. *J Pharmacol Sci, 98*, 319-322.

Oldford SA, Haidl ID, Howatt MA, Leiva CA, Johnston B, Marshall JS. (2010). A critical role for mast cells and mast cell-derived IL-6 in TLR2-mediated inhibition of tumor growth. *J Immunol, 185*, 7067-76.

Ohri, C.M., Shikotra, A., Green, R.H., et al. (2010). Tumour necrosis factor-alpha expression in tumour islets confers a survival advantage in non-small cell lung cancer. *BMC Cancer, 10*, 323.

Pal, S.k., Childs, B.H., Pegram, M, (2011). Triple-Negative Breast Cancer: unmet medical needs. *Breast Cancer Res, 125*(3), 627-636.

Parshad, R., Hazrah, P., Kumar, S., et al. (2005). Effect of preoperative short course famotidine on TILs and survival in breast cancer. *Ind J Cancer, 42*(4), 185-190.

Petit-Bertron, A.F., Machavoine, F., Defresne, M.P., et al. (2009). H_4 histamine receptors mediate cell cycle arrest in growth factor-induced murine and human hematopoietic progenitor cells. *PLoS ONE, 4*, e6504.

Pós, Z., Hegyesi, H., & Rivera, E. (2004). Histamine and cell proliferation. In A. Falus (Ed), *Histamine Biology and Medical Aspects* (pp. 199-217). Hungary: SpringMed Publishing.

Pós, Z., Sáfrány, G., Müller, K., et al. (2005). Phenotypic profiling of engineered mouse melanomas with manipulated histamine production identifies histamine H2 receptor and rho-C as histamine-regulated melanoma progression markers. *Cancer Res, 65*(10), 4458-4466.

Pozzoli, C., Adami, M., Smits, R.A., et al. (2009). Effect of histamine H_4 receptor ligands on cholinergic neurotransmission of the rat duodenum. *Inflamm Res, 58*, Suppl. 1, 59-60.

Raithel, M., Matek, M., Baenkler, H.W., et al. (1995). Mucosal histamine content and histamine secretion in Crohn's disease, ulcerative colitis and allergy enteropathy. *Int Arch Allergy Immunol, 108*, 127-133.

Rajendra, S., Mulcahy, H., Patchett, S., & Kumar, P. (2004). The effect of H2 antagonists on proliferation and apoptosis in human colorectal cancer cell lines. *Dig Dis Sci, 49,* 1634-1640.

Redaelli, A., Lee, JM., Stephens, JM., et al. (2003). Epidemiology and clinical burden of acute myeloid leukemia. *Exp Rev Anticancer Ther, 3,* 695-671

Reynolds, JL., Akhter, J, Adams, W.J, & Morris, D.L. (1997). Histamine content in colorectal cancer: Are there sufficient levels of histamine to affect lymphocyte function? *Eur J Surg. Oncol, 23,* 224-227.

Reynolds, JL., Akhter, JA., Magarey, C.J, et al. (1998). Histamine in human breast cancer. *Br J Surg, 85*(4), 538-541.

Rivera, E., Davio, C., Cricco, G., et al. (1993). Histamine regulation of tumour growth. Role of H1 and H2 receptors. In M. Garcia-Caballero, L. Brandes & S. Hosoda (Eds), *Histamine in normal and cancer cell proliferation* (pp. 299-317). Oxford: Adv. in Bioscience, Pergamon Press.

Rivera, E.S., Cricco, G.P., Engel, N.I., et al. (2000). Histamine as an autocrine growth factor: an unusual role for a widespread mediator. *Sem Cancer Biol, 10,* 15-23.

Rosethorne, E.M., & Charlton, S.J (2011). Agonist-biased signaling at the histamine H4 receptor: JNJ7777120 recruits β-arrestin without activating G proteins. *Mol Pharmacol, 79,* 749-757.

Rydning, A., Lying, O., Adamsen, B.L., et al. (2001). Mast cells are involved in the gastric hyperemic response to acid back diffusion *via* release of histamine. *Am J Physiol Gastrointest Liver Physiol, 280,* G1061-G1069.

Samoszuk, M., Kanakubo, E., & Chan, JK. (2005). Degranulating mast cells in fibrotic regions of human tumors and evidence that mast cell heparin interferes with the growth of tumor cells through a mechanism involving fibroblasts. *BMC Cancer,* 5, 121.

Sander, L.E., Lorentz, A., Sellge, G., et al. (2006). Selective expression of histamine receptors H1R, H2R, and H4R, but not H3R, in the human intestinal tract. *Gut, 55,* 498-504.

Sander, K., Kottke, T., Tanrikulu, Y., et al. (2009). 2,4-Diaminopyridines as histamine H4 receptor ligands-scaffold optimization and pharmacological characterization. *Bioorg Med Chem, 17,* 7186-7196.

Santos, J, Alonso, C., Guilarte, M., et al. (2006). Targeting mast cells in the treatment of functional gastrointestinal diseases. *Curr Opin Pharmacol, 6*, 541-546.

Schnell, D., Brunskole, I., Ladova, K., et al. (2011). Expression and functional properties of canine, rat, and murine histamine H4 receptors in Sf9 insect cells. *Naunyn-Schmiedeberg's Arch Pharmacol, 383*, 457-470.

Schwörer, H., Reimann, A., Ramadori, G., et al. (1994). Characterization of H3 receptors inhibiting 5-HT release from porcine enterochromaffin cells: further evidence for H3 receptor heterogeneity. *Naunyn Schmiedebergs Arch Pharmacol, 350*, 375-379.

Seifert, R., Schneider, E.H., Dove, S., et al. (2011). Paradoxical stimulatory effects of the "standard" histamine H4-receptor antagonist JNJ7777120: the H4 receptor joins the club of 7 transmembrane domain receptors exhibiting functional selectivity. *Mol Pharmacol, 79*, 631-638.

Sieja, K., Stanosz, S., Von Mach-Szczypinski, J, et al. (2005). Concentration of histamine in serum and tissues of the primary ductal breast cancer in women. *Breast, 14*(3), 236-241.

Sirica, A.E. (2005). Cholangiocarcinoma: molecular targeting strategies for chemoprevention and therapy. *Hepatology, 41*, 5-15.

Sorbo, J, Jakobsson A., & Norrby, K. (1994). Mast-cell histamine is angiogenic through receptors for histamine1 and histamine2. *Int J Exp Pathol, 75*, 43-50.

Sotiriou, C., Neo, S., McShane, L.M., et al. (2003). Breast cancer classification and prognosis based on gene expression profiles from a population-based study. *PNAS, 100*, 10393-10398.

Soucek, L., Lawlor, E.R., Soto, D., et al. (2007). Mast cells are required for angiogenesis and macroscopic expansion of Myc-induced pancreatic islet tumors. *Nature Med, 13*(10), 1211-1218.

Stack, W.A., Keely, S.J, O'Donoghue, D.P., et al. (1995). Immune regulation of human colon electrolyte transport in vitro. *Gut, 36*, 395-400.

Starkey JR, Crowle PK, Taubenberger S. (1988). Mast-cell-deficient W/Wv mice exhibit a decreased rate of tumor angiogenesis. *Int J Cancer, 42*, 48-52.

Stegaev, V., Sillat, T., Porola, P., et al. (2012). Brief Report: First identification of H4 histamine receptor in healthy salivary glands and in focal sialadenitis in Sjogren's syndrome. *Arthritis Rheum., 64*, 2663-2668.

Stoyanov, E., Uddin, M., Mankuta, D., et al. (2012). Mast cells and histamine enhance the proliferation of non-small cell lung cancer cells. *Lung Cancer, 75*(1), 38-44.

Strakhova, M.I., Cuff, C.A., Manelli, A.M., et al. (2009a). In vitro and in vivo characterization of A-940894: a potent histamine H4 receptor antagonist with anti-inflammatory properties. *Br J Pharmacol, 157*, 44-54.

Strakhova, M.I., Nikkel, A.L., Manelli, A.M., et al. (2009b). Localization of histamine H4 receptors in the central nervous system of human and rat. *Brain Res, 1250*, 41-48.

Suonio, E., Tuomisto, L., & Alhava, E. (1994). Effects of histamine, H1, H2 and Hic receptor antagonists and alpha-fluoromethylhistidine on the growth of human colorectal cancer in the subrenal capsule assay. *Agents Actions, 41*, 118-120.

Sutton, T.L., Zhao, A., Madden K.B., et al. (2008). Antiinflammatory mechanisms of enteric *Heligmosomoides polygyrus* infection against trinitrobenzene sulfonic acid-induced colitis in a murine model. *Infect Immun., 76*, 4772-4782.

Szincsák, N., Hegyesi, H., Hunyadi, J, et al. (2002). Different h2 receptor antihistamines dissimilarly retard the growth of xenografted human melanoma cells in immunodeficient mice. *Cell Biol Int, 26*(9), 833-836.

Takagaki, K., Osawa, S., Horio, Y., et al. (2009). Cytokine responses of intraepithelial lymphocytes are regulated by histamine H2 receptor. *J Gastroenterol, 44*, 285-296.

Takanami, I., Takeuchi, K., Naruke, M. (2000). Mast cell density is associated with angiogenesis and poor prognosis in pulmonary adenocarcinoma. *Cancer 88(12)*, 2686-2692.

Takeshita, K., Sakai, K., Bacon, K.B., & Gantner, F. (2003). Critical role of histamine H_4 receptor in leukotriene B_4 production and mast cell-dependent neutrophil recruitment induced by zymosan in vivo. *J Pharmacol Exp Ther, 307*, 1072-1078.

Tataroğlu, C., Kargi, A., Ozkal, S., et al. (2004). Association of macrophages, mast cells and eosinophil leukocytes with angiogenesis and tumor stage in non-small cell lung carcinomas (NSCLC). *Lung Cancer 43(1)*, 47-54.

Theoharides T.C., & Conti, P. (2004). Mast cells: the Jekyll and Hyde of tumor growth. *Trends in Immunology, 25,* 235–241.

Thurmond, R.L., Desai, P.J, Dunford, P.J, et al. (2004). A potent and selective histamine H4 receptor antagonist with anti-inflammatory properties. *J Pharmacol Exp Ther, 309,* 404-413.

Tiligada, E., Zampeli, E., Sander, K., et al. (2009). Histamine H3 and H4 receptors as novel drug targets. *Exp Opin Invest Drugs, 18,* 1519–1531.

Tilly, B.C., Tertoolen, L.G.J, Remorie, R., et al. (1990). Histamine as a growth factor and chemoattractant for human carcinoma and melanoma cells: action through Ca2+-mobilizing H1 receptors. *J Cell Biol, 110,* 1211-1215.

Tomita, M., Matsuzaki, Y., Onitsuka, T. (2000). Effect of mast cells on tumor angiogenesis in lung cancer. *Ann Thorac Surg 69(6),* 1686-1690.

Tomita, K., & Okabe, S. (2005). Exogenous histamine stimulates colorectal cancer implant growth via immunosuppression in mice. *J Pharmacol Sci, 97*(1), 116-123.

Tonnesen H, Bulow S, Fischerman K., et al. (1988). Effect of cimetidine on survival after gastric cancer. *Lancet,* ii, 990-992.

Tsunada, S., Fujimoto, K., Gotoh, Y., et al. (1994). Role of histamine receptors in intestinal repair after ischemia-reperfusion in rats. *Gastroenterology, 107,* 1297-1304.

Uçar, K. (1991). The effects of histamine H2 receptor antagonists on melanogenesis and cellular proliferation in melanoma cells in culture. *Biochem Biophys Res Commun., 177*(1), 545-550.

van Rijn, R.M., van Marle, A., Chazot, P.L., et al. (2008). Cloning and characterization of dominant negative splice variants of the human histamine H4 receptor. *Biochem. J, 414*(1), 121-131.

Varga, C. Horvath, K. Berko, A., et al. (2005). Inhibitory effects of histamine H_4 receptor antagonists on experimental colitis in the rat. *Eur J Pharmacol, 522,* 130-138.

Vera, I., Rezende, L., Molina, V., & Sanz-Fuentenebro, J (2012). Clozapine as treatment of first choice in first psychotic episodes. What do we know? *Actas Españolas de Psiquiatría, 40*(5), 281-289.

von Mach-Szczypiński, J, Stanosz, S., Sieja, K., et al. (2009). Metabolism of histamine in tissues of primary ductal breast cancer. *Metabolism, 58*(6), 867-870.

Wagner, W., Ichikawa, A., Tanaka, S., et al. (2003). Mouse mammary epithelial histamine system. *J Physiol Pharmacol, 54*(2), 211-223.

Wang, L.D., Hoeltzel, M., Butler, K., et al. (1997). Activation of the human histamine H2 receptor is linked to cell proliferation and c-fos gene transcription. *Am J Physiol, 273,* C2037-C2045.

Wang, K.Y., Arima, N., Higuchi, S., Shimajiri, S., et al. (2000). Switch of histamine receptor expression from H2 to H1 during differentiation of monocytes into macrophages. *FEBS Lett, 473*(3), 345-348.

Wang, JJ, & Mak, O.T. (2011). Induction of apoptosis in non-small cell lung carcinoma A549 cells by PGD_2 metabolite, 15d-PGJ_2. *Cell Biol Int, 35*(11), 1089-1096.

Warzecha, Z., Dembinski, A., Brzozowski, T., et al. (2000). Gastroprotective effect of histamine and acid secretion on ammonia-induced gastric lesions in rats. *Scand J Gastroenterol, 35,* 916-924.

Watson, S.A., Wilkinson, L.J, Robertson, JF.R., & Hardcastle, JD. (1993). Effect of histamine on the growth of human gastrointestinal tumours: reversal by cimetidine. *Gut, 34,* 1091-1096.

Welsh, T.J., Green, R.H., Richardson, D., et al. (2005). Macrophage and mast-cell invasion of tumor cell islets confers a marked survival advantage in non-small-cell lung cancer. *J Clin Oncol 23(35),* 8959-8967.

Wood, JD. (2006). Histamine, mast cells, and the enteric nervous system in the irritable bowel syndrome, enteritis, and food allergies. *Gut, 55,* 445-447.

Yang, X.D., Ai, W., Asfaha, S., et al. (2011). Histamine deficiency promotes inflammation-associated carcinogenesis through reduced myeloid maturation and accumulation of CD11b+Ly6G+ immature myeloid cells. *Nature Med, 17*(1), 87-95.

Yang, L.P. & Perry, C.M. (2011). Histamine dihydrochloride: in the management of acute myeloid leukaemia. *Drugs, 71*(1), 109-122.

Yu, S., Stahl, E., Li, Q., et al. (2008). Antigen inhalation induces mast cells and eosinophils infiltration in the guinea pig esophageal epithelium involving histamine-mediated pathway. *Life Sci, 82,* 324-330.

Yu, F., Wolin, R.L., Wei, J, Desai, P.J, et al. (2010). Pharmacological characterization of oxime agonists of the histamine H4 receptor. *J Recept, Ligand and Channel Research, 3,* 37-49.

Zakharova, O.P., Karmazanovsky, G.G., & Egorov, V.I. (2012). Pancreatic adenocarcinoma: outstanding problems. *World J Gastrointest Surg, 4,* 104-113.

Zampeli, E., & Tiligada, E. (2009). The role of histamine H4 receptor in immune and inflammatory disorders. *Br J Pharmacol, 157*(1), 24-33.

Zhang, C., Xiong, Y., Li, J, et al. (2012). Deletion and down-regulation of HRH4 gene in gastric carcinomas: a potential correlation with tumor progression. *PloS ONE, 7,* e31207.

Chapter 9

Non Receptor-mediated Histamine Actions

Dimitrios A. Kyriakidis[1*] and Ekaterini Tiligada[2]

[1]Laboratory of Biochemistry, Department of Chemistry, Aristotle University of Thessaloniki

[2]Department of Pharmacology, Medical School University of Athens, Greece

*Corresponding author:
Dimitrios A. Kyriakidis, Laboratory of Biochemistry, Department of Chemistry, Aristotle University of Thessaloniki, 54124 Thessaloniki, Greece, e-mail: kyr@chem.auth.gr

9.1. Introduction

Histamine [2-(1*H*-imidazol-4-yl)ethanamine], a decarboxylation product of histidine that is produced through the catalytic activity of the rate-limiting enzyme histidine decarboxylase (HDC) (Ohtsu, 2010; Figure 9.1), is a short-acting biogenic amine that elicits decisive roles in mammalian (patho)physiological processes, many of which underlie diseases such as inflammation and allergy (Tiligada, 2012; Zampeli & Tiligada, 2009). Related studies performed over the last 80 years have mainly focused on the mechanisms of actions mediated through the binding of histamine to four subtypes of G-protein-coupled receptors, designated H_1, H_2, H_4 and H_4 (Parsons & Ganellin, 2006). However, interactions of histamine with other biologically active components and its contribution in governing vital cellular processes in lower eukaryotic and prokaryotic organisms have been largely overlooked. For instance, histamine induces the adaptive phenotype in the unicellular eukaryote *Saccharomyces cerevisiae* by mechanisms that may involve the differential expression of heat shock proteins (Hsps) and tubulin (Delitheos et al., 2010; Papamichael et al., 2013). In prokaryotic organisms, the experimental data provide evidence for histamine synthesis by some bacteria (Landete et al., 2008; Lucas et al., 2005; Wauters et al., 2004) and for the involvement of membrane transporters driving histidine/histamine exchange (Kimura et al., 2009; Lucas et al., 2005). As opposed to biosynthesis, more species appear to degrade the amine (Leuschner et al., 1998; Reed et al., 2010). In *Escherichia coli*, histamine has been implicated

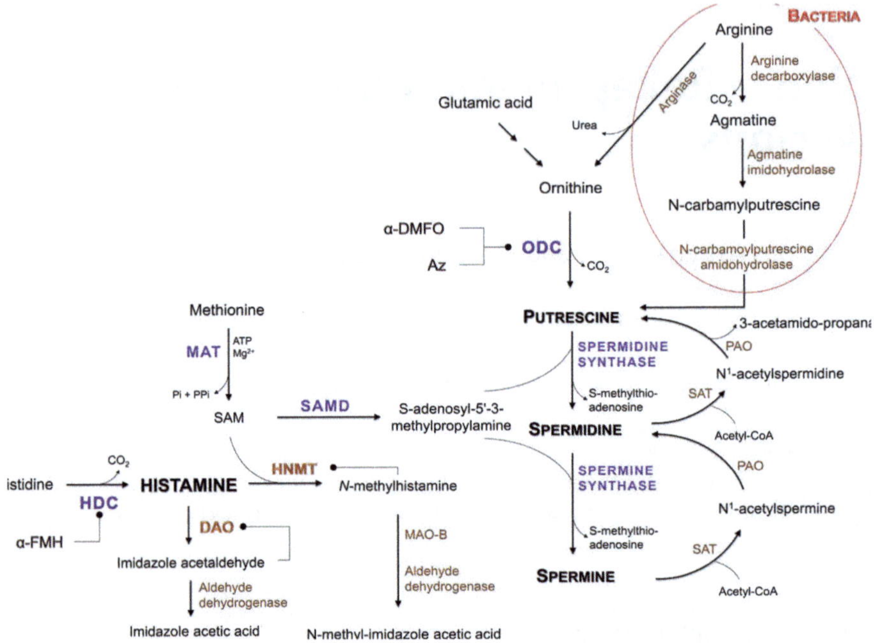

Figure 9.1 Interplay of histamine and polyamine metabolism. Histamine is synthesized from the amino acid histidine through oxidative decarboxylation by histidine-decarboxylase (HDC), a highly conserved pyridoxal 5'-phosphate-dependent enzyme. HDC activity can be inhibited by α-fluoromethylhistidine (α-FMH), a suicide substrate leading to reduction of histamine levels. Inactivation of histamine in the extracellular space is achieved by methylation through histamine N-methyltransferase (HNMT), requiring S-adenosyl-methionine (SAM) as the methyl donor. *N*-methylhistamine undergoes oxidative deamination through a monoamine oxidase (MAO-B). The main histamine-degrading enzyme in peripheral tissues and in invertebrates is diamine oxidase (DAO), which directly converts histamine into imidazoleacetic acid. On the other hand, multiple enzymes catalyse the biosynthesis and interconversion of polyamines, including methionine S-adenosyltransferase (MAT), S-adenosylmethionine decarboxylase (SAMD), ornithine decarboxylase (ODC), the spermidine and spermine synthases, spermidine/spermine N^1-acetyltransferase (SAT) and polyamine oxidase (PAO). ODC is the first rate-limiting enzyme in polyamine biosynthesis, generates putrescine and is inhibited by 2-difluoromethylornithine (DMFO) and the endogenous inhibitor antizyme (Az).

in vital endogenous biosynthetic pathways (Kyriakidis et al., 2008), in Ca^{2+}-mediated signals (Theodorou et al., 2009) and in its chemotactic phenotype (Theodorou et al., 2008).

On the other hand, the structurally related polyamines putrescine, spermidine and spermine are essential components of nearly all living cells, from bacteria and plants to humans (Canellakis et al., 1979; Canellakis et al., 1993). Although their specific functions have not been fully elucidated, the pivotal role of these biogenic amines in cellular physiology is indicated by the severe reductions in cell growth and in some cases by cell death following polyamine depletion (Tabor & Tabor, 1984; Tabor & Tabor, 1985). Regulation of polyamine synthesis

(Figure 9.1) is complex and the key biosynthetic enzyme ornithine decarboxylase (ODC) is probably one of the most highly regulated enzymes in nature (Pegg, 1986). It has been shown that the levels and/or the activity of ODC can be modulated at the transcriptional, translational and post-translational levels (Davis, 1992) through interaction with antizymes, endogenous non-competitive protein ODC inhibitors (Canellakis & Hayashi, 1989; Hayashi & Canellakis, 1989), with the synthesis of these entities being induced by polyamines (Fong et al., 1976; Heller et al., 1976; Kyriakidis et al., 1978; Murakami et al., 1992). Interestingly, antizyme in *E. coli* is also a transcriptional factor of AtoC, the response regulator of the two component system (TCS) AtoSC (Kyriakidis et al., 2012).

A holistic approach integrating current knowledge on biogenic amine interactions and actions from bacteria to humans would challenge the future exploration of amine-modulated pathways in, for example, inflammatory disorders and host-bacteria interactions. To date, histamine receptor homologues have not been identified in prokaryotic cells and reports on the existence of proteins interacting with histamine in microorganisms are limited and inconclusive. Complementary to classical approaches, the recent technological burst in biomedical sciences is expected to add valuable information in this area of research. The progress in metagenomics that incorporates the sequencing of ecosystems, including inflammation-linked 'enterotypes' (Arumugan et al., 2011; Hildebrand et al., 2013; Huttenhower et al., 2012) is an illustrative example of the strength of this field and despite the many layers of high data complexity, it is anticipated to advance our knowledge on the putative intracellular role of histamine.

9.2. Metabolic Interplay of Histamine and Other Biogenic Amines

Although histamine and polyamines represent two different classes of biogenic amines, being synthesized by two independent pathways (Fig.1), they share many characteristics. The comparable equilibrium binding of these molecules to DNA has been known for some time (Ruiz-Chica et al., 2006; Winkle & Crooks, 1988). Polyamines are essential for cell growth as they regulate the intracellular pH and interact with DNA, tRNA, ribosomes and glutathione (Canellakis et al., 1979; Tabor & Tabor, 1985). They also stabilize the conformation of nucleic acids by protecting the DNA from strand breaking induced by external agents (Ha et al., 1997; Ha et al., 1998; Sy et al., 1999) and regulate DNA, RNA and protein biosynthesis, as well as inducing the biosynthesis of the protein inhibitor antizyme, which in turn regulates their levels though a feedback mechanism (Canellakis et al., 1979). Polyamines are converted to acetyl-derivatives by chromatin-bound acetylases (Fig. 1; Canellakis et al., 1979), while several physiological events are associated with their transglutamination (Agostinelli, 2012). Transglutaminases modify proteins post-translationally through a number of methods, including the incorporation of polyamines by transferring the 4-aminobutyl moiety of spermidine to the ε-amino group of a specific lysine residue to form ε-(γ-glutamyl)lysine crosslinks (Folk et al., 1980). Interestingly, the eukaryotic initiation factor 5A (eIF5A), involved in translation, elongation and stimulating peptide bond formation, is the only known protein to contain the unusual amino acid hypusine [*hy*droxy*pu*trescine ly*sine*; N (ε)-(4-amino-2-hydroxybutyl)-lysine], which is formed by post-translational modification in two catalytic steps (hypusination). In the first step, deoxyhypusine synthase catalyzes the cleavage of spermidine and transfer of its 4-aminobutyl moiety to the ε-amino group of a specific lysine residue of the eIF-5A precursor to form deoxyhypusine and 1,3-diaminopropane. Thus, polyamines are essential components for the hypusination of eIF-5A, which is a key protein in lymphocyte blastogenesis and receptor-mediated endocytosis (Park & Wolff, 1988), as well as in the pathogenicity of different diseases, including diabetes, several human cancers, malaria and HIV-1 infections (Kaiser, 2012).

At physiological pH, 96% of histamine is in monocationic form, 3% in dication form and 1% neutral. At low pH, the predominant form appears to be the dication (Fig. 2) and at high pH the amino group loses a proton and appears to be uncharged (Ramirez et al., 2003; Shahid et al., 2009). Studies on the biological activity of histamine and polyamines showed an antagonistic relationship of their metabolism (Garcia-Faroldi et al., 2009b). Polyamines and histamine levels followed opposite profiles in bone marrow cell cultures. α-Difluoromethylornithine (α-DFMO), a suicide inhibitor of ODC, induced polyamine depletion and upregulation of HDC expression and activity along with increased histamine levels, particularly during the active synthetic process in the early stages of cell growth (Garcia-Faroldi et al., 2009a). In contrast, α-DFMO did not induce any effect in either HDC activity or histamine levels of differentiated bone marrow-derived mast cells. Sequence-specific DNA methylation analysis

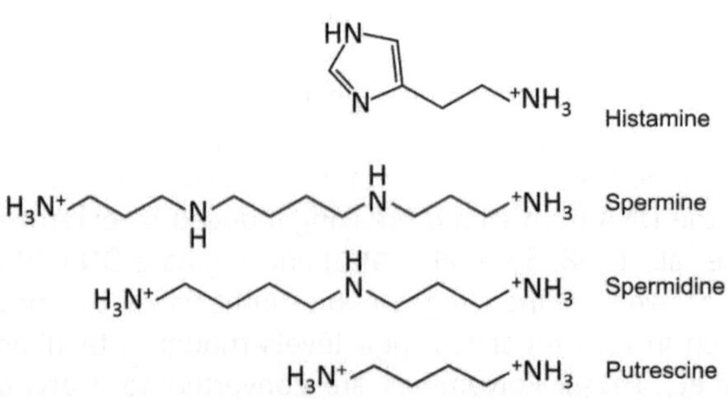

Figure 9.2 Structural similarities between histamine and polyamines.

revealed that the α-DFMO-induced HDC mRNA upregulation, observed in early bone marrow cell cultures, could not be attributed to the demethylation of the gene promoter caused by polyamine depletion. Furthermore, inactivation of histamine in the extracellular space, particularly but not exclusively in the central nervous system (Haas et al., 2008), is achieved by methylation through histamine *N*-methyltransferase (HNMT) (Nishibori et al., 2000; Fig. 1). Histamine methylation requires *S*-adenosyl-methionine (SAM) as the methyl donor to produce *N*-methylhistamine (also known as tele-methylhistamine) and decreased histamine levels (Green et al., 1987). On the other hand, SAM is the substrate for the rate-limiting enzyme in polyamine synthesis, S-adenosylmethionine decarboxylase (Fig. 1), providing the n-propylamine donor for the synthesis of spermidine and spermine from putrescine in mammals, plants and many other species (Pegg, 2009). The metabolic interplay between polyamines and histamine has been confirmed in mast cells (Fajardo et al., 2001a; Fajardo et al., 2001b) and has been linked to chronic inflammation-related processes and cancer development (Garcia-Faroldi et al., 2009b), however, studies exploring the exact contribution of this interplay in clinical settings are lacking.

Taken together, the data supports the cross-talk between histamine and polyamine metabolism and suggests that the resulting regulation depends on the availability of substrates and the expression levels of participating enzymes in various tissues and cells. Therefore, the concept that the metabolic inter-relationship of polyamine and histidine depends on the metabolic state of individual cells, the existing environment and the pathological state of the whole organism deserves careful consideration.

9.3. Actions of Histamine Beyond Histamine Receptor Binding

9.3.1. Histamine Transporters

In mammals, histamine is synthesized in the cytosol and stored in secretory vesicles after transportation by specific transport systems (Riley & West, 1953). Unlike other monoamines, such as serotonin, dopamine and norepinephrine, a specific membrane transporter for histamine has not been reported. The vesicular monoamine transporters (VMATs) are primarily located on the membranes of intracellular vesicles and preferentially transfer monoamines from the cytoplasm to storage vesicles, from which they can be subsequently released on demand (Erickson & Varoqui, 2000; Shahid, 2009). The two subtypes of monoamine transporters have been designated as VMAT1 and VMAT2. Furthermore, organic cation transporters (OCTs) mediate sodium independent electrogenic transport of small organic cations with different molecular structures and play a key role in the clearance of monoamines from the blood stream (Eisenhofer, 2001). The expression of OCT1 was found to be distributed in the liver, kidney and intestine, OCT2 in the brain and kidney and OCT3, also designated as extra-neuronal monoamine transporter (EMT), showed a broad tissue distribution and participates in the cellular uptake and elimination of various cationic substrates, including therapeutically important agents, as well as in the inactivation of biogenic amines (Eisenhofer, 2001; Martel & Azevedo 2003).

The uptake of histamine into the professional histamine-synthesising basophils and mast cells involves transport through two distinct membranes, the plasma membrane and vesicular membrane (Kakavas et al., 2006). Without excluding the contribution of amine metabolism (Ogasawara et al., 2006), the vesicular membrane transporting system appears to be more efficient than the plasma membrane transporting system, as the amount of histamine in the granular fraction is usually higher than the cytosolic fraction (Ohtsu, 2008). OCT1 has not been observed to transport histamine into the cell, in contrast to OCT2 and OCT3 for which histamine has been shown to be a suitable substrate (Gründemann et al., 1999; Martel & Azevedo, 2003; Ogasawara et al., 2006; Schneider et al., 2005). The bidirectional OCT3 appeared to account for the capacity of mast cells and basophils to take up the mediator from the extracellular environment (Ogasawara et al., 2006) and was shown to be involved in the uptake of amines into mast cells and basophil granules (Schneider et al., 2005). OCT3 participates in the control of basophil functions, since exogenous histamine can inhibit not only its own synthesis but also those of interleukin (IL)-4, IL-6, and IL-13 via this transporter (Ohtsu, 2008). Histamine uptake also involves VMAT2, which is responsible for the transport of histamine into secretory granules of enterochromaffin-like cells in the gut (Watson et al., 1999) and gene expression of VMAT2 was found to be modulated positively by TGF-α or negatively by IL-1 and TNF-α (Kazumori et al., 2004). Moreover, the expression of functional VMAT-2 transporter with high affinity for histamine was shown in primary cultures of endometrial cells (Noskova et al., 2006). Therefore, the VMAT2 seems to be essential to insure the passage of histamine across the plasma membrane and crossing of the vesicular membrane.

9.3.2. Histamine Binding to DNA

Histamine is involved in many complex biological processes related to intracellular communication, defense and cellular proliferation (Kyriakidis et al., 2012). The older method of the standard filter assay used to investigate histamine binding to DNA relied on measuring the binding of radioactive histamine to DNA, followed by phase partitioning (Krugh et al., 1981) and an equilibrium binding assay (Winkle & Crooks, 1988). This method incorporates an organic-soluble anion to enhance the partitioning of positively charged radiolabeled compounds between non-polar organic solvents and an aqueous buffer solution. More recently, the interaction between histamine and calf-thymus DNA was analyzed by applying Fourier Transform-Raman spectroscopy, thus opening new perspectives for the molecular understanding of the control of several cell responses (Ruiz-Chica et al., 2006). Being cationic, histamine was shown to bind to salmon sperm DNA with higher affinity to poly(dGdC) sequences (Winkle & Crooks, 1988). Histamine binding was weaker than spermine, probably due to charge differences (Fig. 2) under physiological pH and sequence specificity. Polyamines were observed to compete for histamine binding to DNA, with spermidine affinity being higher than putrescine and lower than spermine. Additionally, the 1,4-diamine H_1 receptor antagonist chlorpheniramine has been reported to interact with DNA and induce conformational changes in the nucleic acid by affecting both phosphodiester bonds and bases (Medina et al., 1998).

9.3.3. Histaminylation of Glutamine Residues in G-proteins

Post-translational modifications are essential for protein activity. Monoaminylation can be described as the transfer of biogenic monoamines such as serotonin, dopamine and norepinephrine to proteins in a transglutaminase-dependent manner with elusive biological consequences (Hummerich et al., 2012). Even though histamine incorporation into proteins has been described almost 30 years ago (Fesus et al., 1985), histamine was recently discovered to be incorporated into mastocytoma proteins via a histaminylation process that functions as a regulatory post-translational modification (Vowinckel et al., 2012). Protein incorporation of histamine catalyzed by transglutaminase II was also shown in mast cells. In this study, glutamine residues modified by histamine attachment were identified in the catalytic core of the small, heterotrimeric G proteins Gαq, Gαo1 and Cdc42 where this modification was shown to lead to constitutive protein activation (Vowinckel et al., 2012). This recent report on the direct involvement of histamine in post-translational modifications provides evidence for a novel histamine function in G protein signaling, guiding the way to an emerging regulatory role in mammalian (patho)physiology.

9.3.4. Histamine Interaction with Heme-containing Proteins

The binding of intracellular histamine to cytochrome P_{450} was proposed by LaBella and Brandes (2000). Cytochrome P_{450} constitutes a family of microsomal enzymes that are present in all cells but are most abundant in the liver (Brandes et al., 2002). These enzymes are involved in the metabolism of xenobiotics, drugs and natural substrates, including hormones that modulate gene function and cell growth (Mahnke et al., 1996; Morgan, 2001; Nebert & Russell, 2002). The microsomal and nuclear histamine sites, designated as H_{IC}, have been implicated in the action of histamine as an intracellular mediator of platelet aggregation and lymphocyte mitogenesis (Bencsath et al., 2002; Labella & Brandes, 2000). Polyamines and hormones, such as estrogen, testosterone and progesterone as well as drugs such as tamoxifen, flutamide, various antidepressants and antihistamines have been reported to inhibit histamine binding to P_{450} (LaBella & Brandes, 2000). In addition to histamine, melatonin and other biogenic amines have also been shown to bind cytochrome P_{450} isozymes and cytochrome C (LaBella & Brandes, 2000). Several histamine antagonists, particularly thioperamide, clobenproprit and ciproxyfan (Yang et al., 2002) bind to the heme moiety of cytochrome P_{450} (Hamelin et al., 1998; Kishimoto et al., 1997), providing a possible explanation for the effects of these antagonists when used at high doses. Thus, heme-containing enzymes may represent common targets for the interaction of multiple bioamines, hormones and drugs for influencing cell function and growth.

Nitrophorins are nitric oxide (NO) transport heme-containing proteins in the saliva of blood-feeding insects, which act as vasodilators and anti-platelet agents and have been associated with histamine (Andersen et al., 1998; Champagne et al., 1995; Ribeiro et al., 1993). *Rhodnius prolixus*, an insect that carries the trypanosome causing Chagas' disease, releases NO-loaded nitrophorins during blood feeding, whereupon the ligand is released into the bloodstream or surrounding tissue of the host (Kirchhoff, 1993). When isolated, these proteins are in the ferric state and contain NO ligated to the heme iron (Champagne et al., 1995; Ribeiro et al., 1993). The NO is released from this complex on dilution at neutral pH, leaving an uncoordinated ferric heme protein that can bind histamine released in response to tissue injury (Ribeiro, 1982; Ribeiro et al., 1994; Weichsel et al., 1998). Recombinant nitrophorin 4 expressed in *E. coli* reconstituted with heme, was found to bind NO and histamine in a manner similar to that of the natural protein (Andersen et al., 1998). The crystal structure of nitrophorin 4 revealed a lipocalin-like eight-stranded β-barrel with heme inserted into one end of the barrel. A detailed comparison with other lipocalins suggests that nitrophorin 4 is closely related to the biliverdin-binding proteins from insects (Andersen et al., 1998).

9.3.5. Histamine Binding to Lipocalins

Some blood-sucking parasites secrete proteins into the feeding site, thus suppressing inflammation and facilitating feeding (Weichsel et al., 1998). The cattle-feeding tick *Rhipicephalus appendiculatus* produces histamine-binding proteins, one of them (Ra2) possessing a lipocalin-fold (Paesen et al., 1999; Paesen et al., 2000). A related protein secreted by the rodent-feeding tick *Dermacentor reticulatus* was reported to bind both histamine and serotonin (Sangamnatdej et al., 2002). Unlike other members of the lipocalin superfamily, which contain a single central cavity to carry hydrophobic ligands within their - barrel structure tick histamine-binding proteins harbour two internal binding sites (Paesen et al., 1999). The well-conserved H site binds histamine with high affinity, while the residue changes in the L site are consistent with weak histamine interaction and the binding of a bulkier serotonin molecule (Sangamnatdej et al., 2002). As histamine is a key inflammatory mediator in animals, the diversification of biogenic amine interactions with parasite proteins may reflect their important role in host adaptation.

9.4. The Role of Histamine in Prokaryotic Physiology

Histamine anabolic and catabolic pathways have been identified in prokaryotic organisms. Studies in Gram-positive bacteria revealed the existence of pyridoxal phosphate-dependent HDCs in the nosocomial pathogen *Enterobacter aerogenes* but not in the commensal potential pathogen *E. coli* (Wauters et al., 2004) and that pyruvoyl-dependent HDCs are associated with Gram-positive bacteria, including some lactic acid species and Staphylococci (Landete et al., 2008). Biogenic primary amines are usually inactivated by oxidative deamination to aldehydes, hydrogen peroxide and ammonia (Leuschner et al., 1998), whereas in *Nocardioides simplex*, histamine is metabolised to imidazole acetaldehyde by histamine dehydrogenase (Reed et al., 2010) (Fig. 1). Histamine metabolism in *E. coli* remains elusive, yet exogenous histamine has been shown to induce proliferation and biomass accumulation during the late lag and the early exponential phases of *E. coli* K-12 growth (Anuchin et al., 2008).

9.4.1. Histamine in AtoSC-regulated Biosynthesis of Poly-R-3-hydroxybutyrate

Exogenous histamine has also been associated with the biosynthesis of the complexed poly-R-3-hydroxybutyrate (cPHB), which is directly regulated by the AtoSC TCS (Theodorou et al., 2006; Theodorou et al., 2007). The abundant, naturally occurring cPHB is a ubiquitous constituent of prokaryotic and eukaryotic cells and elicits fundamental responses in cellular physiology, including Ca^{2+} homeostasis, competence for genetic transformation, protection of complexed proteins from proteolysis and DNA organization (Reusch et al., 1995; Reusch et al., 2002). On the other hand, TCSs modulate many biological processes in bacterial signalling by employing phosphorylation as the means of signal transduction (Kyriakidis & Tiligada, 2009). In particular, the AtoSC TCS has been shown to play a pivotal role in *E. coli* by being implicated to play a role in bacterial homeostasis and pathogenicity (Kyriakidis et al., 2012). AtoSC consists of the AtoS sensor histidine kinase and the cytoplasmic cognate response regulator AtoC (Gao et al., 2007; Parkinson & Kofoid, 1992; West & Stock, 2001) and in the *E. coli* genome, the *atoS* and *atoC* genes are located upstream of the ato operon genes *atoD, atoA, atoE* and *atoB (atoDAEB)* which encode proteins involved in short-chain fatty acid metabolism (Berlyn, 1998).

AtoSC-mediated signal transduction has been associated with acetoacetate (Theodorou et al., 2006), spermidine (Theodorou et al., 2007) or intermediate metabolic compounds of the small fatty acid catabolic pathway (Theodorou et al., 2006). Interestingly, AtoSC has been shown to respond to the pro-inflammatory mediators histamine (Kyriakidis et al. 2008) and platelet activating factor (Theodorou et al., unpublished data), as well as to the basic polyamine compound 48/80 (C48/80) (Kyriakidis et al., 2008) most likely in an extracellular Ca^{2+}-dependent manner (Theodorou et al., 2009). Histamine elicited a suppressive effect on cPHB biosynthesis and counteracted the effect induced by C48/80 synthesis, irrespective of the phase of bacterial growth. In addition to the inhibition of cPHB biosynthesis, the selective histamine-induced activation of *atoC* (Kyriakidis et al., 2008; Theodorou et al., 2011), which also encodes antizyme (Canellakis et al., 1993; Kyriakidis et al., 1978), points to the interplay between histamine and polyamine metabolism (Medina et al., 2003), thus eliciting multiple effects in prokaryotic cells.

9.4.2. Histamine Signalling and Chemotaxis

Host-derived histamine may play a role in bacterial behavior in humans, affecting the microbiota chemotactic response and adherence in the induction of infection and during inflammatory response, at least in part by interacting with biosynthetic and metabolic pathways. This hypothesis is supported by the contribution of the AtoSC TCS in the regulation of *E. coli* motility and chemotaxis (Oshima et al., 2002; Theodorou et al., 2008) and the biphasic histamine-induced alterations in the motility and chemotactic phenotypes of AtoSC-expressing bacteria, where low concentrations enhance *E. coli* motility and higher concentrations are inhibitory (Theodorou et al., 2008).

However, the existing literature does not provide sufficient evidence to support the specificity of histamine effects on bacterial signalling. Candidate cellular components that may contribute to the observed actions of histamine include members of the basic amino acid/polyamine antiporter family such as the putrescine/ornithine antiporter PotE or the histidine/histamine exchanger

HdcP, which could mediate downstream signalling (Lucas et al., 2005). The pH-dependent putative histidine/histamine antiporter HdcT, which transfers histidine into the cytoplasm and takes histamine out of the cell, has been characterized in the opportunistic pathogen *Photobacterium damselae* subsp. *damselae* using information on PotE in *E. coli* (Kimura et al., 2009). Interestingly, Gram-negative bacteria utilize histamine to constitute the iron-sequestering siderophores which are implicated in their survival and protection against host defense mechanisms (Crosa & Walsh, 2002).

Taken together, the scarce data on the putative histamine effects in human microbiota imply that it may directly or indirectly interact with yet unidentified targets to complement its widely accepted implications in mammalian immunoregulatory machinery (Zampeli & Tiligada, 2009) and in the modulation of host defense against bacterial infection (Krämer et al., 2008; Oshima et al., 2002; Zampeli et al., 2009).

9.5. The Role of Histamine in Lower Eukaryotic Organisms

The conservation of the basic cellular and molecular mechanisms from yeasts to higher eukaryotes made the yeast *S. cerevisiae* an established model for the investigation of the cellular stress response (Mager and Ferreira, 1993; Miligkos et al., 2000; Tiligada et al., 2002). Like any other cell, yeasts often encounter and adapt to different types of environmental and microenvironmental stresses and produce or assimilate biologically active molecules, including histamine (Besancon et al., 1992; Gardini et al., 2006). However, relatively limited data are available on the role of histamine in lower eukaryotic organisms (Novak & Falus, 1997; Isik et al., 2009; Rinnerthaler et al., 2006) and on the cross-talk between inflammatory mediators and cellular stress response (van Eden et al. 2007).

However, histamine has been demonstrated to induce cellular stress response in *S. cerevisiae*, leading to Hsp and tubulin overexpression that contribute to the modulation of the adaptive processes (Delitheos et al., 2010). Furthermore, the H_1 receptor antagonist dimethindene elicited a dose-related antifungal action, while the $H_{3/4}$ and H_4 receptor ligands thioperamide and JNJ7777120, respectively, induced the adaptive phenotype in yeast through *de novo* protein synthesis, likely directed towards alterations in Hsp70 and Hsp104 expression (Papamichael et al., 2013). Since yeasts lack known histamine receptors, these findings are indicative of the occurrence of histamine receptor-independent signalling pathways in lower eukaryotic cells that may have interesting implications for understanding the off-target effects of histamine receptor ligands.

9.6. A Working Hypothesis for the Function of Histamine

The broad functions of histamine may be distinguished as 'supportive' or 'specific' (Figure 9.3). 'Supportive' functions can be defined as those in which histamine appears to be replaceable by other cationic molecules, such as polyamines, dopamine, serotonin or noradrenaline and may interact with numerous anionic endogenous components. The non-replaceable functions of histamine are defined as 'specific'. In the latter case, the anime could either be covalently linked to specific macromolecules or engaged in high-affinity linkages. Obviously, the boundaries between 'supportive' and 'specific' functions may be modified as more data become available.

The environment to which the tissues or cells are exposed, either physiologically or experimentally, may modify the observed response. The prior history of the cell defines its 'metabolic state', which establishes a certain level of response within the cell, such as the regulation of the level of expression of metabolic enzymes. The environment in which the tissues or cells are placed at any given time defines the 'external setting', which can qualitatively and quantitatively modify the response of the cell through a cascade of reactions that are not normally elicited by a given stimulus. The experimental manipulations may generate the unique "metabolic state" of the cell. The "external setting" then brings about a cascade of reactions that are not normally elicited by the

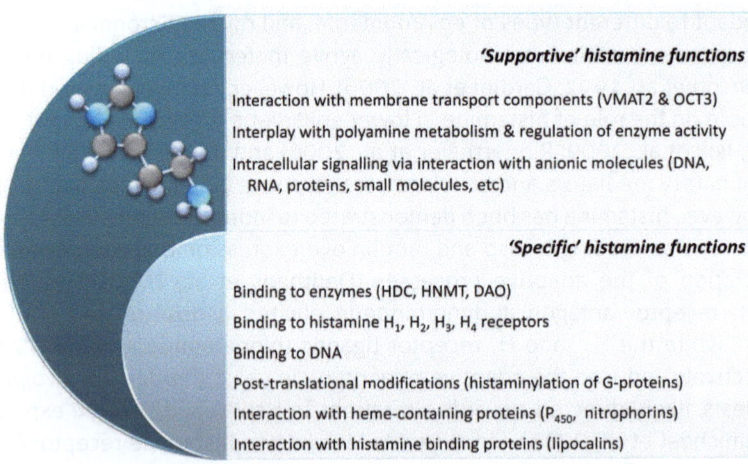

Figure 9.3 Putative '*specific*' and '*supportive*' functions of histamine (see text for details). DAO: diamine oxidase, HDC: histidine-decarboxylase, HNMT: N-methyltransferase; OCT: organic cation transporter, P_{450}: cytochrome P_{450}; VMAT: vesicular monoamine transporter

applied stimulus. As a consequence, the 'potential' of the biological setting may be under investigation rather than its physiological function. Thus, the putative functions of histamine, beyond interaction with the $H_{1\text{-}4}$ receptors are illustrated in this chapter. It is clear that additional information in the field is required in order to elucidate the actual intracellular role of histamine and the related function(s) and regulation.

9.7. Conclusions and Perspectives

In this chapter, the current available information on a number of significant actions of histamine have been described and evaluated. This chapter aimed to highlight the regulatory mechanisms in which histamine may play a role inside the cell, either alone or through cross-talk with other cellular components (Figure 9.4). Besides the histamine responses elicited through binding to its

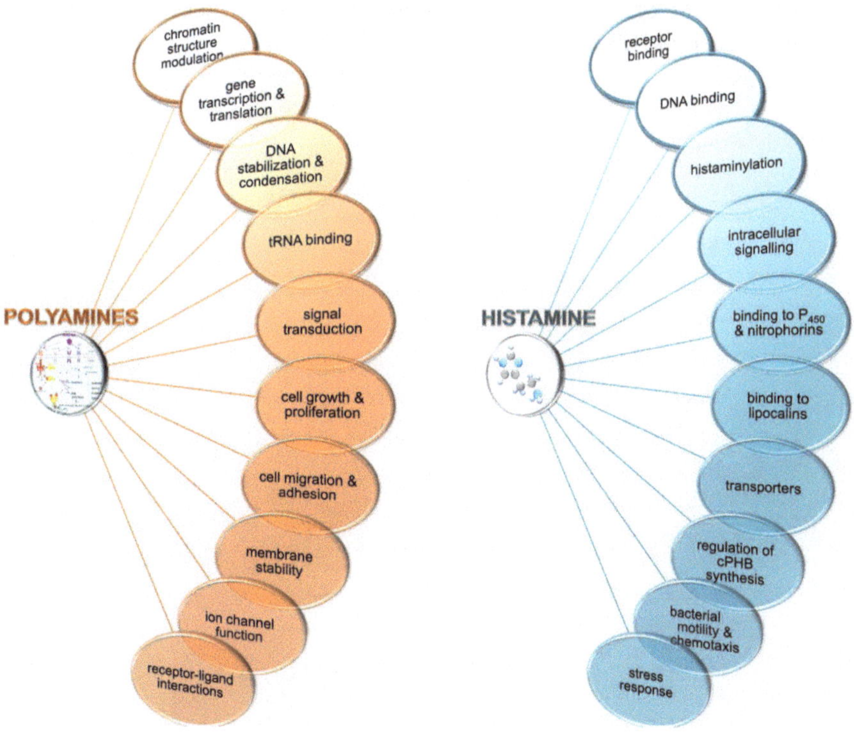

Figure 9.4 Major functions of polyamines and histamine as evidenced by the available data in the literature (see text for details). P_{450}: cytochrome P_{450}; cPHB: complexed poly-R-3-hydroxybutyrate.

currently known membrane receptors, evidence points to the interaction of histamine with nucleic acids and influx/efflux mechanisms, its regulatory role in intracellular signalling (such as through post-translational histaminylation reactions), the amine contribution to xenobiotic metabolism through binding to cytochrome P_{450} and to the response to tissue injury by interacting with nitrophorins and lipocalins. The role histamine in vital adaptive and survival processes including regulation of metabolite biosynthesis, bacterial motility and the cellular stress response has been implied. A comparison with structurally related polyamines was also attempted (Figure 9.4). Our understanding of the functions of histamine in the signalling cascades will expose novel transmission signals in numerous biological processes, thus improving the perspectives for their potential exploitation in health and disease.

Acknowledgments

The work is part of the EU FP7 COST Action BM0806: Recent advances in histamine receptor H_4R research.

References

Agostinelli, E. (2012). Role of polyamines, their analogs and transglutaminases in biological and clinical perspectives. *Amino Acids, 42*, 397-409.

Andersen, J.F., Weichsel, A., Balfour, C. A., et al. (1998). The crystal structure of nitrophorin 4 at 1.5 Å resolution: transport of nitric oxide by a lipocalin-based heme protein. *Structure, 6*, 1315-1327.

Anuchin, A.M., Chuvelev, D.I., Kirovskaia, T.A., et al. (2008). Effects of monoamine neuromediators on the growth-related variables of *Escherichia coli* K-12. *Mikrobiologia, 77*, 758-765.

Arumugam, M., Raes, J., Pelletier, E., et al. (2011). Enterotypes of the human gut microbiome. *Nature, 473*, 174–180.

Bencsath, M., Gidali, J., Brandes, L.J., et al. (2002). Murine and human hematopoietic colony formation: a possible regulatory role for intracellular histamine. *Acta Biol Hung, 53*, 299-306.

Berlyn, M.K.B. (1998). Linkage map of *Escherichia coli* K-12, edition 10: the traditional map. *Microbiol Mol Biol Rev, 62*, 814-984.

Besancon, X., Smet, C., Chabalier, C., et al. (1992). Study of surface yeast flora of Roquefort cheese. *Int J Food Microbiol, 17*, 9–18.

Brandes, L.J., Queen, G.M., LaBella, F.S. (2002). Displacement of histamine from liver cells and cell components by ligands for cytochromes P450. *J Cell Biochem, 85*, 820-824.

Canellakis, E.S., Hayashi, S.I. (1989). The role of antizyme in the regulation of ornithine decarboxylase activity in eukaryotic cells, in U. Bachrach, Y. M. Heimer (Eds.), The Physiology of Polyamines (315-330). FL: CRC, Boca Raton

Canellakis, E.S., Paterakis, A.A., Huang, S-C., et al. (1993). Identification, cloning, and nucleotide sequencing of the ornithine decarboxylase antizyme gene of *Escherichia coli*. *Proc Natl Acad Sci USA, 90*, 7129-7133.

Canellakis, E.S., Viceps-Madore, D., Kyriakidis, D.A., et al. (1979). The regulation and function of ornithine decarboxylase and of the polyamines. In B. L. Horecker & E. R. Stadtman (Eds.), *Curr Topics Cell Reg*, 15, (pp. 155–202). New York, Academic Press Inc.

Champagne, D.E., Nussenzvieg, R.H., Ribeiro, J.M.C. (1995). Purification, partial characterization and cloning of nitric oxide-carrying heme proteins (nitrophorins) from salivary glands of the blood-sucking insect *Rhodnius prolixus*. *J Biol Chem, 270*, 8691-8695.

Crosa, J.H., Walsh, C.T. (2002). Genetics and assembly line enzymology of siderophore biosynthesis in bacteria. *Microbiol Mol Biol Rev, 66*, 223-249.

Davis, R.H., Morris, D.R., Coffino, P. (1992). Sequestered end products and enzyme regulation: the case of ornithine decarboxylase, *Microbiol Rev*, 56, 280-290.

Delitheos, B., Papamichael, K., Tiligada, E. (2010). Histamine modulates the cellular stress response in yeast. *Amino Acids*, 38, 1219-1226.

Eisenhofer, G. (2001). The role of neuronal and extraneuronal plasma membrane transporters in the activation of peripheral catecholamines. *Pharmacol Ther, 91*, 35-62.

Erickson, J.D., Varoqui, H. (2000). Molecular analysis of vesicular amine transporter function and targeting to secretory organelles. *FASEB J, 14*, 2450-2458.

Fajardo, I., Urdiales, J.L., Medina, M.A., et al. (2001a). Effects of phorbol ester and dexamethasone treatment on histidine decarboxylase and ornithine decarboxylase in basophilic cells. *Biochem Pharmacol, 61*, 1101-1106.

Fajardo, I., Urdiales, J.L., Paz, J.C., et al. (2001b). Histamine prevents polyamine accumulation in mouse C57.1 mast cell cultures. *Eur J Biochem, 268*, 768-773.

Fesus, L., Szucs, E.F., Barrett, K.E., et al. (1985). Activation of transglutaminase and production of protein-bound gamma-glutamylhistamine in stimulated mouse mast cells. *J. Biol. Chem., 260*, 13771–13778

Folk, J.E., Park, M.H., Chung, S., et al. (1980). Polyamines as physiological substrates for transglutaminases. *J Biol Chem, 255*, 3695-3700.

Fong, W.F., Heller, J.S., Canellakis, E.S. (1976). The appearance of an ornithine decarboxylase inhibitory protein upon the addition of putrescine to cell cultures. *Biochim Biophys Acta, 428*, 456-465.

Gao, R., Mack, T.R., Stock, A.M. (2007). Bacterial response regulators: versatile regulatory strategies from common domains. *Trends Biochem Sci, 32*, 225-234.

Garcia-Faroldi, G., Correa-Fiz, F., Abrighach, H., et al. (2009a). Polyamines affect histamine synthesis during early stages of IL-3-induced bone marrow cell differentiation. *J Cell Biochem, 108*, 261-271.

Garcia-Faroldi, G., Sanchez-Jimenez, F., Fajardo, I. (2009b). The polyamine and histamine metabolic interplay in cancer and chronic inflammation. *Curr Opin Clin Nutrition Metabolic Care, 12*, 59-65.

Gardini, F., Tofalo, R., Belletti, N., et al. (2006). Characterization of yeasts involved in the ripening of Pecorino Crotonese cheese. *Food Microbiol, 23*, 641-648.

Green, J.P., Prell, G.D., Khandelwal, J.K., et al. (1987). Aspects of histamine metabolism. *Agents and Actions, 22*, 1-15.

Gründemann, D., Liebich, G., Kiefer, N., et al E. (1999). Selective substrates for non-neuronal monoamine transporters. *Mol Pharmacol, 56*, 1-10.

Ha, H.C., Woster, P.M., Yager, J.D., et al. (1997). The role of polyamine catabolism in polyamine analogue-induced programmed cell death. *PNAS USA, 94*, 11557-11562.

Ha, H. C., Yager, J. D., Woster, P. A., et al. (1998). Structural specificity of polyamines and polyamine analogues in the protection of DNA from strand breaks induced by reactive oxygen species. *Biochem. Biophys. Res. Commun., 244*, 298-303.

Haas, H.L., Sergeeva, O.A., Selbach, O. (2008). Histamine in the nervous system. *Physiol Revi, 88*, 1183-1241.

Hamelin, B.A., Gouayad, A., Drolet, B., et al. (1998). *In vitro* characterization of cytochrome P450 2D6 inhibition by classic histamine H_1 receptor antagonists. *Drug Metab Dispos, 26*, 536-539.

Hayashi, S.-I., Canellakis, E.S. (1989). Ornithine decarboxylase antizymes. In S.-I. Hayashi (Ed.), *Ornithine Decarboxylase: Biology, Enzymology and Molecular Genetics* (47-58). New York, NY: Pergamon.

Heller, J.S., Fong, W.F., Canellakis, E.S. (1976). Induction of a protein inhibitor to ornithine decarboxylase by the end products of its reaction. *Proceedings of the National Academy of Sciences USA*, 73, 1858-1862.

Hildebrand, F., Nguyen, A.T., Brinkman, B., et al. (2013). Inflammation-associated enterotypes, host genotype, cage and inter-individual effects drive gut microbiota variation in common laboratory mice. *Genome Biol, 14*, R4.

Hummerich, R., Thumfart, J.O., Findeisen, P., et al. (2012). Transglutaminase-mediated transamidation of serotonin, dopamine and noradrenaline to fibronectin: evidence for a general mechanism of monoaminylation. *FEBS Lett, 586*, 3421-3488.

Huttenhower, C., Gevers, D., & the Human Microbiome Project Consortium. (2012). Structure, function and diversity of the healthy human microbiome. *Nature, 486*, 207-214.

Isik, S., Kockar, F., Aydin, M., et al. (2009). Carbonic anhydrase activators: activation of the b-carbonic anhydrase Nce103 from the yeast *Saccharomyces cerevisiae* with amines and amino acids. *Bioorg Med Chem Lett, 19*, 1662–1665.

Kaiser, A. (2012). Translational control of eIF5A in various diseases. *Amino Acids, 42*, 679-684.

Kakavas, S., Zampeli, E., Papamichael, K., et al. (2006). The mast cell pathway to inflammation and homeostasis: pharmacological insights. *Anti-Inflamm Anti-Allergy Agents Med Chem, 5*, 323-334.

Kazumori, H., Ishihara, S., Rumi, M.A., et al. (2004). Transforming growth factor alpha directly augments histidine decarboxylase and vesicular monoamine transporter 2 production in rat enterochromaffin-like cells. *Am J Physiol, 286*, 508-514.

Kimura, B., Takahashi, H., Hokimoto, S., et al. (2009). Induction of the histidine decarboxylase genes of *Photobacterium damselae* subsp. *damselae* (formally *P. histaminum*) at low pH. *J Appl Microbiol, 107*, 485-497.

Kirchhoff, L.V. (1993). American trypanosomiasis (Chagas' Disease) - A tropical disease now in the United States. *New Engl J Med, 329*, 639-644.

Kishimoto, W., Hiroi, T., Sakai, K., et al. (1997). Metabolism of epinastine, a histamine H_1 receptor antagonist, in human liver microsomes in comparison with that of terfenatine. *Res Commun Chem Pathol Pharmacol, 98*, 273-292.

Krämer, S., Sellge, G., Lorentz, A., et al. (2008). Selective activation of human intestinal mast cells by *Escherichia coli* hemolysin. *J Immunol, 181*, 1438-1445.

Krugh, T.R., Winkle, S.A., Graves, D.E. (1981). Solute enhanced partition analysis-a novel method for measuring the binding of drugs to DNA. *Biochem Biophys Res Commun, 98*, 317-323.

Kyriakidis, D.A., Tiligada, E. (2009). Signal transduction and adaptive regulation through bacterial two-component systems: the *Escherichia coli* AtoSC paradigm. *Amino Acids, 37*, 443-458.

Kyriakidis, D.A., Heller, J.S., Canellakis, E.S. (1978). Modulation of ornithine decarboxylase activity in *Escherichia coli* by positive and negative effectors. *Proc Natl Acad Sci USA, 75*, 4699-4703.

Kyriakidis, D.A., Theodorou, M.C., Tiligada E. (2012). Histamine in two component system-mediated bacterial signaling. *Front Biosci, 17*, 1108-1119.

Kyriakidis, D.A., Theodorou, M.C., Filippou, P.S., et al. (2008). Effect of histamine on the signal transduction of the AtoSC two component system and involvement in poly-(R)-3-hydroxybutyrate biosynthesis in *Escherichia coli. Amino Acids, 35*, 45-52.

LaBella, F.S., Brandes, L.J. (2000). Interaction of histamine and other bioamines with cytochromes P450: implications for cell growth modulation and chemopotentiation by drugs. *Sem Cancer Biol, 10*, 47-53.

Landete, J.M., De las Rivas, B., Marcobal, A., et al. (2008). Updated molecular knowledge about histamine biosynthesis by bacteria. *Crit Rev Food Sci Nutrit, 48*, 697-714.

Leuschner, R.G., Heidel, M., Hammes, W.P. (1998). Histamine and tyramine degradation by food fermenting microorganisms. *Int J Food Microbiol, 39*, 1-10.

Lucas, P.M., Wolken, W.A.M., Claisse, O., et al. (2005). Histamine-producing pathway encoded on an unstable plasmid in *Lactobacillus hilgardii* 0006. *Appl Environ Microbiol, 71,* 1417-1424.

Mager, W.H., Ferreira, P.M. (1993). Stress response of yeast. *Biochem J, 290,* 1-13.

Mahnke, A., Roos, P.H., Hanstein, W.G., et al. (1996). *In vivo* induction of cytochrome P450 CYP3A expression in rat leukocytes using various inducers. *Biochem Pharmacol, 51,* 1579-1582.

Martel, F., Azevedo, I. (2003). An update on the extraneuronal monoamine transporter (EMT): characteristics, distribution and regulation. *Curr Drug Metabol, 4,* 313-318.

Medina, M.A., Ramírez, F.J., Ruiz-Chica, J., et al. (1998). DNA-chlorpheniramine interaction studied by spectroscopic techniques. *Biochim Biophys Acta, 1379,* 129-133.

Medina, M.A., Urdiales, J.L., Rodríguez-Caso, C., et al. (2003). Biogenic amines and polyamines: similar biochemistry for different physiological missions and biomedical applications. *Crit Revi Biochem Mol Biol, 38,* 23-59.

Miligkos, V., Tiligada, E., Papamichael, K., et al. (2000). Anticancer drugs as inducers of thermotolerance in yeast. *Folia Microbiol (Praha), 45,* 339-342.

Morgan, E.T. (2001). Regulation of cytochrome P450 by inflammatory mediators: why and how. *Drug Metabol Dispos, 29,* 207-212.

Murakami, Y., Matsufuji, S., Miyazaki, Y. et al. (1992). Antizyme, a protein induced by polyamines, accelerates the degradation of ornithine decarboxylase in Chinese-hamster ovary-cell extracts. *J Biol Chem, 267,* 13138-13141.

Nebert, D.W., Russell, D.W. (2002). Clinical importance of the cytochrome P450. *Lancet, 360,* 1155-1162.

Nishibori, M., Tahara, A., Sawada, K., et al. (2000). Neuronal and vascular localization of histamine N-methyltransferase in the bovine central nervous system. *Eur J Neurosci, 12,* 415-424.

Noskova, V., Bottalico, B., Olsson, H., et al. (2006). Histamine uptake by human endometrial cells expressing the organic cation transporter EMT and the vesicular monoamine transporter-2. *Mol Human Reprod, 12,* 483-489.

Novak, I., Falus, A. (1997). Molecular biology and role of histamine in physiological and pathological reactions. A review. *Acta Biol Hungar, 48*, 385-394.

Ogasawara, M., Yamauchi, K., Satoh, Y., et al. (2006). Recent advances in molecular pharmacology of the histamine systems: organic cation transporters as a histamine transporter and histamine metabolism. *J Pharmacol Sci, 101*, 24-30.

Ohtsu, H. (2008). Progress in allergy signal research on mast cells: the role of histamine in immunological and cardiovascular disease and the transporting system of histamine in the cell. *J Pharmacol Sci, 106*, 347-353.

Ohtsu, H. (2010). Histamine synthesis and lessons learned from histidine decarboxylase deficient mice. *Adv Exp Med Biol, 709*, 21-31.

Oshima, T., Aiba, H., Masuda, Y., et al. (2002). Transcriptome analysis of all two-component regulatory system mutants of *Escherichia coli* K-12. *Mol Microbiol, 46*, 281-291.

Paesen, G.C., Adams, P.L., Harlos, K., et al. (1999). Tick histamine-binding proteins: isolation, cloning, and three-dimensional structure. *Mol Cell, 3*, 661-671.

Paesen, G.C., Adams, P.L., Nuttall, P.A., et al. (2000). Tick histamine-binding proteins: lipocalins with a second binding cavity. *Biochim Biophys Acta, 1482*, 92-101.

Papamichael, K., Delitheos, B., Tiligada, E. (2013). A subset of histamine receptor ligands improve thermotolerance of the yeast *Saccharomyces cerevisiae. J Appl Microbiol, 114*, 492-501.

Park, M.H., Wolff, E. (1988). Cell-free synthesis of deoxyhypusine: Separation of protein substrate and enzyme and identification of 1,3-diaminopropane as a product of spermidine cleavage. *J Biol Chem, 263*,15264-15269.

Parkinson, J.S., Kofoid, E.C. (1992). Communication modules in bacterial signaling proteins. *Ann Revi Genetics, 26*, 71-112.

Parsons, M.E., Ganellin, R.C. (2006). Histamine and its receptors. *Br J Pharmacol, 147*, S127-135.

Pegg, A.E. (1986). Recent advances in the biochemistry of polyamines in eukaryotes. *Biochem J, 234*, 249-262.

Ramirez, F.J., Tunon, I., Collado, J.A., et al. (2003). Structural and vibrational study of the tautomerism of histamine free-base in solution. *JACS, 125*, 2328-2340.

Reed, T., Lushington, G., Xia, H.Y., et al. (2010). Crystal structure of histamine dehydrogenase from *Nocardioides simplex*. *J Biol Chem, 285*, 25782-25791.

Reusch, R.N., Huang, R., Bramble, L.L. (1995). Poly-3-hydroxybutyrate/polyphosphate complexes form voltage-activated Ca^{2+} channels in the plasma membranes of *Escherichia coli*. *Biophys J, 69*, 754-766.

Reusch, R.N., Shabalin, O., Crumbaugh, A., et al. (2002). Posttranslational modification of *E. coli* histone-like protein H-NS and bovine histones by short-chain poly-(R)-3-hydroxybutyrate (cPHB). *FEBS Lett, 527*, 319-322.

Ribeiro, J.M.C. (1982). The antiserotonin and antihistamine activities of salivary secretion of *Rhodnius prolixus*. *J Insect Physiol, 28*, 69-75.

Ribeiro, J.M.C., Hazzard, J.M.H., Nussenzveig, R.H., et al. (1993). Reversible binding of nitric oxide by a salivary heme protein from a bloodsucking insect. *Science, 260*, 539-541.

Riley, J.F., West, G.B. (1953). The presense of histamine in tissue mast cell. *J Physiol, 120*, 528-537.

Rinnerthaler, M., Jarolim, S., Heeren, G., et al. (2006). MMI1 (YKL056c, TMA19), the yeast orthologue of the translationally controlled tumor protein (TCTP) has apoptotic functions and interacts with both microtubules and mitochondria. *Biochim Biophys Acta, 1757*, 631-638.

Ruiz-Chica, A.J., Soriano, A., Tunon, I., et al. (2006). FT-Raman and QM/MM study of the interaction between histamine and DNA. *Chem Phys, 324*, 579–590.

Sangamnatdej, S., Paesen, G.C., Slovak, M., et al. (2002). A high affinity serotonin- and histamine-binding lipocalin from tick saliva. *Insect Mol Biol, 11*, 79-86.

Schneider, E., Machavoine, F., Pléau, J.M., et al. (2005). Organic cation transporter 3 modulates murine basophil functions by controlling intracellular histamine levels. *J Exp Med, 202*, 387-393.

Shahid, M., Tripathi, T., Sobia, F., et al. (2009). Histamine, histamine receptors, and their role in immunomodulation: An updated systematic review. *Open Immunol J, 2*, 9-41.

Sy, D., Hugot, S., Savoye, C., et al. (1999). Radioprotection of DNA by spermine: A molecular modelling approach. *Int J Rad Biol, 75*, 953-961.

Tabor, C.W., Tabor, H. (1984). Polyamines. *Annual Review of Biochemistry*, 53, 749-790.

Tabor, C.W., Tabor, H. (1985). Polyamines in microorganisms. *Microbiol Rev, 49*, 81–99.

Theodorou, E.C., Theodorou, M.C., Samali, M.N., et al. (2011). Activation of the AtoSC two-component system in the absence of the AtoC N-terminal receiver domain in *E. coli. Amino Acids, 40*, 421-430.

Theodorou, M.C., Panagiotidis, C.A., Panagiotidis, C.H., et al. (2006). Involvement of the AtoS-AtoC signal transduction system in poly-(R)-3-hydroxybutyrate biosynthesis in *Escherichia coli. Biochim Biophys Acta, 1760*, 896-906.

Theodorou, M.C., Theodorou, E.C., Panagiotidis, C.A., et al. (2007). Spermidine triggering effect to the signal transduction through the AtoS-AtoC/Az two-component system in *Escherichia coli. Biochim Biophys Acta, 1770*, 1104-1114.

Theodorou, M.C., Tiligada, E., Kyriakidis, D.A. (2008). The involvement of AtoSC two-component system in *E. coli* chemotaxis. *FEBS J, 275* (Suppl 1), 290.

Theodorou, M.C., Tiligada, E., Kyriakidis, D.A. (2009). Extracellular Ca^{2+} transients affect poly-(R)-3-hydroxybutyrate regulation by the AtoSC system in *E. coli. Biochem J, 417*, 667-672.

Tiligada, E. (2012). Editorial: Is histamine the missing link in chronic inflammation? *Journal of Leukoc Biol, 92*, 4-6.

Tiligada, E., Miligkos, V., Delitheos, A. (2002). Cross-talk between cellular stress, cell cycle and anticancer agents: mechanistic aspects. *Curr Med Chem, 2*, 553-566.

van Eden, W., Wick, G., Albani, S., et al. (2007). Stress, heat shock proteins, and autoimmunity: how immune responses to heat shock proteins are to be used for the control of chronic inflammatory diseases. *Ann NY Acad Sci, 1113*, 217-237.

Vowinckel, J., Stahlberg, S., Paulmann, N., et al. (2012). Histaminylation of glutamine residues is a novel posttranslational modification implicated in G-protein signaling. *FEBS Lett, 586*, 3819-3824.

Watson, F., Deavall, D.G., MacRo, J.A., et al. (1999). Transcriptional activation of vesicular monoamine transporter 2 in the pre-B cell line Ea3.123. *Biochem J, 337*, 193-199.

Wauters, G., Avesani, V., Charlier, J., et al. (2004). Histidine decarboxylase in *Enterobacteriaceae* revisited. *J Clin Microbiol, 42*, 5923-5924.

Weichsel, A., Andersen, J.F., Champagne, D.E., et al. (1998). Crystal structures of a nitric oxide transport protein from a blood-sucking insect. *Nature Struct Biol, 5*, 304-309.

West, A.H., Stock, A.M. (2001). Histidine kinases and response regulator proteins in two-component signaling systems. *Trends Biochem Sci, 26*, 369-376.

Winkle, S.A., Crooks, P.A. (1988). Equilibrium binding of spermine and histamine to salmon sperm DNA and poly(dGdC). *J Pharmacy Pharm, 40*, 809-811.

Yang, R., Hey, J.A., Aslanian, R., et al. (2002). Coordination of histamine H_3 receptor antagonists with human adrenal cytochrome P450 enzymes. *Pharmacology, 66*, 128-135.

Zampeli, E., Tiligada, E. (2009). The role of histamine H_4 receptor in immune and inflammatory disorders. *Br J Pharmacol, 157*, 24-33.

Zampeli, E., Pitychoutis, P., Papadopoulou-Daifoti, Z., et al. (2009). Systemic challenge with lipopolysaccharide increases histamine levels in the conjunctiva and cartilage, but not hypothalamus of Sprague-Dawley rats. *Inflamm Res, 58*(Suppl 1), 49-50.

Chapter 10

Perspectives in H_4R Research and Therapeutic Exploitation

Ekaterini Tiligada

Department of Pharmacology, Medical School University of Athens, Greece

Address for correspondence: Department of Pharmacology, Medical School University of Athens, M. Asias 75, GR-11527 Athens, Greece, E-mail: aityliga@med.uoa.gr

10.1. Introduction

Histamine [2-(1*H*-imidazol-4-yl)ethanamine] is an endogenous short-acting biogenic amine synthesised from the basic amino acid L-histidine through the catalytic activity of the rate-limiting enzyme histidine decarboxylase that is widely distributed throughout the body (Ohtsu, 2010). The classical source of histamine is the pluripotent heterogeneous mast cell, where it is stored in granules and released in response to various immunological and non-immunological stimuli (Kakavas *et al.*, 2006; Riley & West, 1952). Histamine is also produced by a number of other cell types, including gastric enterochromaffin-like cells (Prinz *et al.*, 2003), various types of blood cells, such as basophils (Falcone *et al.*, 2006), macrophages, dendritic cells, lymphocytes (Zwadlo-Klarwasser *et al.*, 1994) and platelets (Masini *et al.*, 1998), chondrocytes (Maslinska *et al.*, 2004), neurons (Lethbridge & Chazot, 2010; Tiligada *et al.*, 2011) and tumour cells (Falus *et al.*, 2001).

Since its discovery at the beginning of the 20th century (Dale & Laidlaw, 1910), histamine has been one of the most studied biological molecules in medicine (Parsons & Ganellin, 2006; Figure 10.1). It is widely accepted to elicit a spectrum of actions, largely, but not exclusively (Delitheos *et al.*, 2010; Kyriakidis *et al.*, 2012; Papamichael *et al.*, 2013), through binding to four currently known G protein-coupled receptors (GPCRs), designated as H_1, H_2, H_3, and H_4 (H_1R-H_4R) that are differentially expressed in various cell types (Akdis & Simons, 2006; Bongers *et al.*, 2010; Parsons & Ganellin, 2006). Histamine receptor diversity is supported by pharmacological (Hill *et al.*, 1997; Lim *et al.*, 2005; Lim *et al.*, 2010; Nijmeijer *et al.*, 2013) and biochemical evidence (Leurs *et al.*, 2009; Nijmeijer *et al.*, 2012b) as well as by the low sequence homology among them (Leurs *et al.*, 2009; Nijmeijer

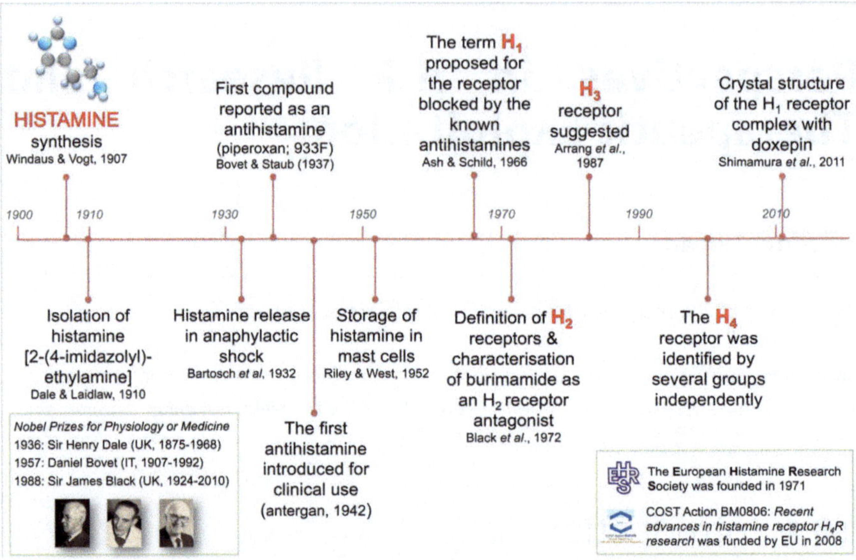

Figure 10.1 Timeline featuring historical highlights related to histamine and histamine receptors (H_1-H_4).

et al., 2012a). Antagonists of the H_1R (antihistamines) and the H_2R have attained blockbuster status for the treatment of allergy and gastrointestinal diseases, respectively (Parsons & Ganellin, 2006), whereas the relatively recent discovery of the H_3R and the H_4R revived the scientific interest in histamine research as attractive perspectives for the potential therapeutic exploitation of these new drug targets (Thurmond *et al.*, 2008; Tiligada *et al.*, 2009).

10.2. The Histamine H_4 Receptor in Disease

The identification and cloning of the H_4R in 2000 by several groups led to intense research over the last decade, which offers renewed hope that histamine may be the 'missing link' in chronic inflammation and even exposes additional roles for the H_1Rs and H_2Rs (Tiligada, 2012). Yet, our understanding of the functional role of histamine in the multiple interconnected immunological responses and inflammatory signals remains largely elusive.

The H_4R is predominately expressed on cells of haematopoietic origin, such as eosinophils, mast cells, basophils, neutrophils, dendritic cells, monocytes and T cells (cf Zampeli & Tiligada, 2009). The H_4R shows ~19% amino acid sequence homology to the classical pro-inflammatory H_1R or the H_2R and ~37% homology with the H_3R. H_3Rs and H_4Rs are most closely related to each other and have a closer phylogenetic relationship with peptide ligand GPCRs, and remotely related to H_1Rs and H_2Rs (Leurs *et al.*, 2009; Nakamura *et al.*, 2000; Nijmeijer *et al.*, 2012a). Although histamine has been implicated in inflammation for over 80 years, H_1R and H_2R ligands are largely ineffective in the treatment of a number of immune-associated disorders characterised by increased numbers of mast cells and histamine levels in various body fluids and tissues, including some allergic disorders, asthma and autoimmune diseases (Ceponis *et al.*, 1998; Church & Maurer, 2012; Frewin *et al.*, 1986; Konttinen *et al.*, 1990; Permin *et al.*, 1981). Interestingly, the use of H_1R antagonists for the treatment of itch that frequently complements many skin disorders is well documented; however, antihistamines targeting the H_1Rs and H_2Rs are of limited therapeutic value in chronic skin diseases, such as atopic dermatitis (Akdis *et al.*, 2006; Gutzmer *et al.*, 2011). This could be attributed to the considerable variations of extracellular histamine levels that result from mast cell degranulation and/or the 100-1000-fold lower rates of *de novo* histamine synthesis by nonprofessional histamine-producing cells compared to mast cells and basophils. Thus, the levels of the amine in many tissues and body fluids may be too low to stimulate the pro-inflammatory H_1Rs (pKi = 4.2) and the H_2Rs (pKi = 4.3), but may be sufficient to activate the high affinity H_4Rs (pKi = 7.8) or the H_3Rs (pKi = 8.0) (Lim *et al.*, 2005; Thurmond *et al.*, 2008). Additionally, H_4R genomic properties, such as single nucleotide polymorphisms (SNPs), splice variants and/or gene copy number variations may influence the course of inflammatory diseases (Mommert *et al.*, 2011; Yu et al., 2010a; Yu et al., 2010b). Therefore, the novel concept that histamine exerts immunomodulatory actions in inflammation through H_4R signalling and the potential therapeutic exploitation of this activity for a range of the major poorly treatable chronic inflammatory diseases (Gschwandtner *et al.*, 2012; Makris *et al.*, 2012; O'Mahony *et al.*, 2011; Thurmond *et al.*, 2008; Tiligada *et al.*, 2009; Zampeli & Tiligada, 2009) and associated central pathologies (Moya-Garcia *et al.*, 2011; Rossbach *et al.*, 2011) is currently being the subject of worldwide investigation. In this respect, the emerging view that combined treatment with drugs targeting the H_4R and other H_xRs may have a significant therapeutic effect on chronic inflammatory disorders, such as dermatitis (Ohsawa & Hirasawa, 2012; Rossbach *et al.*, 2011) and multiple sclerosis (Saligrama *et al.*, 2012) deserves careful consideration.

10.2.1. Preclinical Evidence for the Implication of the H_4R in Disease: Perspectives in H_4R Therapeutic Exploitation

A large body of preclinical evidence derived from *in vivo* and *in vitro* studies using animal models of disease and human biological samples points to the key role of the H_4R in the orchestration of histamine-associated (patho)physiological immune responses (Table 10.1) that are commonly reflected by immune cell

Table 10.1 Indicative recent evidence supporting the contribution of the H_4R in pathological conditions.

Pathological condition	Experimental model/tissue	Findings	Reference
Airway inflammation	$H_4R^{-/-}$ mice sensitized with OVA	Diminished allergic responses; decreased Th2 responses	Dunford *et al.*, 2006; Thurmond *et al.*, 2008
	BALB/c female mice sensitized with OVA	H_4R antagonism ameliorated Th2-driven pathologies	Cowden *et al.*, 2010
	BALB/c mice	H_4R receptor agonist reduced airway inflammatory responses; accumulation of FoxP3$^+$ T cells	Morgan *et al.*, 2007
Dermal inflammation	NC/Nga mice (chronic dermatitis by local application of picryl chloride)	JNJ7777120 attenuated scratching and improved dermatitis; the effects were augmented by combined treatment with olopatadine	Ohsawa & Hirasawa, 2012
	Human atopic dermatitis	H_4R functional expression in key effector cells (T_H & APC)	Mommert *et al.*, 2011
	CD-1, NMRI, BALB/c, C57BL/6J mice (acute croton oil-induced inflammation)	CD-1 mice: reduction by JNJ777720 NMRI, BALB/c or C57BL/6J: no effect of JNJ7777120	Coruzzi *et al.*, 2012b
	Human monocytes & keratinocyte cultures	H_2R- & H_4R-mediated IL-27 down-regulation	Gschwandtner *et al.*, 2012
	BALB/c $H_4R^{-/-}$ mice	Orchestration of skin-relted immunological responses by H_2R & H_4R	Gschwandtner *et al.*, 2012
	Human skin (chronic spontaneous urticaria)	Differential H_4R expression profile in skin biopsies from responsive and unresponsive to therapy patients	Makris *et al.*, 2012
	Human plasmacytoid dendritic cells (psoriasis)	High H_4R expression	Gschwandtner *et al.*, 2011
Pruritus	Mice (acute itch induced by the selective H_3R inverse agonist pitolisant)	Pruritus blocked by combined treatment with cetirizine & JNJ7777120 Expression of H_1R, H_3R & H_4R on skin innervating sensory neurons Histamine induced Ca^{2+} increases in skin-specific sensory neurons via H_1R & H_4R activation & H_3R inhibition	Rossbach *et al.*, 2011
Ocular inflammation	Rat and human conjunctival goblet cells	Expression of all HxRs Stimulation: $H_1R=H_3R>H_4R>H_2R$ Inhibition of histamine-stimulated conjunctival goblet cell secretion by H_2R & H_3R antagonists	Hayashi *et al.*, 2012

APC: antigen presenting cells; EAE: experimental allergic encephalomyelitis; H_4R: histamine H_4 receptor; $H_4R^{-/-}$: H_4R-deficient (knock-out); H_xR: histamine receptor; IL: interleukin; OA: osteoarthritis; OVA: ovalbumin; RA: rheumatoid arthritis; T_H: helper T-cell.

continued Table 10.1 Indicative recent evidence supporting the contribution of the H_4R in pathological conditions.

Pathological condition	Experimental model/tissue	Findings	Reference
	Rat conjunctiva	JNJ7777120 induced increases in tissue histamine levels	Zampeli *et al.*, 2009
	Human vernal keratoconjunctivitis	Increased H_1R, H_2R and H_4R expression	Leonardi *et al.*, 2011
Arthritis	Human RA	H_4R amplifications	Yu *et al.*, 2010a
	ATDC5 chondrocytes (osteoarthritis)	H_4R expression; H_4R-associated differentiation of chondrocytes into hypertrophic cells	Yamaura *et al.*, 2012a
	Human synovial tissue	Expression of H_1R and H_4R lower in RA than OA but comparable H_2R expression	Ikawa *et al.*, 2005; Yamaura *et al.*, 2012b
Systemic lupus erythematosus	Human	Increased H_4R expression	Yu *et al.*, 2010a
Sjögren's syndrome	Human	Low H_4R levels in salivary glands	Stegaev *et al.*, 2012
Multiple sclerosis	H_1H_2RKO & H_3H_4RKO mice (EAE)	H_1R and H_2R: propathogenic H_3R and H_4R: antipathogenic Combined H_xR targeting in disease-modifying therapy	Saligrama *et al.*, 2012
Diabetes-associated renal disease	Rat	H_4R overexpression in diabetic animals; highest immune-positivity in the medulla	Rosa *et al.*, 2012
Gastric ulcer	CD-1 mice (ulcerogenic effect induced by indomethacin & bethanechol)	Strain-related differences in H_4R antagonist-mediated gastroprotection	Adami *et al.*, 2012
Cancer	Human gastric carcinoma	Down-regulation of H_4R gene	Zhang *et al.*, 2012
	Human colorectal carcinoma	Abnormal H_4R expression	Fang *et al.*, 2011
	Human melanoma WM35 & M1/15 cells	H_4R expression in 42% of biopsies	Massari *et al.*, 2011
	MDA-MB-231 and MCF-7 cells (breast cancer)	Functional H_4R expression	Medina *et al.*, 2011

APC: antigen presenting cells; EAE: experimental allergic encephalomyelitis; H_4R: histamine H_4 receptor; $H_4R^{-/-}$: H_4R-deficient (knock-out); H_xR: histamine receptor; IL: interleukin; OA: osteoarthritis; OVA: ovalbumin; RA: rheumatoid arthritis; T_H: helper T-cell.

trafficking and by modifications of the cytokine and chemokine milieu at the sites of inflammation (Dunford *et al.*, 2006; Gutzmer *et al.*, 2011; Thurmond *et al.*, 2004; Tiligada, 2012; Zampeli & Tiligada, 2009).

Accumulating data support the implication of the H_4R in airway inflammation and allergy, largely through autocrine and paracrine processes and the facilitation of inflammatory cell recruitment in asthmatic lungs by direct chemotaxis of additional dendritic cells, eosinophils and mast cells, thus amplifying allergic symptoms and maintaining chronic inflammation (Thurmond *et al.*, 2008). Decreased allergic responses have been reported in H_4R deficient mice or following oral administration of the H_4R antagonist JNJ7777120 (Cowden *et al.*, 2010; Dunford *et al.*, 2006). However, in another study, inhibition of airway resistance and inflammation was observed using the selective H_4R agonist 4-methylhistamine administered intratracheally into the lungs of asthmatic mice (Morgan *et al.*, 2007). Although these discrepancies have been attributed to the route of administration of the compounds (Morgan *et al.*, 2007), the recently described functional selectivity or biased signaling (Seifert *et al.*, 2011) and the marked strain-related differences (Coruzzi *et al.*, 2012b) in H_4R ligand pharmacology have to be considered carefully before translating preclinical data to clinical human settings. Despite these facts, the H_4R antagonists JNJ39758979, PF3893787 and UR63325 have entered clinical trials for allergic diseases and are the first H_4R-targeting candidates with the potential to translate the preclinical data into a therapeutic use, while many pharmaceutical companies report to have H_4R antagonists in their clinical pipeline (Salcedo *et al.*, 2013; Tiligada *et al.*, 2009).

In the field of skin inflammation, the contribution of the H_4R is strongly supported by the available data derived from *in vitro* studies on human immune cells and *in vivo* murine models. The immunomodulatory role of the H_4R on various cell types and its involvement in helper T (T_H) cell polarization links innate and adaptive immune pathways in atopic dermatitis (Mommert *et al.*, 2011). H_4R upregulation has been observed during the differentiation of monocytes into monocyte-derived dendritic cells and in the Th2 to Th1 shift during transition from acute to chronic skin inflammation in human samples (Dijkstra *et al.*, 2007; Gutzmer *et al.*, 2005). In an attempt to identify an essential regulatory pathway that is critical for the pathogenesis of allergic and inflammatory skin diseases, such as chronic eczema and psoriasis, the H_2Rs and H_4Rs have been identified as orchestrators of the immunological responses associated with IL-27 down-regulation in human peripheral monocytes and skin keratinocytes (Gschwandtner *et al.*, 2012; Tiligada, 2012). Furthermore, a differential H_4R expression profile was observed in skin biopsies obtained from groups of patients with chronic spontaneous urticaria who were responsive and unresponsive to drug therapy (Makris *et al.*, 2012). Finally, the neuronal H_4R has been suggested to be implicated in itch phenomena that are frequently associated with dermal inflammation (Rossbach *et al.*, 2011), while the reported mast cell-independent implication of the H_4R in inflammatory pruritus draws attention to the rigorous characterisation of the functional role of the receptor in itch and pain (Dunford *et al.*, 2007; Thurmond *et al.*, 2008). In the context of these multifaceted activities of the H_4R, further research will elucidate whether H_4R agonists or antagonists can yield promising drugs in the treatment of dermal inflammation. Likewise, in parallel to the development of dual-action H_1R/H_4R antagonists offering therapeutic potential in itch and pain (Tiligada *et al.*, 2009), the development of dual-action H_2R/H_4R ligands would provide more accurate tools to assess the potential of histamine receptors to shape pro- or/and anti-inflammatory signals in chronic skin diseases (Gutzmer *et al.*, 2011).

More recently, the functional expression of the H_4R in a variety of non-immume tissues and cells (Table 10.1), such as neurons (Connelly *et al.*, 2009; Lethbridge & Chazot, 2010) and salivary glands (Stegaev *et al.*, 2012) suggests an extensive biological role of the H_4R. Increased H_4R expression levels have been associated with various cancers (Fang *et al.*, 2011; Massari *et al.*, 2011; Medina *et al.*, 2011), systemic lupus erythematosus, arthritis, proteinuria (Yu *et al.*, 2010a) and diabetes (Rosa *et al.*, 2012). In contrast, lower receptor levels were shown in Sjögren's syndrome (Stegaev *et al.*, 2012). The observation that the H_4R is expressed in human submandibular glandular intercalated duct and acinar cells, in a testosterone-dependent manner, implicates the H_4R dysregulation in the pathology of Sjögren's syndrome and provides insights into the sexual dimorphism associated with the disease (Stegaev *et al.*, 2012). Additional evidence supporting the role of the H_4R in autoimmune diseases has been obtained from studies on rheumatoid arthritis (RA). In addition to the presence of H_1Rs and H_2Rs, considerable variations in H_4R expression in human synovial cells could be related to RA severity and duration (Ikawa *et al.*, 2005; Yamaura *et al.*, 2012b). The localisation of H_4 receptors in synovial and vascular wall cells of patients with RA (Grzybowska-Kowalczyk *et al.*, 2007) and osteoarthritis (Grzybowska-Kowalczyk *et al.*, 2008) and the identification of H_4Rs in fibroblast- and macrophage-like cells from RA synovial tissues (Ohki *et al.*, 2007) further support the contribution of the receptor in the pathophysiology of arthritic diseases. Moreover, evidence for the systemic contribution of histamine in the arthritic phenotype was obtained using a rat model of adjuvant arthritis with the functional role of the H_4R in the normal cartilage (Zampeli *et al.*, 2008) and in the inflamed blood vessels (unpublished data) proposed, raising attractive questions regarding the systemic H_4R-mediated mechanisms in arthritis.

Finally, a range of yet elusive new potential roles for the H_4R await further investigation. For instance, the H_4R has been implicated in the putative automodulatory function of histamine in peripheral inflamed tissues (Zampeli *et al.*, 2009), hematopoiesis (Schneider *et al.*, 2009), inflammatory bowel disease and gastroprotection (Adami *et al.*, 2012; Coruzzi *et al.*, 2012a), while the presence of H_4R on human epithelial cells of patients with cystic fibrosis may provide an important route to activating chloride transport.

10.3. Limitations and Perspectives in H_4R Research and Drug Development

Following the deorphanization of the H_4R, a significant number of agonists and antagonists with high affinity for the H_4R and selectivity over the other histamine receptors were successfully designed and synthesized and numerous novel patent applications have been filed covering the new developments in this field (Istyastono *et al.*, 2011). The efforts to develop well-characterized selective H_4R-binding compounds resulted to the description of new classes of H_4R ligands, including (imidazolyl)alkyl derivatives and 2,4-diaminopyrimidines, which were developed by in-house library search and virtual screening/*de novo* design, respectively, and showed both a very good structure-affinity and especially structure-efficacy relationships (Kottke *et al.*, 2011; Łazewska *et al.*, 2009; Sander *et al.*, 2009). Additions to the panel of available selective H_4R ligands include 2-aryl benzimidazole compounds and indolecarboxamides (Engelhardt *et al.*, 2012).

Cell and tissue variability in histamine-mediated signal transduction pathways (Nijmeijer *et al.*, 2013; Reher *et al.* 2012; Rosethorne & Charlton, 2011) as well as the profound intra- and inter-species differences (Coruzzi *et al.*, 2012b) in potency, selectivity and off-target effects (Sugata *et al.*, 2011; Thurmond *et al.*, 2008) of H_4R ligands limits investigations in animal models and more cautious interpretation of ligand effects *in vivo* is required. A large body of evidence shed light on the H_4R intra- & inter-species variations and the structural and pharmacological similarities between H_3Rs and H_4Rs and provided support for the therapeutic opportunities offered through targeting the H_4R in treating inflammatory disorders (Leurs *et al.* 2009; Lim *et al.* 2005; Lim *et al.* 2010). Although JNJ7777120 (1-[(5-chloro-1H-indol-2-yl)carbonyl]-4-methyl-piperazine), the first highly selective H_4R antagonist developed (Jablonowski *et al.*, 2003; Thurmond *et al.*, 2004) has been a very useful tool for dissecting the (patho)physiological function of the H_4R for almost a decade, recent reports evidenced the marked strain-related differences in JNJ7777120 pharmacological activity (Coruzzi *et al.*, 2012b). This is particularly important as it can complicate the interpretation of *in vivo* studies. To add further complication, JNJ7777120 can also activate β-arrestin in a Gi-protein-independent manner and it has been reported to exhibit partial agonist efficacy with respect to Gi-protein activation at certain H_4R species orthologs (Seifert *et al.*, 2011). Thus, the complex pharmacology of the H_4R ligands can be attributed, at least partly, to the functional selectivity or biased signaling, exhibited by many GPCRs, where ligand-specific stabilization of distinct receptor conformations induce distinct biological responses (Nijmeijer *et al.*, 2012b; Nijmeijer *et al.*, 2013; Reher *et al.* 2012; Rosethorne & Charlton, 2011).

10.4. Concluding Remarks

The extensive investigations and the developments in the field of H_4R research during the last decade revealed novel roles for histamine in immune homeostasis. Although a large body of preclinical evidence identifies the H_4R as a key player in the initiation and propagation of immune and inflammatory responses, suitable pharmacological tools as well as appropriate human biological samples and *in vivo* animal models of disease are essential for future investigations to reach solid conclusions on the putative receptor functions (Figure 10.2). Numerous unresolved questions need to be addressed and novel strategies to improve methodologies for H_4R investigation and ligand development are required before reaching beneficial end-points for human (patho)physiology.

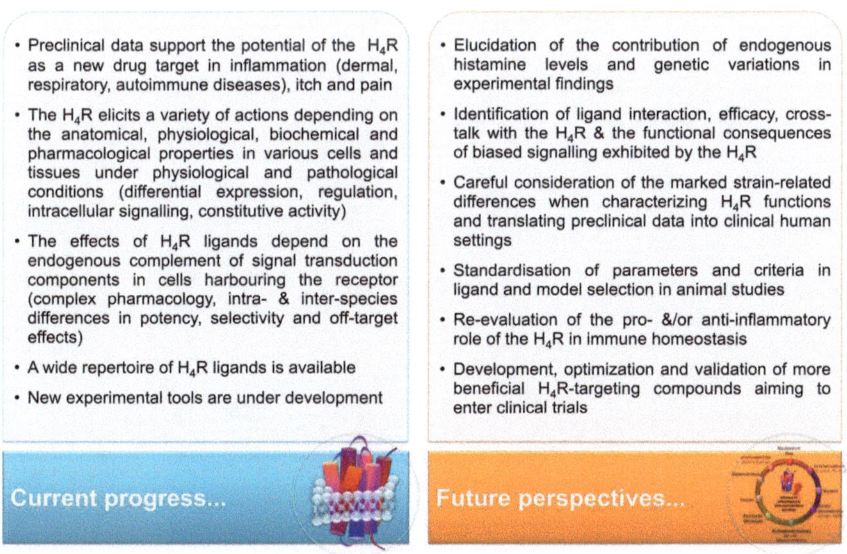

Figure 10.2 Current progress and future perspectives in histamine H_4 receptor (H_4R) research. The advances in the field of H_4R characterisation during the last decade point to the need for the dissection of the role of the H_4R in immunoregulation and the inflammatory response. The underlying mechanisms are clearly more complex than initially anticipated. Given the importance of histamine in human (patho)physiology, continued efforts are directed at the comprehensive elucidation and understanding of the emerging properties of the H_4R and its ligands, thus challenging the potential for their beneficial therapeutic exploitation.

Acknowledgments

This work was supported by the EU FP7 COST Action BM0806: *Recent advances in histamine receptor H4R research.*

References

Adami, M., Pozzoli, C., Menozzi, A., et al. (2012). Effects of histamine H_4 receptor ligands in a mouse model of gastric ulceration. *Pharmacology, 89*, 287-294.

Akdis,C.A,Simons,F.E.(2006).Histaminereceptorsarehotinimmunopharmacology. *Eur J Pharmacol, 533*, 69-76.

Akdis, C.A., Akdis, M., Bieber, T., et al. (2006). Diagnosis and treatment of atopic dermatitis in children and adults: European Academy of Allergology and Clinical Immunology/American Academy of Allergy, Asthma and Immunology/PRACTALL Consensus Report. *Allergy, 61*, 969-987.

Arrang, J.-M., Garbarg, M., Lancelot, J.-C., et al. (1987). Highly potent and selective ligands for histamine H_3-receptors. *Nature, 327*, 117-123.

Ash, A.S.F., Schild H.O. (1966). Receptors mediating some actions of histamine. *Br J Pharmacol, 27*, 427-439.

Bartosch, R., Feldberg, W., Nagel, E. (1932). Das Freiwerden eines histaminhnlichen Stoffes bei der Anaphylaxie des Meerschweinchens. *Archiv für die gesamte Physiologie, 230*, 129.

Black, J.W., Duncan, W.A.M., Durant, G.J., et al. (1972). Definition and antagonism of histamine H_2-receptors. *Nature, 236*, 385-390.

Bongers, G., de Esch, I., Leurs, R. (2010). Molecular pharmacology of the four histamine receptors. *Adv Exp Med Biol, 709*, 11-19.

Bovet, D., Staub, A.-M. (1937). Action protectrice des éthers phénoliques an cours de l'intoxication histaminique. *Comptes Rendus des Seances de la Societe de Biologie (Paris), 124*, 547-549.

Ceponis, A., Konttinen, Y.T., Takagi, M., et al. (1998). Expression of stem cell factor (SCF) and SCF receptor (c-kit) in synovial membrane in arthritis: correlation with synovial mast cell hyperplasia and inflammation. *J Rheumatol, 25,* 2304-2314.

Church, M.K., Maurer, M. (2012). H_1-antihistamines and urticaria: how can we predict the best drug for our patient? *Clin Exp Allergy, 42,* 1423-1429.

Connelly, W.M., Shenton, F.C., Lethbridge, N., et al. (2009). The histamine H_4 receptor is functionally expressed on neurons in the mammalian CNS. *Br J Pharmacol, 157,* 55-63.

Coruzzi, G., Adami, M., Pozzoli, C. (2012a). Role of histamine H_4 receptors in the gastrointestinal tract. *Front Biosci (Schol Ed), 4,* 226-239.

Coruzzi, G., Pozzoli, C., Adami, M., et al. (2012b). Strain-dependent effects of the histamine H_4 receptor antagonist JNJ7777120 in a murine model of acute skin inflammation. *Exp Dermatol, 21,* 32-37.

Cowden, J.M., Riley, J.P., Ma, J.Y, et al. (2010). Histamine H_4 receptor antagonism diminishes existing airway inflammation and dysfunction via modulation of Th2 cytokines. *Resp Res, 11,* 86.

Dale, H.H., Laidlaw, P.P. (1910). The physiological action of β-imidazolylethylamine. *J Physiol, 41,* 318–341.

Delitheos, B., Papamichael, K., Tiligada E. (2010). Histamine modulates the cellular stress response in yeast. *Amino Acids, 38,* 1219-1226.

Dijkstra, D., Leurs, R., Chazot, P., et al. (2007). Histamine downregulates monocyte CCL2 production through the histamine H_4 receptor. *J Allergy Clin Immunol, 120,* 300-307.

Dunford, P.J., O'Donnell, N., Riley, J.P., et al. (2006). The histamine H_4 receptor mediates allergic airway inflammation by regulating the activation of $CD4^+$ T cells. *J Immunol, 176,* 7062-7070.

Dunford, P.J., Williams, K.N., Desai, P.J., et al. (2007). Histamine H_4 receptor antagonists are superior to traditional antihistamines in the attenuation of experimental pruritus. *J Allergy Clin Immunol, 119,* 176-183.

Engelhardt, H., de Esch, I.J., Kuhn, D., et al. (2012). Detailed structure-activity relationship of indolecarboxamides as H_4 receptor ligands. *Eur J Med Chem, 54,* 660-668.

Falcone, F.H., Zillikens, D., Gibbs, B.F. (2006). The 21st century renaissance of the basophil? Current insights into its role in allergic responses and innate immunity. *Exp Dermatol, 15*, 855-864.

Falus, A., Hegyesi, H., Lazar-Molnar, E., et al. (2001). Paracrine and autocrine interactions in melanoma: histamine is a relevant player in local regulation. *Trends Immunol., 22*, 648-652.

Fang, Z., Yao, W., Xiong, Y., et al. (2011). Attenuated expression of HRH4 in colorectal carcinomas: a potential influence on tumor growth and progression. *BMC Cancer, 11*(195), 1-11.

Frewin, D.B., Cleland, L.G., Jonsson, J.R., et al. (1986). Histamine levels in human synovial fluid. *J Rheumatol, 13*, 13-14.

Grzybowska-Kowalczyk, A., Maslinska, D., Wojciechowska, M., et al. (2008). Expression of histamine H_4 receptor in human osteoarthritic synovial tissue. *Inflammation Res, 57*(S1), S63-S64.

Grzybowska-Kowalczyk, A., Wojtecka-Lukasik, E., Maslinska, D., et al. (2007). Distribution pattern of histamine H_4 receptor in human synovial tissue from patients with rheumatoid arthritis. *Inflammation Res, 56*(S1), S59-S60.

Gschwandtner, M., Bunk, H., Köther, B., et al. (2012). Histamine down-regulates IL-27 production in antigen presenting cells. *J Leukoc Biol, 92*, 21-29.

Gschwandtner, M., Mommert, S., Köther, B., et al. (2011). The histamine H_4 receptor is highly expressed on plasmacytoid dendritic cells in psoriasis and histamine regulates their cytokine production and migration. *J Invest Dermatol, 131*, 1668-1676.

Gutzmer, R., Diestel, C., Mommert, S., et al. (2005). Histamine H_4 receptor stimulation suppresses IL-12p70 production and mediates chemotaxis in human monocyte-derived dendritic cells. *J Immunol, 174*, 5224-5232.

Gutzmer, R., Gschwandtner, M., Rossbach, K., et al. (2011). Pathogenetic and therapeutic implications of the histamine H_4 receptor in inflammatory skin diseases and pruritus. *Front Biosci, 3*, 985-994.

Hayashi, D., Li, D., Hayashi, C., et al (2012). Role of histamine and its receptor subtypes in stimulation of conjunctival goblet cell secretion. *Invest Ophthalmol Visual Sci, 53*, 2993-3003.

Hill, S.J., Ganellin, C.R., Timmerman, H., et al. (1997). International Union of Pharmacology. XIII. Classification of histamine receptors. *Pharmacol Rev, 49,* 253-278.

Ikawa, Y., Suzuki, M., Shiono, S., et al. (2005). Histamine H_4 receptor expression in human synovial cells obtained from patients suffering from rheumatoid arthritis. *Biol Pharm Bull, 28,* 2016-2018.

Istyastono, E.P., de Graaf, C., de Esch, I.J., et al. (2011). Molecular determinants of selective agonist and antagonist binding to the histamine H4 receptor. *Curr Topics Med Chem, 11,* 661-679.

Jablonowski, J.A., Grice, C.A., Chai, W., *et al.* (2003). The first potent and selective non-imidazole human histamine H_4 receptor antagonists. *J Med Chem, 46,* 3957–3760.

Kakavas, S., Zampeli, E., Papamichael, K., et al. (2006). The mast cell pathway to inflammation and homeostasis: Pharmacological insights. *Anti-inflammatory & Anti-Allergy Agents in Medicinal Chemistry, 5,* 323-334.

Konttinen, Y.T., Tuominen, S., Segerberg-Konttinen, M., et al. (1990). Mast cells in the labial salivary glands of patients with Sjögren's syndrome: a histochemical, immunohistochemical, and electron microscopical study. *Ann Rheum Dis, 49,* 685–689.

Kottke, T., Sander, K., Weizel, L., et al. (2011). Receptor-specific functional efficacies of alkyl imidazoles as dual histamine H_3/H_4 receptor ligands. *Eur J Pharmacol, 654,* 200-208.

Kyriakidis, D.A., Theodorou, M.C., Tiligada, E. (2012). Histamine in two component system-mediated bacterial signaling. *Front Biosc, 17,* 1108-1119.

Łazewska, D., Wiecek, M., Ligneau, X., et al. (2009). Histamine H_3 and H_4 receptor affinity of branched 3-(^{1}H-imidazol-4-yl)propyl N-alkylcarbamates. *Bioorg Med Chem Lett, 23,* 6682-6685.

Leonardi, A., Di Stefano, A., Vicari, C., et al. (2011). Histamine H_4 receptors in normal conjunctiva and in vernal keratoconjunctivitis. *Allergy, 66,* 1360-1366.

Lethbridge, N., Chazot, P.L. (2010). Immunological identification of the mouse H_4 histamine receptor on spinal cord motor neurons using a novel anti mouse H_4R antibody. *Inflamm Res, 59*(S2), S197-S198.

Leurs, R., Chazot, P.L., Shenton, F.C., et al. (2009). Molecular and biochemical pharmacology of the histamine H_4 receptor. *Br J Pharmacol, 157,* 14-23.

Lim, H.D., de Graaf, C., Jiang, W., et al. (2010). Molecular determinants of ligand binding to H_4R species variants. *Mol Pharmacol, 77,* 734-743.

Lim, H.D., van Rijn, R.M., Ling, P., et al. (2005). Evaluation of histamine H_1-, H_2-, and H_3-receptor ligands at the human histamine H_4 receptor: identification of 4-methylhistamine as the first potent and selective H_4 receptor agonist. *J Pharmacol Exp Ther, 314,* 1310-1321.

Makris, M.P., Lethbridge, N.L., Zampeli, E., et al. (2012). Blood histamine levels and skin H_4R expression in patients with chronic spontaneous urticaria. *Inflamm Res, 61*(S2), S50.

Masini, E., Di Bello, M.G., Raspanti, S., et al. (1998). The role of histamine in platelet aggregation by physiological and immunological stimuli. *Inflamm Res, 47,* 211-220.

Maslinska, D., Gujski, M., Laure-Kamionowska, M., et al. (2004). Subcellular localization of histamine in articular cartilage chondrocytes of rheumatoid arthritis patients. *Inflamm Res, 53,* S35-36.

Massari, N.A., Medina, V.A., Martinel Lamas, D.J., et al. (2011). Role of H_4 receptor in histamine-mediated responses in human melanoma. *Melanoma Res, 21,* 395-404.

Medina, V.A., Brenzoni, P.G., Lamas, D.J., et al (2011). Role of histamine H_4 receptor in breast cancer cell proliferation. *Front Biosci (Elite Ed),* 2011, *3,* 1042-1060.

Mommert, S., Gschwandtner, M., Gutzmer, R., et al (2011). The role of the histamine H_4 receptor in atopic dermatitis. *Curr Allergy Asthma Rep, 11,* 21-28.

Morgan, R.K., McAllister, B., Cross, L., et al. (2007). Histamine 4 receptor activation induces recruitment of $FoxP3^+$ T cells and inhibits allergic asthma in a murine model. *J Immunol, 178,* 8081-8089.

Moya-Garcia, A.A., Rodriguez, C.E., Morilla. I., et al. (2011). The function of histamine receptor H_4R in the brain revealed by interaction partners. *Front Biosci (Schol Ed), 3,* 1058-1066.

Nakamura, T., Itadani, H., Hidaka, Y., et al. (2000). Molecular cloning and characterization of a new human histamine receptor, HH4R. *Biochemical and Biophys Res Commun, 279*, 615-620.

Nijmeijer, S., de Graaf, C., Leurs, R., et al. (2012a). Molecular pharmacology of histamine H_4 receptors. *Front Biosci, 17*, 2089-2106.

Nijmeijer, S., Vischer, H.F., Rosethorne, E.M., et al. (2012b). Analysis of multiple histamine H_4 receptor compound classes uncovers Gαi protein- and β-arrestin2-biased ligands. *Mol Pharmacol, 82*, 1174-1182.

Nijmeijer, S., Vischer, H.F., Sirci, F., et al. (2013). Detailed analysis of biased histamine H_4 receptor signalling by JNJ 7777120 analogues. *Br J Pharmacol*, [Epub ahead of print]

Ohki, E., Suzuki, M., Aoe, T., et al. (2007). Expression of histamine H_4 receptor in synovial cells from rheumatoid arthritic patients. *Biol Pharm Bull, 30*, 2217-2220.

Ohsawa, Y., Hirasawa, N. (2012). The antagonism of histamine H_1 and H_4 receptors ameliorates chronic allergic dermatitis via anti-pruritic and anti-inflammatory effects in NC/Nga mice. *Allergy, 67*, 1014-1022.

Ohtsu, H. (2010). Histamine synthesis and lessons learned from histidine decarboxylase deficient mice. *Adv Exp Med Biol, 709*, 21-31.

O'Mahony, L., Akdis, M., Akdis, C.A. (2011). Regulation of the immune response and inflammation by histamine and histamine receptors. *J Allergy Clin Immunol, 128*, 1153-1162.

Papamichael, K., Delitheos, B., Tiligada, E. (2013). A subset of histamine receptor ligands improve thermotolerance of the yeast *Saccharomyces cerevisiae. J Appl Microbiol, 114*, 492-501.

Parsons, M.E., Ganellin, C.R. (2006). Histamine and its receptors. *Br J Pharmacol, 147*(S1), S127-S135.

Permin, H., Skov, P.S., Norn, S., et al. (1981). Possible role of histamine in rheumatoid arthritis: treatment with cimetidine and mepyramine. *Allergy, 36*, 435-436.

Prinz, C., Zanner, R., Gratzl, M. (2003). Physiology of gastric enterochromaffin-like cells. *Ann Rev Physiol, 65*, 371-382.

Reher, T.M., Neumann, D., Buschauer, A., et al. (2012). Incomplete activation of human eosinophils via the histamine H_4-receptor: evidence for ligand-specific receptor conformations. *Biochem Pharmacol, 84*, 192-203.

Riley, J.F., West, G.B. (1952). Histamine in tissue mast cells. *J Physiol, 117*, 72-73.

Rosa, A.C., Grange, C., Pini, A., et al. (2013). Overexpression of histamine H_4 receptors in the kidney of diabetic rat. *Inflamm Res, 62*, 357-365.

Rosethorne, E.M., Charlton, S.J. (2011) Agonist-biased signaling at the histamine H_4 receptor: JNJ7777120 recruits β-arrestin without activating G proteins. *Mol Pharmacol, 79*, 749–757.

Rossbach, K., Nassenstein, C., Gschwandtner, M., et al. (2011). Histamine H_1, H_3 and H_4 receptors are involved in pruritus. *Neuroscience*, 2011, *190*, 89-102.

Salcedo, C., Pontes, C., Merlos, M. (2013). Is the H_4 receptor a new drug target for allergies and asthma? *Front Biosci (Elite Ed)*, *5*, 178-187.

Saligrama, N., Noubade, R., Case, L.K., et al. (2012). Combinatorial roles for histamine H_1-H_2 and H_3-H_4 receptors in autoimmune inflammatory disease of the central nervous system. *Eur J Immunol, 42*, 1536-1546.

Sander, K., Kottke, T., Tanrikulu, Y., et al. (2009). 2,4-Diaminopyrimidines as histamine H_4 receptor ligands - Scaffold optimization and pharmacological characterization. *Bioorg Med Chem, 17*, 7186-7196.

Schneider, E., Bertron, A.F., Dy, M. (2011). Modulation of hematopoiesis through histamine receptor signaling. *Front Biosci (Schol Ed)*, *3*, 467-473.

Seifert, R., Schneider, E.H., Dove, S., et al. (2011). Paradoxical stimulatory effects of the "standard" histamine H_4-receptor antagonist JNJ7777120: the H_4 receptor joins the club of 7 transmembrane domain receptors exhibiting functional selectivity. *Mol Pharmacol, 79*, 631-638.

Shimamura, T., Shiroishi, M., Weyand, S., et al. (2011). Structure of the human histamine H_1 receptor complex with doxepin. *Nature, 475*, 65-70.

Stegaev, V., Sillat, T., Porola, P., et al. (2012). Brief report: first identification of H_4 histamine receptor in healthy salivary glands and in focal sialadenitis in Sjögren's syndrome. *Arthritis Rheum, 64*, 2663-2668.

Sugata, Y., Okano, M., Fujiwara, T., et al. (2007). Histamine H_4 receptor agonists have more activities than H_4 agonism in antigen-specific human T-cell responses. *Immunology, 121*, 266-275.

Thurmond, R.L., Desai, P.J., Dunford, P.J., et al. (2004). A potent and selective histamine H_4 receptor antagonist with anti-inflammatory properties. *J Pharmacol Exp Ther, 309*, 404-413.

Thurmond, R.L., Gelfand, E.W., Dunford, P.J. (2008) The role of histamine H_1 and H_4 receptors in allergic inflammation: the search for new antihistamines. *Nature Rev Drug Disc, 7*, 41-53.

Tiligada, E. (2012). Editorial: Is histamine the missing link in chronic inflammation? *J Leukoc Biol, 92*, 4-6.

Tiligada, E., Kyriakidis, K., Chazot, P.L., et al. (2011). Histamine pharmacology and new CNS drug targets. *CNS Neurosci Ther, 17*, 620-628.

Tiligada, E., Zampeli, E., Sander, K., et al. (2009). Histamine H_3 and H_4 receptors as novel drug targets. *Expert Opinion on Investigational Drugs, 18*, 1519-1531.

Windaus, A., Vogt, W. (1907). Syntheses des imidazolylathylamines. *Ber Deutsch Chemisch Gesell, 3*, 3691

Yamaura, K., Akiyama, S., Ueno, K. (2012a). Increased expression of the histamine H_4 receptor subtype in hypertrophic differentiation of chondrogenic ATDC5 cells. *J Cell Biochem, 113*, 1054-1060.

Yamaura, K., Oda, M., Suzuki, M., et al. (2012b). Lower expression of histamine H_4 receptor in synovial tissues from patients with rheumatoid arthritis compared to those with osteoarthritis. *Rheumatol Int, 32*, 3309-3313.

Yu, B., Shao, Y., Li, P., et al. (2010a). Copy number variations of the human histamine H_4 receptor gene are associated with systemic lupus erythematosus. *Br J Dermatol, 163*, 935-940.

Yu, B., Shao, Y., Zhang, J., et al. (2010b). Polymorphisms in human histamine receptor H_4 gene are associated with atopic dermatitis. *Br J Dermatol, 162*, 1038-1043.

Zampeli, E., Tiligada, E. (2009). The role of histamine H_4 receptor in immune and inflammatory disorders. *Br J Pharmacol, 157*, 24-33.

Zampeli, E., Thurmond, R.L., Tiligada, E. (2009). The histamine H_4 receptor antagonist JNJ7777120 induces increases in the histamine content of the rat conjunctiva. *Inflamm Res, 58,* 285-291.

Zampeli, E., Thurmond, R.L., Tiligada, E. (2008). Effect of the H_4R antagonist JNJ7777120 on the cartilage histamine content in rats with adjuvant arthritis. *Fund Clin Pharmacol, 22*(S2), 10.

Zhang, C., Xiong, Y., Li, J., et al. (2012). Deletion and down-regulation of HRH4 gene in gastric carcinomas: a potential correlation with tumor progression. *PLoS One, 7,* e31207.

Zwadlo-Klarwasser, G., Braam, U., Mühl-Zürbes, P., et al. (1994). Macrophages and lymphocytes: alternative sources of histamine. *Agents Actions, 41,* C99-100.

List of Figures

binding site as predicted by Moya-García et al., 2008 using quantum mechanics/ molecular mechanics methods. The carbon skeleton of the adduct is represented in magenta and the skeletons of the polypeptide residues are in green. Nitrogen and oxygen atoms are depicted in blue and red, respectively, in both the external aldimine and the protein residues. The depicted residues have been located close to the external aldimine in both rat and human HDCs (Moya-García et al., 2008; Kumori et al., 2012).

Figure 3.6 Histamine transport through plasma and vesicle membranes. Histamine can be passively transported through the plasma membrane by organic cation transporter (OCT) 2 and 3, and by the plasma membrane monoamine transporter (PMAT). If cells are polarised such as epithelial cells, transport proteins are commonly detected at the basolateral side while OCT3 has been found at the apical membrane in bronchial epithelial cells and small intestinal enterocytes. Transport of histamine from the cytoplasm into vesicles is mediated by the vesicular monamine transporter (VMAT) 2, a secondary active antiport using an electrochemical gradient of protons established by a H^+-ATPase. Abbreviations: HA = histamine; SG = secretory granule; TJ = tight junction.

Figure 3.7 Histamine secretion pathways. (A) Histamine secretion may occur constitutively through small vesicular carriers after integration of VMAT2 in those vesicles and subsequent uptake of histamine from the cytoplasm. (B) One of the regulated exocytosis pathways of histamine is "piecemeal degranulation" detected in mast cells and basophils. It is mediated by vesicular transport of granule content from secretory granules to the plasma membrane. (C) This "piecemeal degranulation" is increased if vesicles are forming channels between the granule and the plasma membrane. (D) Regulated exocytosis of histamine by fusion of granules with the cell surface. Abbreviations: ER/TGN = endoplasmic reticulum/trans-Golgi network; SG = secretory granule.

Figure 3.8 Structures of the human HMT and DAO proteins. (A) Structure of monomeric human HMT complexed with histamine and S-adenosyl-L-homocysteine (PDB ID: 1JQD; Horton et al., 2001). (B) Structure of homodimeric human DAO containing 2 Cu^{++} and 4 Ca^{++} ions and 12 N-acetyl-D-glucosamine residues (PDB ID: 3HI7; McGrath et al., 2009). Images were downloaded from the Protein Data Bank website (www.rcsb.org) using standard cartoon style and secondary structure coloring options.

Figure 3.9 Metabolism of N^t-methylhistamine. MAO-B, monoamine oxidase type B; DAO, diamine oxidase; ALDH, aldehyde dehydrogenase; ALO, aldehyde oxidase; XO, xanthine oxidase.

Figure 8.1 Histamine catabolism in MDA-MB-231 and MCF-7 breast cancer cells. (A) Cells were left untreated or were treated with histamine (Ha, 10 μM) for 24 h. (A) HMT (555 bp) and DAO (558 bp) expression was determined by RT-PCR. Lanes: 1, untreated MCF-7 control; 2, histamine-treated MCF-7 cells; 3, untreated MDA-MB-231 cells; 4, histamine-treated MDA-MB-231 cells; 5, positive control (kidney lysate); M, molecular size marker. Glyceraldehyde 3-phosphate dehydrogenase (GAPDH, 576 bp) was used as load control (B) Western blot analysis of HMT (33 kDa). Lanes: 1, untreated MCF-7 control; 2, histamine-treated MCF-7 cells; 3, untreated MDA-MB-231 cells; 4, histamine-treated MDA-MB-231 cells. (C) Western blot analysis of DAO (100 kDa), lanes: 1, untreated MCF-7 control; 2, histamine-treated MCF-7 cells; 3, untreated MDA-MB-231 cells; 4, histamine-treated MDA-MB-231 cells; 5, positive control (kidney lysate); β-actin (42 kDa) was used as load control (D) HMT activity determination by a radiometric assay based on the transmethylation of histamine with S-adenosyl-L-[methyl-^{14}C] methyonine. Error bars represent the means ± SEM. (***P<0.0001, vs. Control (C); t-Test). (E) Histamine intracellular content was determined by immunofluorescence analysis and flow cytometry. Error bars represent the means ± SEM. (F) Immunohistochemistry of HMT, DAO and histamine on the triple-negative MDA-MB-231 xenografted tumour induced in nude mice. Representative immunohistochemical images in paraffin-embedded tumour specimens of untreated and histamine (1 mg/kg, *s.c.*) treated animals. Original magnification, x630. Scale bar = 20 μm.

Figure 8.2 Correlations between H_4R and PCNA protein expression determined by immunohistochemistry in human nevi and melanoma biopsies. (A) Representative immunohistochemical images in paraffin-embedded specimens. Original magnification, x630. Scale bar = 20 μm. (B) Contingency table: includes total immunohistochemical analysis of nevi and melanomas. 16/19 nevus samples were positive for the expression of H_4R and 18/19 were negative for PCNA. 8/19 melanoma samples were positive for the expression of H_4R and 5/19 were negative for PCNA. Chi square Test ***p= 0.0002 and Fisher exact Test ***p= 0.0003.

Figure 8.3 LLC model in mast cell deficient Kit^{W-sh} mice. LLC development in mast cell-deficient and wild type mice was checked 14 days after the induction of the disease. (A) The number of metastatic colonies (empty bars), lung area and lung/total weight ratio (black bars and dotted line) were compared. Data are the mean ± SD from mast cell-deficient and wild type mice. (B) Lung sections of LLC-injected mast cell-deficient (I) and wild type (II) mice. H&E staining of normal looking areas (a) and cancerous areas (b) of the lung tissue sections (10×) and (c) toluidine blue staining of cancerous lung tissue sections (40×). Arrows indicate mast cells. Representative pictures obtained from mast cell-deficient and wild

are clearly more complex than initially anticipated. Given the importance of histamine in human (patho)physiology, continued efforts are directed at the comprehensive elucidation and understanding of the emerging properties of the H_4R and its ligands, thus challenging the potential for their beneficial therapeutic exploitation.

List of Tables

Index

T

U

V

X

Z

Bei Fragen zur Produktsicherheit wenden Sie sich bitte an:
If you have any questions regarding product safety,
please contact:

Walter de Gruyter GmbH
Genthiner Straße 13
10785 Berlin
productsafety@degruyterbrill.com